STRENGTH SECOND EDITION
OF MATERIALS

◆ John N. Cernica

PROFESSOR AND HEAD, CIVIL ENGINEERING DEPARTMENT

YOUNGSTOWN UNIVERSITY

HOLT, RINEHART AND WINSTON

New York · Chicago · San Francisco · Toronto · London

*To my wife Patricia
and my daughters Kathleen, Jude, Alice,
Johanna, Tricia, and Sarah*

PREFACE TO THE SECOND EDITION

While the basic format, perhaps one feature most responsible for the outstanding success of the first edition, was closely adhered to in this edition, some additions are indeed timely; the inevitable universal adoption of the International System of units (SI) makes the introduction of this item at this time an indespensible added objective.

Most of the example problems have the dual, english and metric, units. Some are treated in the english units with the significant values also given in the metric equivalent; others are treated totally in the metric system. The answers to the problems, given in Appendix C for the even-numbered problems, are likewise in the *dual* system.

The addition of nearly 140 new problems and examples bring the total number in this edition to almost 800; of these about 100 are example problems worked out in detail. This should provide for a rather wide choice of assignments for a long period, for many sections.

Every effort was made to correct some errors detected in the first edition. For this much appreciation is extended to many who brought them to the attention of the author; much appreciation is expressed to those who wish to do so in the future.

The author is most grateful to Dr. R. G. Boggs of U.S. Coast Guard Academy who did a most thorough and exhaustive evaluation of the book, and whose comments were most valuable. Also, thanks is extended to the following persons for their constructive review evaluation and recommendations: Dr. W. J. Lnenicka, Georgia Institute of Technology; Dr. K. Muhlbauer, University of Missouri-Rolla; Prof. C. M. Antony, University of Syracuse; Dr. W. C. Crisman, Le Tourneau College; Dr. D. H. Suchora, Youngstown State University; Dr. J. M. Dalrymple, Michigan Technological University; Mr. Douglas Kearns, Graduate Student, Youngstown State University; to Miss Mary Ann Stasiak and Mrs. Karen Vanderbilt for the typing of the manuscript.

J. N. Cernica

Youngstown, Ohio
June 1976

PREFACE TO THE FIRST EDITION

The study of strength of materials is perhaps the first opportunity for the student to put to use in a constructive way some of the fundamentals presented to him in calculus, physics, and engineering mechanics. Up to this point he has learned fundamentals for use in the future. He cannot, from these fundamentals alone, tie in the *analysis* of a structure to the *design* of that structure; that is, he cannot intelligently select a material and specify size and other essential properties to meet some specific need. *Strength of Materials* develops the fundamental relationship between applied forces and internal effects in a systematic and basic manner, and thus sets the foundation for the transition to *application*. Furthermore, it permits — and in some instances compels — the student to take up where theory leaves off, and in so doing, provides an avenue for development of sound judgment (we might prefer to call it engineering judgment) and for applying theory in a practical and realistic way.

It is hoped that this book's content, and the fashion used in presenting this content, will form the bridge the student may cross to broader design fields. More specifically, the book has the following objectives:

1. To give the student fundamentals, which once mastered, can serve as tools for him to build with.
2. To emphasize an orderly and systematic format for solutions to problems.
3. To emphasize the need for good judgment in *defining* a problem, making the *assumptions* needed to solve the problem, *selecting* the material, and considering the *design* of the component part of a structure — be it a machine part, a beam for a bridge, or some other item.

To achieve the first objective, the author has taken great care in explaining the basic concepts involved, and in establishing fundamental expressions which, although rigorously derived, would be easily understood and used by the student, with a full knowledge of their physical meaning and of the assumptions and limitations on which their derivations are based. Chapter 1 covers the concepts of stress and strain in a broad sense. Although an attempt was made to keep the presentation concise, no effort was spared to elaborate on various phases of these basic concepts where needed. The remainder of the book puts these concepts into use, always deriving theoretical expressions from fundamentals.

To attain the second objective, the author has presented numerous example problems in *component form: Given; To Find; Solution.* The separation of these components into explicit forms, and the systematic procedure followed to get an answer, makes for an efficient and easy-to-understand system for solving problems. Once in the habit of using such a method to solve problems, the student should benefit, not only by decreasing his errors of computation, but also by acquiring a desirable form for presentation to others who must check, approve, or just look, at his work.

The final objective of the text is to emphasize common sense, rather than unquestioned use of available formulas, methods, or specifications.

In presenting this material, the author has drawn upon material presented in one form or another by many authors and many publications, and upon the author's own notes from teaching the subject for many years. The author wishes to acknowledge his indebtedness to Dr. Frank D'Isa and Professor Robert Sorokach for their constructive criticism, and to the Misses Gerri Sfara and Patricia M. Olinik for their typing of the manuscript.

J. N. Cernica

Youngstown, Ohio
January 1966

CONTENTS

viii • Contents

SYMBOLS AND ABBREVIATIONS

A area

c distance from neutral axis to extreme fiber

d diameter

E modulus of elasticity

e eccentricity

F force

ft   foot or feet

ft-lb   foot-pound

G shear modulus or modulus of rigidity

g gravitational acceleration constant

h height; depth of a beam

hp   horsepower

I moment of inertia of area

i radius of gyration

in.   inch or inches

in.-lb   inch-pound

J polar moment of inertia of area

K stress concentration factor

k symbol for p/EI; spring constant; factor

kN   kilonewtons

kip   kilopound (1000 lb)

ksi   kilopounds (or kips) per square inch

l or L length

lb   pound

M bending moment

max   maximum

min   minimum

N normal force; Newtons

n cycles; number

P force; concentrated load

Pa   pascals

p pressure per unit area

psf   pounds per square foot

psi   pounds per square inch

Q force; statical moment of area

q load per unit length; shear flow

R reaction; radius; resultant force

r radius; radius of gyration

rad   radian

rpm   revolutions per minute

rps   revolutions per second

s arc length; distance

T torque; temperature

t thickness

U strain energy

u strain energy per unit volume

V shearing force; volume

v velocity

W weight; total load

w load per unit length; weight per unit volume

x, y, z coordinates

Z section modulus

α temperature coefficient of expansion; angle

β angle

δ deflection; total elongation

γ shearing strain

ϵ tensile or compressive strain

θ slope of elastic line; angle of twist per unit length

μ Poisson's ratio

ρ radius of curvature

σ normal stress

τ shearing stress

ϕ angle of twist; angular coordinate

ω angular velocity

∂ specific weight

BASE SI UNITS

Multiplication Factor	Prefix	SI Symbol
1 000 000 000	giga	G
1 000 000	mega	M
1 000	kilo	k
0.001	milli	m
0.000 001	micro	μ
0.000 000 001	nano	n

Quantity	Unit	Symbol
length	meter	m
mass	kilogram	kg
force	newton	N
time	second	s

DERIVED SI UNITS

Quantity	Derived SI unit	Name	Symbol
area	square meter	—	m^2
volume	cubic meter	—	m^3
density	kilogram per cubic meter	—	kg/m^3
force	kilogram-meter per second squared	newton	N
moment of force	newton-meter	—	N-m
pressure	newton per meter squared	pascal	Pa
stress	newton per meter squared	pascal	Pa or N/m^2
work, energy	newton-meter	joule	J
power	joule per second	watt	W

kilopound-force (kip)	kilonewtons (kN)	4.45
kilopound-force/sq in.	meganewtons/meter^2 (MN/m^2)	6.895
one kilogram force (kgf)	newtons (N)	9.81
pounds per square foot (psf)	newtons per square meter (N/m^2)	47.9
pounds per square inch (psi)	kilonewtons per square meter (kN/m^2)	6.9
inch-pound force (ft-lbf)	newton-meter (N-m)	0.113
foot-pound force (ft-lbf)	newton-meter (N-m)	1.356
horsepower (hp = 550 ft-lbf/sec)	newton-meter/sec (N-m/sec)	745.7

To convert	To	Multiply by
inches (in.)	millimeters (mm)	25.40
inches (in.)	centimeters (cm)	2.540
inches (in.)	meters (m)	0.0254
feet (ft)	meters (m)	0.305
miles (miles)	kilometers (km)	1.61
yards (yd)	meteres (m)	0.91
square inches (sq. in.)	square centimeters (cm^2)	6.45
square feet (sq ft)	square meters (m^2)	0.093
square yards (sq yd)	square meters (m^2)	0.836
acres (acre)	square meters (m^2)	4047
square miles (sq miles)	square kilometers (km^2)	2.59
cubic inches (cu in.)	cubic centimeteres (cm^3)	16.4
cubic feet (cu ft)	cubic meters (m^3)	0.028
cubic yards (cu yd)	cubic meters (m^3)	0.765
pounds (lb)	kilograms (kg)	0.453
tons (ton)	kilograms (kg)	907.2
kilopound-force (kip)	kilonewtons (kN0	4.45
kilopound-force/sq in.	meganewtons/meter^2 (MN/m^2)	6.895
one kilogram force (kgf)	newtons (N)	9.81
pounds per square foot (psf)	newtons per square meter (N/m^2)	47.9
pounds per square inch (psi)	kilonewtons per square	6.9
gallons (gal)	cubic meters (m^3)	0.0038
acre-feet (acre-ft)	cubic meters (m^3)	1233
gallons per minute (gal/min)	cubic meters per minute (m^3/min)	0.0038
newtons per square meter (N/m^2)	pascals (Pa)	1.00
inch-pound force (ft-lbf)	newton-meter (N-m)	0.113
foot-pound force (ft-lbf)	newton-meter (n-m)	1.356
horsepower (hp = 550 ft-lbf/sec)	newton-meter/sec)	745.7

Chapter 1 ◆ FORCES AND DEFORMATIONS

1—1 Introduction

Safety and economy, in that order, are perhaps the most important aspects of structural design. A designer must choose a size and type of material strong enough to carry the forces to which it is subjected and rigid enough not to deform excessively, but economical and serviceable enough to dictate its choice over some other material. Sometimes economy—but never safety—is sacrificed for aesthetic values, durability, or low cost of maintenance.

Strength-of-material analysis develops systematically the relationship between externally applied loads and the resulting internal effects, loads, and deformations induced in the body subjected to these loads. This forms a basis for understanding design problems and their solution in a fundamental way, and supplies the background for safe and economical design work. Once a student masters these fundamentals, he can expand and grow; without these fundamentals, he is forever limited.

For example, a carpenter knows from experience that a floor joist does not serve well when it lies with its deeper side on the support; the joist deflects excessively and might perhaps rupture in bending. He draws on the experience of others and his own personal observations, and with this insight he manages to solve this problem by placing the joist with its deeper side vertical. The size of the joist necessary to span a certain distance, and the spacing of these joists, are problems he again solves in a crude way. He has no systematic way of attacking the problem. He could not span other structures of any general type because he lacks fundamental knowledge. He is limited to something he saw or was handed down to him by his predecessors, and he is therefore limited in growth.

The problems encountered in this book generally fall in the category of *design* or *analysis*. If the problem is to select the size and material to build a

1

machine or a structure to satisfy a certain function under given or assumed conditions, it is a design problem. If the capacity of a given completed structure to carry a certain load is to be investigated, the problem becomes one of analysis.

Seldom, however, is either an *analysis* or a *design* problem clear-cut. Its solution depends on conditions that are *variable*. Almost always, the approach requires reasonable and basic assumptions, utilizing practical considerations and available experimental and theoretical information, before a systematic solution to the problem can be attempted. For example, again assume that a joist supporting a floor is to be *analyzed* for deflection. Load, span, size, and strength of the wood are important factors in the solution of the problem. Although the size can be measured, the strength characteristics are obtained from either testing in the laboratory, or accepting typical test results from others. Furthermore, if plywood sheets are nailed on the top of the joists, should these sheets be assumed effective in reducing the deflection, and, if so, how much? Only a few nails may make the effect of the plywood insignificant, and many nails might mean just the reverse. Likewise, is the load evenly distributed over the span, and is the span the same as the length of the joists, or just the distance between the inside faces of the two walls, or somewhere in between?

In the same manner, in designs for assumed conditions of load, many sizes may fit a certain restriction, such as deflection or strength. But a comprehensive consideration of all the restrictions may eliminate many choices. Thus, the process of selection and elimination to reach a satisfactory design is a process of analysis. The conditions of load and function to be served, as well as the actual selection of suitable components, are idealized in a systematic way, and are proven to be reliable and reasonably basic in insuring satisfactory results.

Much is still to be done in perfecting and refining the approaches to the solution of many problems. But much has been done and much is to be gained by understanding these accomplishments. The presentation of these concepts and the understanding of the basis on which they exist is the undertaking of this book.

1—2 Loads and Reactions

The forces that *act* on a member are referred to as *loads*. The forces *counteracting* the effect of these loads are called *reactions*. Loads and reactions may be generally categorized as external or internal, and static or dynamic. They may be further defined with respect to the area subjected to contact. It is important that definitions of various types be established now in order to eliminate confusion later on.

A *static* load is a force applied gradually and slowly, and not repeated many times.

A *sustained* load is a force acting for a long period of time. The dead weight of a structure is an example of such a load. For some materials under certain conditions of temperature and load, this force has an appreciable effect on the structure, such as permanent deflection of a beam or permanent shortening of a column.

A *repeated* load is a force applied thousands or perhaps millions of times; for example, the forces in various parts of a running engine.

An *impact* load is a force applied within a short period of time, such as a hammer hitting a nail or a weight dropping to the floor. Sometimes this type of load is described as an *energy* load.

An *axial* load is a force or force system whose resultant passes through the centroid of the section on which the force acts.

A *concentrated* load is a force applied at a point.

A *distributed* load is a force or force system spread over an area, either uniformly or nonuniformly.

An *external* load is a force acting on the outside of a structure set in equilibrium. The loads P and the reactions R and H in Figure 1–1 are examples of such loads.

An *internal* load is a force effect within the entity or any component part of a structure set in equilibrium. The forces in members U_1L_1, U_2L_2, U_3L_3 in Figure 1–1 are examples of such loads.

To analyze *external* reactions, the loads acting on a body may be replaced by their resultant, frequently a mathematically convenient simplification. *External* reactions are not altered by the application of a load to *any* point along its line of action. *Internal* effects, however, are altered by these

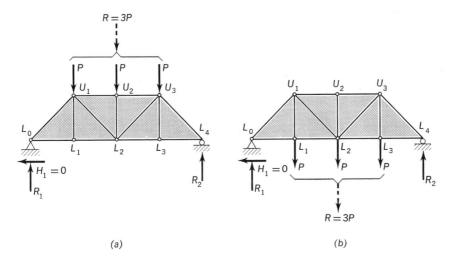

(a)   (b)

Figure 1–1 A truss symmetrically loaded with three equal loads.

loading substitutions. Figure 1–1 illustrates this. Whether the loads act at points U_1, U_2, and U_3, as shown in Figure 1–1a, or at points L_1, L_2, and L_3, as shown in Figure 1–1b, the reactions are the same; likewise, the reactions would be the same if the loads were replaced by their resultants (represented by the dashed arrows).

Now let us investigate the internal loads, such as the forces in the vertical members. In Figure 1–1a, the forces in the three vertical members, represented by the two capital letters at the joints or connections of these members (a common notation), are $U_1L_1 = 0$, $U_2L_2 = P$, $U_3L_3 = 0$. In Figure 1–1b the forces in these members are different, that is, $U_1L_1 = P$, $U_2L_2 = 0$, and $U_3L_3 = P$. This is an important aspect to keep in mind, because repeatedly problems are solved by first finding the external effects, and then the internal forces.

Example 1–1:

Given: The cantilever beam shown in Figure *a*.

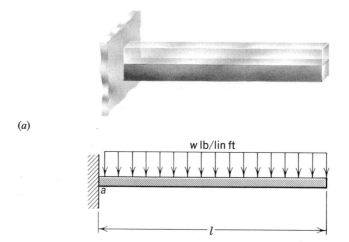

(*a*)

Find: The external reactions acting *on* the beam.
Solution: The *uniform* load applied to the beam is usually so much larger than the unit weight of the beam that the beam's weight is neglected in the calculations. When and if to neglect the beam's weight is an engineering judgment applied to each situation as it arises. For our beam we shall assume that the weight of the beam is incorporated in *w*. We proceed by setting the beam in equilibrium, as shown in Figure *b*, that is, only the beam *outside* the wall. (A sign convention: Forces acting upward and moments or bending effects creating tension on the bottom surface of the beam are assumed positive, as is explained in Chapter 4. Thus, *R* being upward, the sign is positive; the negative sign

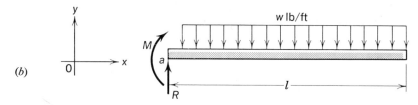

$\sum F_y = 0$; therefore, $R - wl = 0$. $R = wl$ ft, as shown.

$\sum M_a = 0$; therefore, $+M + \dfrac{wl^2}{2} = 0$. $M = \dfrac{-wl^2}{2}$ ft-lb.

for M indicates that the orientation of M should be reversed, that is, counterclockwise.)

Example 1–2:

Given: The cantilever truss shown in Figure *a*.

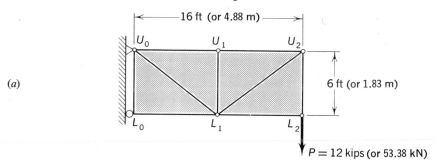

Find: The external reactions and internal forces in members U_0L_1, U_2L_2.

Procedure: The roller support provides only a horizontal support; the other provides a vertical and a horizontal support, as shown below in the free-body diagram *b*.

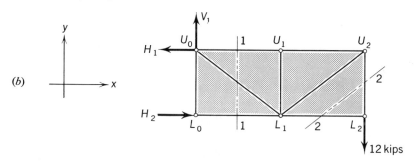

$\sum M_{U_0} = 0$; therefore, $12 \times 16 - H_2 \times 6 = 0$; $H_2 = 32$ kips (or 142.34 kN)
$\sum F_X = 0$; $|H_1| = |H_2| = 32$ kips (or 142.34 kN)
$\sum F_y = 0$; $V_1 = 12$ kips (or 53.38 kN)

To determine the internal forces, we may proceed by isolating (by an imaginary cut) a complete part of the truss, as shown in Figure *b* by the cuts 1–1 and 2–2. This is generally called the *method of sections*. Of course, the member for which the load is to be found is one of those separated by this imaginary cut. Caution: If the cut has more than three unknowns, the forces cannot be determined by statics equations alone. Also remember that a truss member is assumed to carry only axial load, tension, or compression. The following procedure illustrates this. The *direction* of the forces are assumed and corrected later if the result is a negative sign.

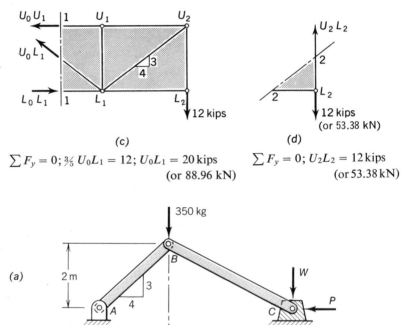

(c)

$\sum F_y = 0; \; \tfrac{3}{5} \, U_0 L_1 = 12; \; U_0 L_1 = 20 \, \text{kips}$
(or 88.96 kN)

(d)

$\sum F_y = 0; \; U_2 L_2 = 12 \, \text{kips}$
(or 53.38 kN)

(a)

Example 1–3:

Given: The frame shown in Figure *a*. Assume points *A*, *B*, and *C* to be frictionless hinges. If $\dot{W} = 70$ kg and the coefficient of friction between *W* and the plane is 0.3.

Find: (*a*) The axial force in bar *BC*. (*b*) The magnitude of *P* necessary to keep the block from sliding to the right.

Procedure: The forces on the block *W* are shown in Figure *b*.

(b)

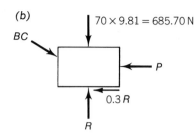

From Figure b,

$$\sum F_y: \quad 685.70 \text{ N} + \frac{2}{\sqrt{20}} BC = R \quad (1)$$

$$\sum F_x: \quad 0.3R + P = \frac{4}{\sqrt{20}} BC \quad (2)$$

By isolating joint B, as shown in Figure c, we can find force BC:

(c)

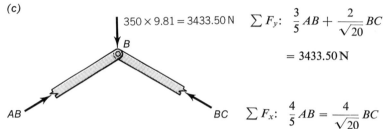

$$\sum F_y: \quad \frac{3}{5} AB + \frac{2}{\sqrt{20}} BC$$
$$= 3433.50 \text{ N} \quad (3)$$

$$\sum F_x: \quad \frac{4}{5} AB = \frac{4}{\sqrt{20}} BC \quad (4)$$

From Equation (4),

$$AB = \frac{5}{4} \left(\frac{4}{\sqrt{20}} BC \right)$$

Substituting in Equation (3), we have

$$\frac{3}{\sqrt{20}} BC + \frac{2}{\sqrt{20}} BC = 3433.50 \text{ N}$$

or

$$BC = \frac{\sqrt{20}}{5} (3433.50 \text{ N}) = \textbf{3071.02 N comp.}$$

Substituting in Equations (1) and (2), we have

$$685.70 \text{ N} + \tfrac{2}{5}(3433.50 \text{ N}) = R \quad (1')$$

and

$$0.3 [685.70 \text{ N} + \tfrac{2}{5} (3433.50 \text{ N})] + P = \tfrac{4}{5} (3433.50 \text{ N}) \quad (2')$$

Solving for P, we have

$$P = \textbf{1657.89 N}$$

PROBLEMS

1–1 In Figure P 1–1, assume pin connections for *A, B, C,* and *D.* Neglecting the weight of the members, determine (*a*) the magnitude and directions of the reactions at points *A* and *B*, (*b*) the force in cable between *CD*, and (*c*) the force in member *BC*.

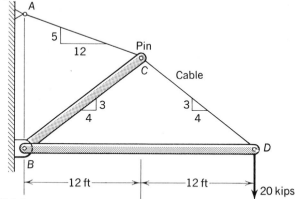

Figure P 1–1

1–2 In Figure P 1–2, assume that the weight of the members is negligible. For the given uniform load, determine (*a*) the reaction at *A*, (*b*) the tension in linkage tie *BC*, and (*c*) the reaction at *D*.

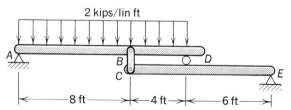

Figure P 1–2

1–3 What is the force in the turnbuckle tie of Figure P 1–3 when *P* = 9 kips?

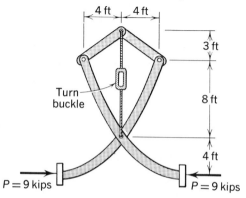

Figure P 1–3

1-4 In Figure P 1–4 the arch is hinged at points *A, B,* and *C.* Draw a free-body diagram of portions *AB* and *BC* and show all the forces (magnitudes and directions) acting on these members. Neglect the weight of the members.

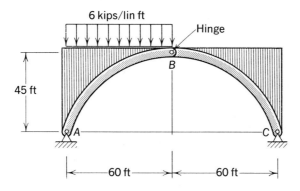

Figure P 1–4

1-5 In Figure P 1–5 the beam is fixed against rotation in the wall. Joints *B* and *C* are rigid; that is, they can take bending. Assuming the weight of the members to be negligible, (*a*) determine the reactions at *A,* and (*b*) set beam *AB* in equilibrium and show all the forces acting on it. [Hint: At point *B,* there is a *moment* and a *transverse force* acting.]

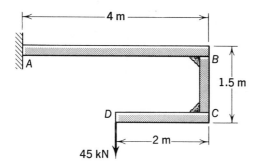

Figure P 1–5

1-6 The hoist in Figure P 1-6 consists of two compression members assumed to be pin connected at points A, B, and C, and a pulley at point B. Draw free-body diagrams and show all the forces acting on AB, BC, and the pulley system when the weight W is 150 lb.

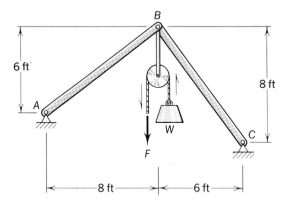

Figure P 1-6

1-7 For the truss shown in Figure P 1-7, determine (a) the reaction at points L_0 and U_4, and (b) the forces in members L_1L_2 and U_2L_2.

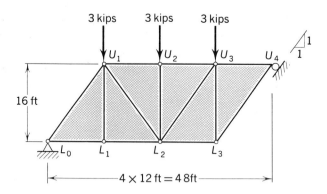

Figure P 1-7

1-8 Determine the reactions at points A and B in terms of P for the structure shown in Figure P 1-8.

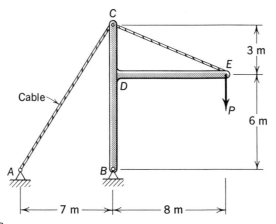

Figure P 1–8

1–9 For the system in Figure P 1–9, assume points A and B to be frictionless hinges. If $W = 100$ lb and the coefficient of friction between W and the plane is 0.3, determine (a) the range of magnitude of P for equilibrium when the mechanism is in the position shown, (b) the forces in members AB and BC in the shown position, and (c) pin reaction at points A and C.

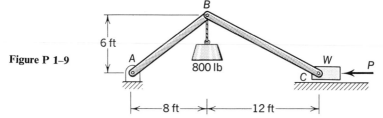

Figure P 1–9

1–10 In Figure P 1–10, a 90-kg man walks to the left on a rigid plank as shown. At what point x will the 14-kg mass begin to move upward? What are the reactions at points A and B when impending motion takes place?

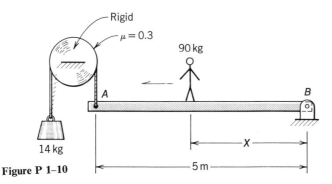

Figure P 1–10

1–11 Determine the reactions at points A and B in Figure P 1–11.

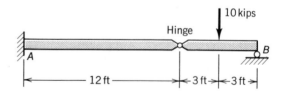

Figure P 1–11

1–12 For the frame shown in Figure P 1–12, determine (a) the reactions at A and E, and (b) the axial forces in the members AC, DC, and EB.

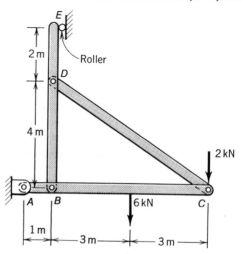

Figure P 1–12

1–13 In Figure P 1–13 a man lifts the 150-lb weight via a frictionless pulley, as shown. Determine the axial forces in AB and BC.

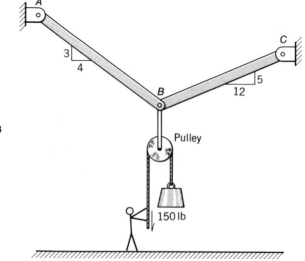

Figure P 1–13

1–14 For the load shown in Figure P 1–14, determine (*a*) the axial force in the tie, and (*b*) the reactions at points *A* and *C*.

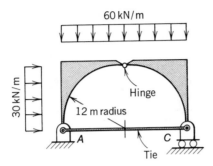

60 kN/m

30 kN/m

Hinge

12 m radius

A

C

Tie

Figure P 1–14

1–15 Determine the reactions at points *A* and *B* in Figure P 1–15.

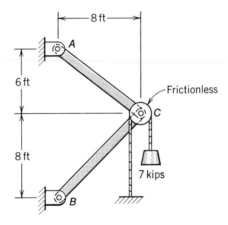

8 ft

A

6 ft

Frictionless

8 ft

C

7 kips

B

Figure P 1–15

1—3 Stress

Stress is a basic concept used to denote the intensity of internal force. *Stress* is defined as the *force per unit area*. It is a convenient basis for analyzing the internal resistance of a structure subjected to load, and for selecting the most appropriate material and size for a design.

To formulate the concept of stress, assume that the link of Figure 1–2a is cut at section 1–1 perpendicular to its longitudinal axis, as shown in Figure 1–2b. For equilibrium, the axial load P (neglecting the weight of the link) must be resisted by the sum of internal forces, composed of the stress σ

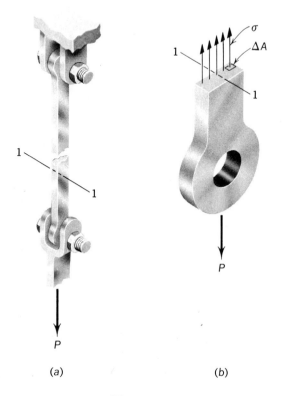

(a)   (b)

Figure 1–2

and the area ΔA on which the stress acts. If the cut shown is reasonably far away from load P, the stress may be assumed uniform over the cross section (an assumption commonly made in the analysis of such members[1]). Then, by summing forces in the vertical direction, we have

$$P = \sigma A \quad \text{or} \quad \sigma = P/A$$

This will be taken as the definition for simple stress in *equation form*, that is, it defines normal (tension or compression) or shear stresses. As we shall see in later chapters, there are other stresses, such as *bending, torsion,* and *combined* stresses. However, these can be reduced to variations of compression, tension, or shear.

---

[1] The portion of the link near the load is subjected to localized stress distribution, discussed in Section 1–14.

A more exact and general definition of stress is obtained by taking an increment ΔP of a load P and dividing it by an increment ΔA of the area A on which P acts. This condition is shown in Figure 1–3. The body is subjected to a system of forces P_1, P_2, and P_3, and is held in equilibrium by the resultant R, which is comprised of the sum of the increment of forces ΔP acting on the small element of area ΔA, whose plane is assumed parallel to the $x–y$ plane shown. Then, by resolving ΔP into two components, one perpendicular and the other parallel to the plane as shown in Figure 1–3, we obtain a general expression for the *normal* stress

$$\sigma = \lim_{\Delta A \to 0} \frac{\Delta P_n}{\Delta A} = \frac{dP_n}{dA} \tag{1–1}$$

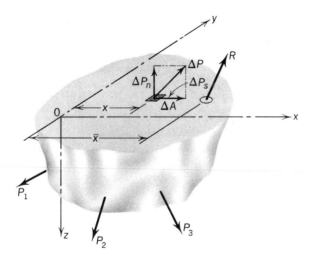

Figure 1–3

Again, $\sigma = P/A$ is a special case of Equation (1–1) when the stress is uniform over the cross-sectional area. For the stress to be uniform, the external resultant of the applied loads P_1, P_2, and P_3 must have a line of action passing through the centroid of the cross section. Of course, the internal resultant R must pass through the same line of action (centroid); otherwise, there would be a rotational effect on the structure.

The proof for the above statement is as follows: Let $\bar{x}$ represent the x coordinate of the point of application of the resultant R, and let x_c represent the coordinate of the centroid c of the cross section. Then

$$R = \int_A \sigma \, dA$$

Also, from taking moments about the y axis,

$$R\bar{x} = \int_A x\,(\sigma\,dA)$$

But because σ is a constant (uniform stress) over the whole cross section, it can be taken outside the integrals. Therefore, we have

$$R = \sigma \int_A dA \qquad \text{and} \qquad R\bar{x} = \sigma \int_A x\,dA$$

from which, when substituting for R, we obtain

$$\left(\sigma \int_A dA\right)\bar{x} = \sigma \int_A x\,dA$$

$$A\bar{x} = \int_A x\,dA = A\cdot x_c \qquad \text{or} \qquad \bar{x} = x_c$$

The stress created by the horizontal component of P_s (parallel to the cut plane in Figure 1–3) is called a *shear stress*, denoted by the Greek letter τ. The general expression for it is

$$\tau = \lim_{\Delta A \to 0} \frac{\Delta P_s}{\Delta A} \tag{1–2}$$

The pin in Figure 1–2a is subjected to a shear stress. Figure 1–4a shows the connection unit as a free body. Figure 1–4b shows a magnified version

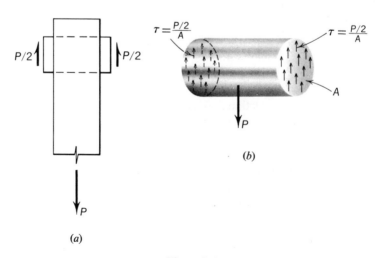

(a)

(b)

Figure 1–4

of the bolt subjected to a force P and resisted by the forces on the two faces where shear failure might occur. Again, the common assumption for the analysis of such a stress is that its magnitude is uniform over the whole area. Actually, this is at best only a rough approximation of the actual stress distribution. Figure 1–5 shows what might be the load distribution on this pin. From this we note that the pin or bolt tends to bend as well as shear, because the resultant load and reactions are not collinear, and the farther apart they are, the greater the bending effect. This is pointed out to make us aware of the qualitative fallacy in the assumption, but not to suggest that it be abandoned. With further theoretical and experimental research, such a substitution might be warranted; at present, it is not.

For a long time the unit of stress most commonly used in the United States is *psi*, for pounds per square inch. Another is *ksi*, representing kips (1000 lb) per square inch. Rapidly gaining in popularity, however, is kilonewtons /meter² (kN/m^2). It is important that the student distinguish between force (lb, N, and so on) and stress (psi, kN/m^2, and so on) and bear this difference in mind.

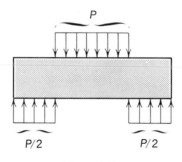

Figure 1–5

Example 1–4:

Given: A cargo ship loads cargo from the dock by lifting it with a boom crane and then, with a horizontal cable, guiding the weight to a location on the ship (shown in dotted lines). Schematically, this is shown in Figure *a*. The area of each cable is 0.90 in.² (or 5.81 cm²) and the allowable stress is 15,000 psi (or 35.47 MN /m²).

Find: The maximum weight the crane can lift at any one time.

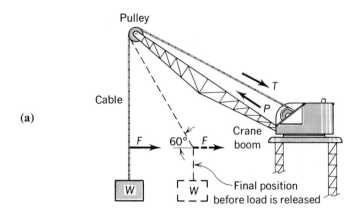

(a)

Pulley

Cable

F 60° F

W

W

Crane boom

T

P

Final position before load is released

(b)

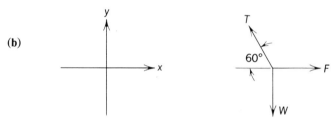

Procedure:

$$T = \sigma \cdot A = (15{,}000 \text{ lb/in.}^2)(0.90 \text{ in.}^2)$$
$$T = 13{,}500 \text{ lb}$$

From $\Sigma F_y = 0$

$$T \sin 60° = W$$

Therefore

$$W = (13{,}500)(0.866) = \textbf{11,700 lb (or 52 kN)}$$

From $\Sigma F_x = 0$

$$F = T \cdot \cos 60° = (13{,}500)(0.5) = 6750 \text{ lb}$$

Therefore, the stress in the horizontal cable is

$$\sigma = \frac{F}{A} = \frac{6750}{0.9} = 7500 \text{ psi (or 51.71 MN/m}^2)$$

Example 1-5:

Given: Two rotating shafts are coupled together by a flange coupling. Four ¾-in. (or 19 mm)-diam bolts carry the load across the connection shown in the figure.

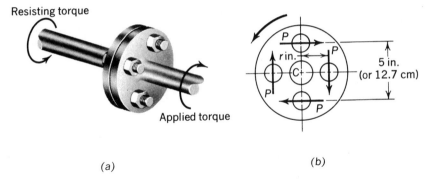

(a)

(b)

Find: The maximum safe torque the coupling can transmit if the allowable shear stress in the bolts is 10,000 psi (or 68.95 MN/m²).

Solution: Area of 1 bolt = $(\pi/4)(3/4)^2$ = 0.442 in.²

Transverse load that 1 bolt can safely carry = $P = \tau \cdot A$ = 4420 lb. At this point we shall assume that each bolt is in pure shear (no bending) and that each one takes one-fourth of the total torque transmitted across the joint.

From $\sum M_c = 0$, we have

Torque = $T = P \cdot r$ ·number of bolts

$$T = (4420)(5/2)(4) = \textbf{44,200 in.-lb (or 59.93 kN-m)}$$

PROBLEMS

1–16 A steel wire is suspended vertically from one end, supporting only its own weight. It has a cross-sectional area A, a length L, and a specific weight γ. Determine (*a*) the maximum axial stress in the wire in terms of A, L, and γ, and (*b*) the maximum length that the wire may have if the maximum permissible stress in the wire is 27,000 psi. Take the specific weight of 490 lb/cu ft.

1–17 A *round* bar in Figure P 1–17 is made up of steel, aluminum, and copper. Determine the radius of each portion necessary in order to carry the given load, if the allowable stresses in the steel, aluminum, and copper are 20,000, 10,000, and 5000 psi, respectively.

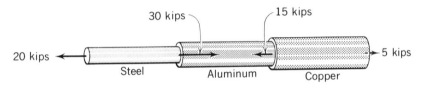

Figure P 1–17

1-18 Determine the minimum cross section the cable in Figure P 1-1 must have if the allowable stress in the cable is 10,000 psi. Check both lengths AC and CD and let the most critical portion control the size for the whole length ACD.

1-19 Consider the loads of Figure P 1-19 to be axially applied at sections A, B, and C to the steel bar which has a cross-sectional area of 3 in.² Determine the axial stress in the bar on (a) a section 10 in. to the right of A, (b) a section 60 in. to the right of A.

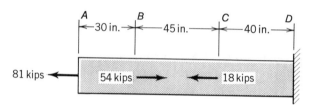

Figure P 1-19

1-20 A short, hollow steel cylinder is to carry an axial compressive load of 50 tons, at a stress of 18,000 psi. Determine the wall thickness necessary for a ratio of outside diameter to wall thickness of 3π to 1.

1-21 Refer to Figure P 1-21. Determine the stress in the truss bar AB ($A = 6 \times 10^{-4} \text{m}^2$).

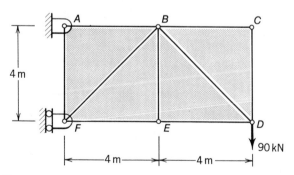

Figure P 1-21

1–22 Figure P 1–22 shows a typical link connection. Neglecting any bending in the pin, determine (a) the diameter of the link with an allowable tensile stress of 20,000 psi, and (b) the diameter of the bolt if
the allowable stress in shear is 10,000 psi.

Figure P 1–22

1–23 Figure P 1–23 represents a typical connection of a mechanism joint.
All the parts are made of steel. If the allowable stresses in tension,
compression, and shear are 22,000, 15,000, and 10,000 psi, respectively,
determine the appropriate size of bars and pins when the load in the
turnbuckle rod is 13 kips.

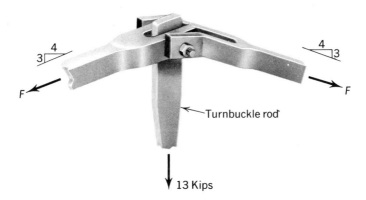

Figure P 1–23

1–24 In Figure P 1–24, a man lifts a 150-lb weight, as shown. Determine the stress in each rod, if they are of a ½-in. diameter.

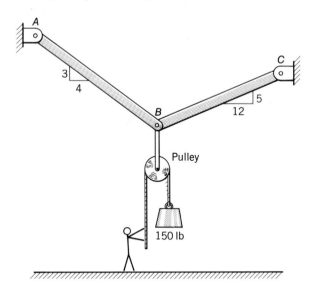

Figure P 1–24

1–25 Determine the percentage variation in stress in the members of the bottom chord of the truss in Figure P 1–7 (that is, members L_0L_1, L_1L_2, and L_2L_3) if the cross-sectional area for each of these members is 0.90 in.^{2}. Use member L_0L_1 as the basis of comparison.

1–26 Two sheets of plywood are spliced together as shown in Figure P 1–26. If the allowable shear stress in the glue is 100 psi, determine the length L of the splice pieces in order to carry the 4800–lb load, assuming uniform shear-stress distribution.

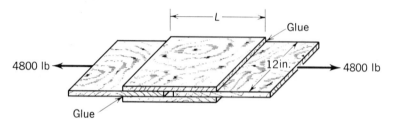

Figure P 1–26

1–27 The torque supplied by the two forces P in Figure P 1–27 is transmitted to the shaft by means of a shear key. Determine the maximum
value of P if the stress in the shear key is not to exceed 10,000 psi.

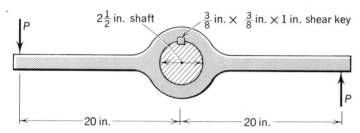

Figure P 1–27

1–28 Determine a size for the pin shown in Figure P 1–28 to resist the
torque created by the forces P when $(a) P = 40$ lb and $\tau = 10,000$
psi in the pin, and $(b) P = 40$ lb and $\tau = 6000$ psi in the pin.

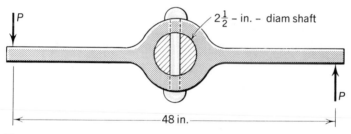

Figure P 1–28

1–29 Figure P 1–29 shows a timber truss. Assuming the allowable compression and shearing stresses to be 1000 and 200 psi, respectively,
determine the dimensions a and b if $P = 10$ kips. Assume that the
top joint is a hinge.

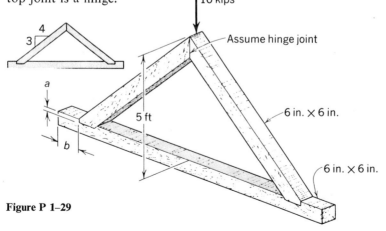

Figure P 1–29

1–30 A 7000-lb weight is supported by means of a pulley as shown in Figure P 1–30. The pulley is supported by the frame *ABC*. Find the required cross-sectional areas for *AC* and *BC* if the allowable stress in tension is 20,000 psi and in compression is 14,000 psi.

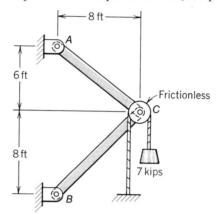

Figure P 1–30

1–31 In Figure P 1–31 determine the cross-sectional area of the cable and of the pulley support if the allowable stress in each is 140 MN/m².

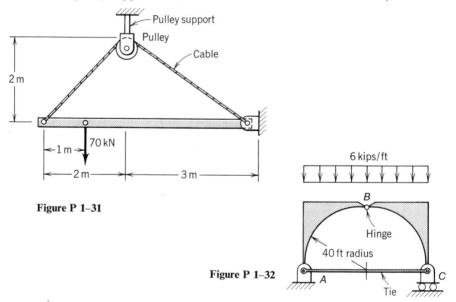

Figure P 1–31

Figure P 1–32

1–32 Refer to Figure P 1–32. If the stress in the steel "tie" is not to exceed 20,000 psi determine the cross-section of area needed for tie.

1—4 Strain

All bodies undergo deformation when subjected to load or temperature changes. The deformation may be a constant for a given load or temperature, or it may be progressive and cumulative, depending upon the magnitude of load, the temperature change, or the temperature at the time the load is applied to the material itself, and perhaps upon the manner in which the load is applied (such as rate of application, for example). Temperature effects are discussed in Section 1–10. For now we shall limit our discussion to deformations caused by loads only. For example, a steel bar of uniform cross section, subjected to an axial load P as shown in Figure 1–6, elongates an amount δ, which increases as the load is increased.

It is reasonable to conclude that this bar would elongate twice as much under the same load if it were twice as long. Therefore, we need a basic unit to represent the deformation in a relative sense. *Strain* is such a unit, defined as the *deformation per unit length*. The strain in Figure 1–6 is in the axial direction, or normal to the cross section of the bar. This strain is thus referred to as *axial* or *normal* strain, symbolically represented by ϵ. It may be represented by

$$\epsilon_x = \lim_{\Delta x \to 0} \frac{\Delta \delta_x}{\Delta x} = \frac{d\delta_x}{dx} \tag{1–3}$$

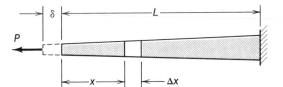

Figure 1–6

The average normal strain for the bar in Figure 1–6 is

$$\epsilon_x = \frac{\delta}{L}$$

In Figure 1–6, the strain is caused by a tensile load. This type is generally considered *positive;* the strain caused by a compressive force is *negative*. This will be the convention adopted in this text.

The deformations discussed thus far are only those normal to cross section in the direction of applied load. Accompanying such axial strains are strains perpendicular to the direction of the load. For example, the bar in Figure 1–6 not only elongates but also reduces in diameter. If the load were a *compressive* type, the deformation would be a shortening (compressive strain) in the axial direction and an increase in the lateral dimension.

We will not discuss this behavior in detail at this time,[2] but note that these deformations represent *changes* in *linear dimensions* (length or width). In addition, there are deformations representing a change of geometry. This is illustrated by Figure 1–7.

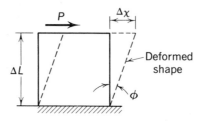

Figure 1–7

If the block in Figure 1–7 is subjected to a *shear* load P as shown, the block will deform, as exaggerated by the dotted outline. The actual deformation is quite small, and therefore

$$\tan \phi = \frac{\Delta x}{\Delta L} \approx \phi$$

This unit angular deformation is defined as the *shearing strain*; in equation form it is

$$\gamma = \frac{\Delta x}{\Delta L} \approx \phi \tag{1–4}$$

We again note that, as in the case of the normal strain, the shearing strain represents the ratio of two quantities of the same dimensions and is therefore a dimensionless quantity. However, the units make the concept meaningful and therefore it will be employed in defining strains, either normal or shearing.

Both normal and shear strains are useful tools in the analysis of structural components. In addition to this, the magnitudes of these strains can be governing criteria in the functioning of a machine part, or some other structure. Let us take the one-panel truss in Figure 1–8 as an example.

Figure 1–8

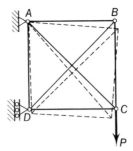

_____

[2] Section 1–8 discusses this phenomenon in greater detail.

As it stands, the forces in the truss members cannot be determined from the equations of simple statics alone. By considering the axial deformations of these bars, however, a solution is possible (Section 1–9 covers this topic). Thus, deformation here serves as a tool for analysis. On the other hand, the magnitude of deflection of, say, point C on the truss might be restricted for perhaps a functional or aesthetic reason. Thus, we must know what this deflection is for a given load. The magnitudes of deformations in the various members now assume importance for this deflection.

Example 1–6:

Given: The three wires shown in Figure *a* are of the same material and have the same cross section. Assume the bar to be rigid, that is, no bending.

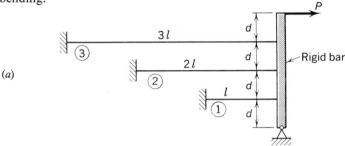

(a)

Find: The ratio of the strains of the three wires, using the short wire as the basis of comparison.

Procedure: Because the bar does not bend, the deformation varies linearly from the hinge on out. This is shown below:

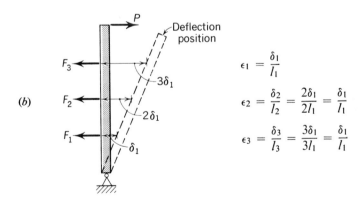

(b)

$$\epsilon_1 = \frac{\delta_1}{l_1}$$

$$\epsilon_2 = \frac{\delta_2}{l_2} = \frac{2\delta_1}{2l_1} = \frac{\delta_1}{l_1}$$

$$\epsilon_3 = \frac{\delta_3}{l_3} = \frac{3\delta_1}{3l_1} = \frac{\delta_1}{l_1}$$

Therefore

$$\frac{\epsilon_2}{\epsilon_1} = \frac{\epsilon_2}{\epsilon_3} = \frac{\epsilon_3}{\epsilon_1} = 1, \cdots$$

Example 1-7:

Given: The strains in the steel rod and wood beam are 1×10^{-3} in./in. and 3.33×10^{-4} in./in., respectively.

Find: The *vertical* deflection of point B. Use a graphical approach.

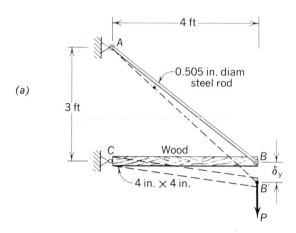

(a)

Procedure: In Figure a, the length of the steel rod is (5 ft) (12 in./ft) = 60 in. The wood beam is (4 ft)(12 in./ft) = 48 in. long. Let δ_{St} = total elongation of the steel rod, and δ_{Wd} = total elongation of the wood beam.

$$\delta_{St} = 1 \times 10^{-3} \times 60 = 6 \times 10^{-2} \text{ in. (or 1.52 mm)}$$

$$\delta_{Wd} = 3.33 \times 10^{-4} \times 48 = 1.6 \times 10^{-2} \text{ in. (or 0.41 mm)}$$

The final length of the steel rod is thus $(60 + 6 \times 10^{-2})$ in. The wood beam has a final length of $(48 - 1.6 \times 10^{-2})$ in.

Because the rod has point A as its center of rotation and the beam has point C as its center of rotation, plotting the new dimensions on a large enough scale would give the deflected location of point B. However, this is not practical, because the magnitude of the chosen scale necessary to reflect the effect of the strains would be too large. Therefore, graphically, only the *changes of lengths* are plotted instead of the lengths themselves. Starting with point B, the rod is assumed to deform down-

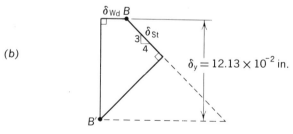

(b)

ward along the slope of the rod an amount δ_{St}. The wood is assumed to contract to the left, relative to point B, by an amount δ_{Wd}. The two ends of the rod and beam must meet at the intersection of the arcs that would be formed by the lengths. (This point may be approximated with perpendicular lines as shown in Figure b). The arbitrarily chosen scale for sketching the deformations is 1 in. $= 10 \times 10^{-2}$ in. Therefore,

$$\delta_y = \mathbf{12.13} \times \mathbf{10^{-2}} \textbf{ in.} \text{ (measured)}$$

PROBLEMS

1–33 A surveyor's chain, 100-ft standard length, is 0.24 in. longer because of an increase in temperature. Determine the average thermal strain for this chain.

1–34 A standard tensile steel specimen, 0.505 in. in diameter, shown in Figure P 1–34, is marked in ½-in. increments, as shown in Figure P 1–34a before the test. After the test, these increments are a little longer than ½-in., as shown in Figure P 1–34b. Determine the strain for each ½-in. length and the average strain for the 2-in. gage length.

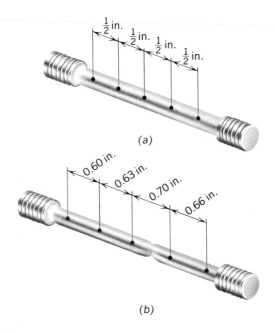

(a)

(b)

Figure P 1–34

1-35 A hard-drawn copper sleeve fits over a steel bolt, as shown in Figure P 1–35, so that when the bolt is tightened, the bolt elongates and the sleeve contracts. The cross-sectional areas are such that the steel elongates only half as much as the copper tube contracts. The positive (tensile) strain in the steel is not to exceed 1×10^{-3} in./in., and the negative (compressive) strain in the copper is not to exceed 1.75×10^{-3} in./in. Determine the maximum number of turns the nut can make if the bolt has 32 threads per in. Assume the deformations in the nut and bolt head to be negligible.

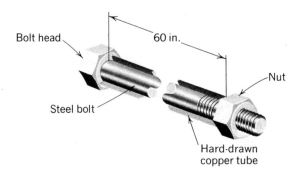

Figure P 1–35

1-36 The force P in Figure P 1–36 lowers the *rigid* arm, which is hinged at the point A and suspended by a rod BC as shown. Points D and E are frictionless pin hinges. When the rigid beam ACD is horizontal the gap at point F is 0.1 in. Determine the strain in rod BC when the gap is 0.2 in.

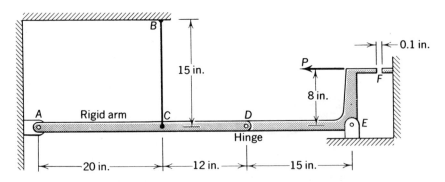

Figure P 1–36

1-37 The horizontal bar of Figure P 1–37 is assumed to be rigid. The two wires supporting the bar are tightened so that when the strain in wire *DE* is 1000 × 10⁻⁶ in./in., the strain in *BC* is 100 × 10⁻⁶ in./in. and the bar is horizontal. Then the load *P* is added and the strain in *DE* becomes 2000 × 10⁻⁶ in./in. Determine the new strain in *BC*.

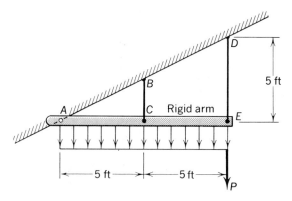

Figure P 1–37

1–38 The handbrakes on a bicycle consist of two hard-rubber bars fixed to the frame of the bike that press against the wheel when actuated for stopping. Figure P 1–38 shows a schematic plan view of this arrangement, and the dotted lines are the form the rubber might take when the brakes are applied. Determine the shearing strain in the rubber if the deformation is assumed linear from the fixed plane.

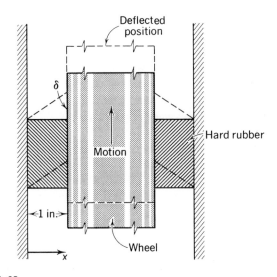

Figure P 1–38

1–39 Figure P 1–39 shows the schematic of a wheel-brake arrangement. Assume that the shearing strain in the wheel varies linearly from zero at r_1 (shaft radius) to γ_2 at r_2. Determine the strain at any radius on the wheel.

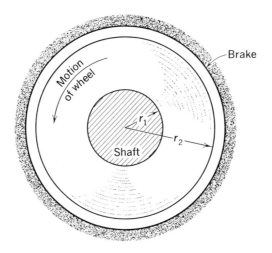

Figure P 1–39

1—5 Elastic Deformation and Hooke's Law

All materials deform when subjected to load.[3] For most materials a *change in load* results in a *corresponding*, but not necessarily linear, *change in deformation*. Furthermore, most materials that we are concerned with, such as metal, wood, or concrete, tend to regain their original shape at normal temperatures after the removal of a load, if the load is not excessive. If, upon the removal of the load, a body returns to its original size and shape, the body has undergone *elastic deformation*. The ability of a body to regain its original shape is known as *elasticity*. If the body does not completely recover its original shape, it is said to be *partially elastic;* it is *perfectly* elastic when full recovery takes place.

In his 1678 treatise *Ut tensio sic vis*, Robert Hooke (1635–1703) recognized a definite relationship between elastic deformation and load. Generalized in its simplest form, Hooke's law says that *stress is proportional to strain*. In 1807, Thomas Young (1773–1829) introduced a constant of proportionality, *the modulus of elasticity, or Young's modulus*, which is a measure of stiffness of the material. It represents the ratio of stress to

---

[3] Although this holds true for solids, liquids, or gases, in this book we are concerned almost exclusively with solids. Therefore, unless specified otherwise, the word *material* will mean solid materials only.

strain. For tension or compression, the modulus of elasticity, designated by E, is expressed in equation form as

$$E = \frac{\sigma}{\epsilon} \qquad (1\text{--}5a)$$

For shear stresses and strains, the proportionality constant is G (sometimes called the modulus of rigidity, or occasionally the torsional modulus). In equation form it is

$$G = \frac{\tau}{\gamma} \qquad (1\text{--}5b)$$

There is an important relation between the constants E and G, and another constant not yet discussed, Poisson's ratio μ. The relation is

$$G = \frac{E}{2(1 + \mu)} \qquad (1\text{--}6)$$

In Equation (1–6),[4] μ is always positive and ranges between 0 and ½.[5] Therefore, from Equation (1–6), we see that G is anywhere from one-third to one-half the magnitude of E. This agrees well with experimental findings.

For most of the metallic materials we encounter in design, the E and G are almost constant, regardless of the quality of the material. For example, the modulus of elasticity E is about 30×10^6 psi (or 200×10^6 kN/m²) for both structural-grade steel (ultimate strength 60,000 psi or 414 MN/m²) and high-strength steel (ultimate strength up to 300,000 psi or 2060 MN/m²). Similarly, the value of G for steel is about 12×10^6 psi (or 83×10^6 kN/m²), regardless of grade.

For concrete, however, this is not true. The E value varies with strength and the material content and even within the elastic range of one sample, because the relation of stress to strain is not linear. For such cases, the value of E is approximated at some specified stress level where the strain is known.[6]

From Equation (1–5a) we derive a rather important expression for axial deformation: δ_a. From Equation (1–3), $d\delta_x = \epsilon_x\,dx$. Also, from Equation (1–5a), $\epsilon_x = \sigma_x/E$. Therefore, $d\delta_x = (\sigma_x/E)\,dx$, and over a length L,

$$\delta_x = \int_0^L \frac{\sigma_x}{E}\,dx \qquad (1\text{--}7)$$

If P is a constant throughout the length L, and if A and E are also constant, the total deformation is

$$\delta_x = \frac{PL}{AE} \qquad (1\text{--}7a)$$

---

[4] This equation is derived in Section 1–18.
[5] Poisson's ratio is discussed in Section 1–8.
[6] This is further discussed in Section 1–7.

where P = axial load

L = length of member being loaded

A = cross-sectional area of member

E = modulus of elasticity of material of member

Similarly, the relationship between the shearing deformation δ_s and shearing force P (frequently represented by V) can be derived from Equation (1–5b). *Assuming constant P, L, A,* we have

$$G = \frac{\tau}{\gamma} = \frac{P/A}{\delta_s/L} = \frac{PL}{\delta_s A}$$

Rearranging terms, we get

$$\delta_s = \frac{PL}{AG} \tag{1–8}$$

where P = shearing force

A = area of element subjected to shear force

L = length of element perpendicular to the direction of shear force

G = modulus of rigidity

The association of these expressions with schematic diagrams is shown in Figures 1–9a and b.

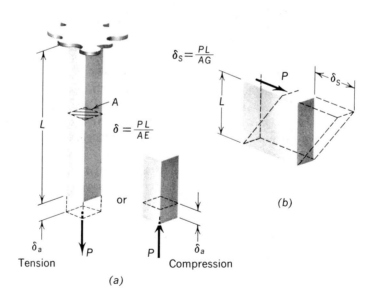

Figure 1–9 Forces and deformations.

Example 1—8:

Given: A steel wire of uniform cross-sectional area A hangs vertically under its own weight from one end, as shown in the figure.
Find: The total elongation δ of the wire.

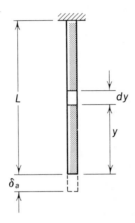

Solution: The stress in the wire is obviously caused by the weight of the wire only, since there is no external load at the lower end. Thus, at the lower end, the stress is zero and varies to a maximum at the support. At some distance y, the force is $wA \cdot y$, where w represents the weight per unit volume. For a small element of length dy, the elongation $d\delta$ can be obtained from Equation (1–7).

$$d\delta = \frac{P(dy)}{A \cdot E} = \frac{(wAy)(dy)}{AE} = \frac{w \cdot y \cdot dy}{E}$$

Therefore, the total elongation δ is

$$\delta = \int_0^l \left(\frac{w}{E}\right)ydy = \frac{w}{E}\int_0^l ydy = \frac{wl^2}{2E}$$

Recognizing that $w \cdot A \cdot l$ represents the total weight W of the wire, the above deformation may be written as

$$\delta = \frac{(wAl)l}{2AE} = \frac{Wl}{2AE}$$

Example 1—9:

Given: A concrete column 18-ft (or 5.49-m) high supports a 50-ton (or 444.82 kN) load. Its square dimension at the top is 14 in. $\times$ 14 in. and at the bottom 20 in. $\times$ 20 in. as shown in Figure *a*.
Find: The total compressive deformation in the column, neglecting the weight of the column. Assume that $E_C = 3 \times 10^6$ psi (or 20×10^6 kN/m²).

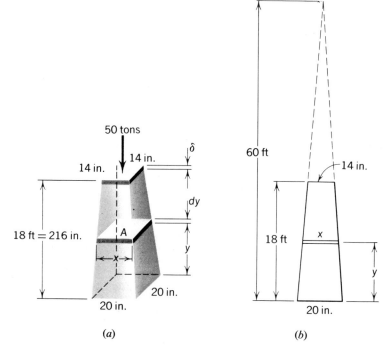

(a)　　　　　　　　　　　(b)

Solution: For an element of length dy,

$$d\delta = \frac{P\,dy}{AE}$$

where A varies from 14 in. × 14 in. = 196 in.² at the top to 20 in. × 20 in. = 400 in.² at the bottom. The area at the y level is x^2. From similar triangles,

$$\frac{20\text{ in.}}{60\text{ ft}} = \frac{x\text{ in.}}{(60 - y)\text{ ft}}$$

$$x = (20 - y/3)\text{ in.}$$

$$A = x^2 = (20 - y/3)^2$$

Therefore

$$d\delta = \frac{P\,dy}{(20 - y/3)^2 E} = \frac{100{,}000\,dy}{3 \times 10^6 \times (20 - y/3)^2}$$

$$\delta = \frac{9 \times 100{,}000}{3 \times 10^6} \int_0^{18\text{ ft}} \frac{dy}{(60 - y)^2}$$

$$\delta = 3 \times 10^{-1}\left[\frac{1}{(60 - y)}\right]_0^{18\text{ ft}} = 3 \times 10^{-1}\left[\frac{1}{(60 - 18)} - \frac{1}{(60 - 0)}\right]$$

$$\delta = 3 \times 10^{-1}[0.0238 - 0.0167] \times 12\text{ in.}/\text{ft} = \mathbf{0.0258\ in.\ (or\ 0.65\ mm)}$$

Example 1-10:

Given: The 20-ft (or 6.1-m) bar is made up of one aluminum and one steel length, connected and loaded as shown in Figure *a*.

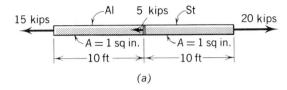

(a)

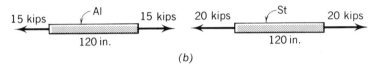

(b)

Find: (a) The elongation of the 20-ft length, and (b) the ratio of elongation of the aluminum to that of steel ($E_{Al} = 10 \times 10^6$ psi, $E_{St} = 30 \times 10^6$ psi).

Solution: Each section may be cut and set in equilibrium.

$$\delta_{Al} = \frac{15,000 \times 120}{1 \times 10 \times 10^6} \qquad \delta_{St} = \frac{20,000 \times 120}{1 \times 30 \times 10^6}$$

$$\delta_{Al} = 18 \times 10^{-2} \text{ in.} \qquad \delta_{St} = 8 \times 10^{-2} \text{ in.}$$

Because the deformations are both positive (elongation), the deformation of the 20-ft length is

(a) $\delta_{tot} = (18 + 8) \times 10^{-2} = \textbf{26} \times \textbf{10}^{-2}$ **in. (or 6.64 mm)**

(b) Ratio: $\dfrac{\delta_{Al}}{\delta_{St}} = \dfrac{18 \times 10^{-2}}{8 \times 10^{-2}} = \textbf{2.25}$

PROBLEMS

1-40 The steel specimen in Figure P 1-40 is used in a tension test. At what load P will the mechanical gage read 0.002-in. deformation? Assume $E = 30 \times 10^6$ psi.

Figure P 1-40

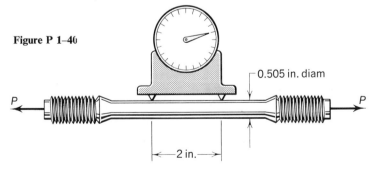

1–41 An aluminum bar, 2-in. × ½-in. cross section, is subjected to a tensile load of 16,000 lb. At this load, the axial strain in the bar is 1650×10^{-6} in./in. Assuming elastic deformation, determine the modulus of elasticity in tension (E) for the aluminum specimen.

1–42 The two rods in Figure P 1–42 support a uniform load of 1000 lb/ft, including the weight of the beam and a concentrated load P. If the cross section is the same for both, derive expressions for x, the position of P, in terms of P and l, assuming the beam to remain horizontal, if (a) the two bars are made of steel, and (b) the longer is steel, and the other aluminum.

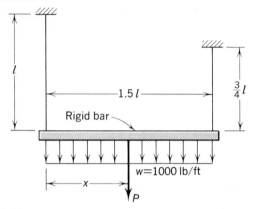

Figure P 1–42

1–43 The vertical bar in Figure P 1–43 is made of aluminum and steel. Determine the ratio of the cross-sectional area of steel to aluminum if the two lengths are to deform equally. Neglect the weight of the bars.

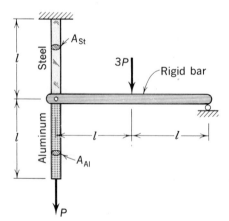

Figure P 1–43

1–44 The steel bar in Figure P 1–44 has a circular cross section, 1 in. diameter at the top, ½ in. at the bottom. (*a*) Determine the value of *P* for which the maximum stress is 15,000 psi. Assume the unit weight of steel of 490 lb/cu ft. (*b*) Determine the total deformation of the bar when the load *P* is 16,000 lb.

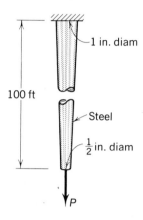

1 in. diam

100 ft

Steel

$\frac{1}{2}$ in. diam

P

Figure P 1–44

1–45 Determine the vertical distance that the 45-kN load in Figure P 1–45 travels if the arm *CD* was originally horizontal. Assume the member *BCD* to be rigid.

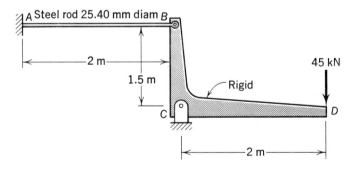

A Steel rod 25.40 mm diam B

2 m

1.5 m

45 kN

Rigid

C

D

2 m

Figure P 1–45

1–46 Determine the stress and deformation in each bar of the cantilever truss shown in Figure P 1–46. Assume the modulus of elasticity of the bars to be $E = 30 \times 10^6$ psi, and the cross-sectional area of each bar to be 4 sq in.

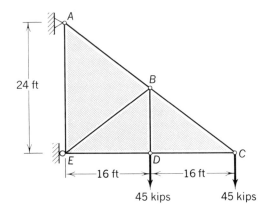

Figure P 1–46

1–47 A post is guyed by a steel wire, as shown in Figure P 1–47. Assuming the post to be rigid (no bending) and neglecting the axial deformation of the post, determine the horizontal sway of the top of the post caused by the uniform load as shown. For the steel wire, $E = 30 \times 10^6$ psi.

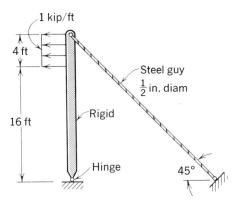

Figure P 1–47

1–48 How far apart will the supports A and B in Figure P 1–48 spread when the 20-kip load is lifted with the pulley arrangement shown? Neglect the bending of bar AC and BC.

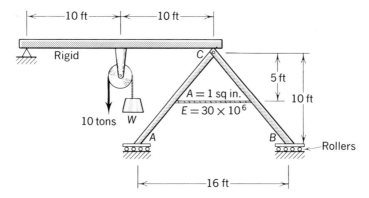

Figure P 1–48

1–49 A homogeneous slender rod of uniform cross-sectional area A rotates with a constant angular velocity ω rad/sec about a fixed vertical axis through one end. This is shown in Figure P 1–49. If γ is the weight per unit volume of rod, determine the total elongation of the rod.

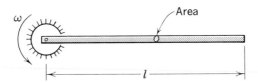

Figure P 1–49

1–50 For Figure P 1–50, determine the deflection of point C. Assume the beam to be rigid.

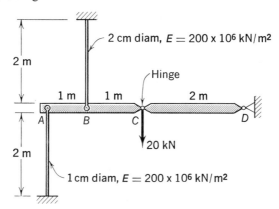

Figure P 1–50

1–51 Determine the size (diameter) of the steel tie shown in Figure P 1–51 so that (a) the stress in the tie is 15,000 psi, and (b) the rotation of *AB* about *A* is 0.05 rad.

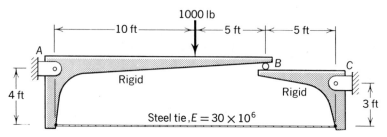

1000 lb

10 ft — 5 ft — 5 ft

A

Rigid

4 ft

B

C

Rigid

3 ft

Steel tie, $E = 30 \times 10^6$

Figure P 1–51

1–52 Determine the vertical deflection of point *B* in Figure P 1–52. Neglect the weight of the members. (See Example 1–7.)

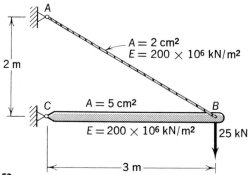

A

$A = 2$ cm^2
$E = 200 \times 10^6$ kN/m^2

2 m

C $A = 5$ cm^2 B

$E = 200 \times 10^6$ kN/m^2 25 kN

— 3 m —

Figure P 1–52

1–53 Steel bands are used to tie wooden staves to form a barrel, as shown in Figure P 1–53. Neglecting the compressive deformation in the staves, determine the stress in the steel bands as a band is forced down from a spot where the diameter is 5.000 ft to one where the diameter is 5.005 ft.

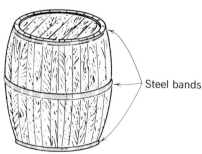

Steel bands

Figure P 1–53

1-54 Determine the deflection of point C, Figure P 1–54, as a result of the deformation of the two supports. Assume the horizontal beam to be rigid (no bending).

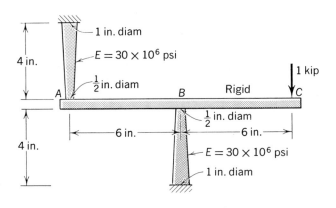

Figure P 1-54

1-55 Through how many turns must the nut at point C (Figure P 1–55) be turned to create a stress of 18,000 psi in rod AB? The nut has 8 threads per in.; $E = 30 \times 10^6$ psi for all the rods. [*Hint:* $\sum F_x = 0$ and deformation considerations provide the necessary equation for the solution.]

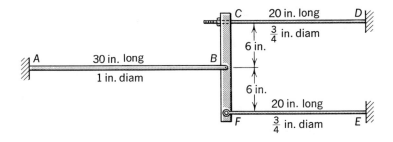

Figure P 1-55

1–56 In Figure P 1–56 determine (a) the *strains* in the two rods (area = 1 in.²), and (b) the total movement of the 400-lb force.

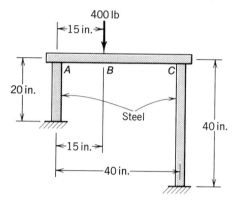

Figure P 1–56

1–57 An axial load P is hung on the end of the bar as shown in Figure P 1–57. This load causes the lower end of the bar to elongate 0.03 in. Determine the load P. Note that $E_{Cu} = 15 \times 10^6$ psi.

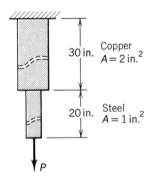

Figure P 1–57

1–58 Determine the vertical travel of point C for the condition given in Figure P 1–58. The beam is rigid (it will not bend) and the *steel* rope (A = 1 in.²) passes over *frictionless* supports as shown. (Approximate the original length of the cable to the nearest foot.)

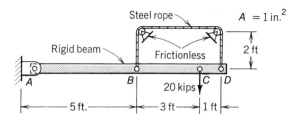

1-59 All members in the frame in Figure P 1–59 have a cross-sectional area of 3 cm² and are steel. Determine how much point C will move to the right when the 1000-kg load is applied as shown.

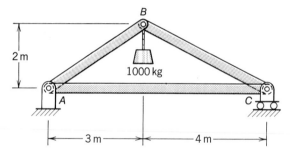

Figure P 1–59

1-60 Referring to Figure P 1-60, and assuming member AC and DE as rigid (DO NOT BEND), determine the elongation of (a) Members BG. and (b) CD as a function of P and AE (the EA is the same for members CD and BG.)

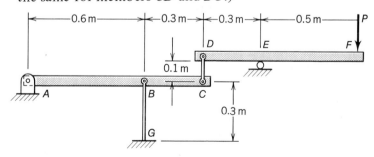

Figure P 1–60

1—6 Plastic Deformation

In the previous section we established that, within the elastic limit of the material, there is no permanent deformation of the material. Also, for most materials with which we most commonly deal, stress is proportional to strain up to a limit, that is, *a proportional limit* of the material. Now we wish to investigate the behavior of a material beyond this limit, in the *inelastic* or *plastic* range.

Figure 1–10 gives the typical shape of the load-deformation (or stress-strain) relationship for many materials at the "early" stages of static load. When first loaded, the specimen deforms somewhat linearly (constant slope) to the proportional limit. However, as the loading is increased, the

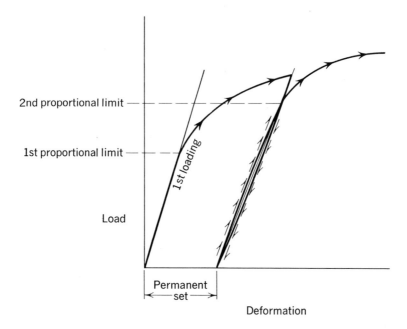

Figure 1–10

deformation increases at a faster rate, decreasing the slope of the curve from that in the linear range. If the load is released, the deformation is not zero; rather a plastic or permanent deformation is retained, as indicated in Figure 1–10. The slope of the load-deformation curve during the unloading and reloading cycle is nearly the same as the slope of the curve during the original loading. For small plastic deformations, the unloading curve is straight and parallel to the "elastic" curve. For large plastic deformations, the unloading curve will be slightly curved, with its slope somewhat smaller than that of the elastic curve. When reloading, however, the opposite curvature is obtained, forming what is commonly referred to as a *hysteresis loop*.

The resulting proportional limit is appreciably higher than that of the elastic segment, a beneficial effect taken into account in the utilization of springs, extrusions, and various cold-working processes.

The amount of plastic deformation before rupture may be used as an indication of the brittleness or ductility of a material. If the plastic deformation at failure is quite small, the material is called *brittle;* it is called *ductile* if the plastic deformation is very large. These terms should not be confused with *stiffness*, the ability of a material to resist deformation. The modulus of elasticity is a measure of stiffness.

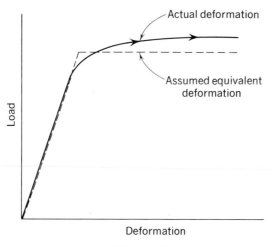

Figure 1–11

Some materials deform large amounts without an appreciable increase in load after the yield point is reached, as shown in Figure 1–11. The actual curve, represented by the solid line, is frequently replaced by the dashed line for the purpose of idealizing, and thus simplifying the plastic analysis approach sometimes used in present-day design.

1—7 Stress-Strain Diagram

A good pictorial description of some useful mechanical properties of a material is given by a graph known as the *stress-strain diagram*. It is obtained by plotting stress versus strain from an axial loading, which is either tension or compression. Dividing the total deformation by the length over which the deformation is measured, we obtain the strain; dividing the load by the cross-sectional area, we obtain the stress. Strain

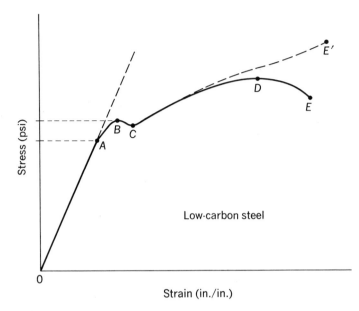

Figure 1–12

is plotted as the abscissa and the corresponding stress as the ordinate. Figure 1–12 gives the shape of a typical stress-strain diagram for a low-carbon steel specimen in tension. This graph should prove helpful in making more meaningful some of the terms used in connection with mechanical properties of materials.

From 0 to *A* the stress is proportional to strain, satisfying Hooke's law quite well within this range. If the load is increased past *A*, we note that the strain increases more rapidly, resulting in a curve that deviates from a straight line. This point of departure is known as the *proportional limit*, the point where the proportionality begins to disappear. A slight increase in load brings us to point *B* at which there is an appreciable increase in strain with no appreciable increase in load. For mild steels (low-carbon steels, structural steels, and so forth) we sometimes can detect a decrease in load even though the elongation keeps increasing. The result is a slight dip in the curve, indicated by point *C*. The region between points *B* and *C* is the *yielding region;* the magnitude of stress at points *B* or *C* is known as the *yield stress, yield strength,* or *yield point.* One may further classify point *B* as the upper yield and point *C* as the lower yield, whenever they exist. For most materials, however, the dip is neither detectable nor existent. Such a case is shown in Figure 1–13. The yield point for such a material is defined as the intersection of the stress-strain curve with line *BC*. Line *BC* is parallel to the straight portion of the curve; the distance 0–*C* is usually specified and varies for different materials and sometimes

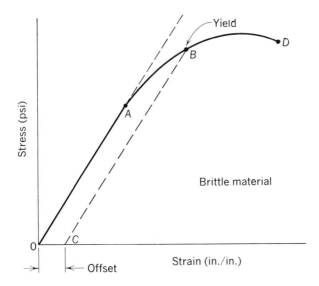

Figure 1–13

even for the same material. For steel, as an example, this offset usually ranges between 0.1–0.2 percent in./in. (that is, 0.001–0.002 in./in.).

The *ultimate strength* is represented by the highest point on the diagram. For low-carbon steels, such as typified by Figure 1–12, this point is appreciably higher than the yield point, perhaps 1½ to 1¾ as high. For the higher carbon steels, the difference is lower.

For brittle metals, the ultimate strength and the *rupture strength* are represented by the same point (see point D in Figure 1–13). For the more ductile material, Figure 1–12 for example, the specimen begins to neck down at some point along its length when the ultimate load is reached (point D, Figure 1–12), resulting in an increase in axial deformation and a decrease in load; hence, we note a dip in the curve before rupture. If we were to measure the actual cross-sectional area once the necking begins at each value of load, we would get the *true stress*, represented by the dashed curve in Figure 1–12. Instead, the solid curve between points D and E represents a *nominal stress*, which is lower than the *actual* or *true stress*.

1—8 Poisson's Ratio

As mentioned previously, axial elongation is always accompanied by a lateral contraction (and vice versa). The ratio of the lateral strain to the axial strain is expressed by a constant known as the *Poisson's ratio,* after the French mathematician, Simeon Poisson (1781–1840), who identified its existence in 1811. Represented by μ, Poisson's ratio is

$$\mu = \frac{-\text{ lateral strain}}{\text{axial strain}} = \frac{-\epsilon \text{ lateral}}{\epsilon \text{ axial}} \qquad (1\text{--}9)$$

For most metals, Poisson's ratio varies from $\frac{1}{4}$ to $\frac{1}{3}$. For the *isotropic* materials, materials that have the same properties in all directions, Poisson designated a value of $\mu = \frac{1}{4}$, although most metals show that the actual measured values are closer to $\frac{1}{3}$ than to $\frac{1}{4}$. Poisson's ratio remains about constant during the elastic straining of the material, and increases as the specimen yields. For steel, as an example, μ varies from about 0.3 in the elastic range up to perhaps 0.5 in the plastic region, values which may be verified experimentally.

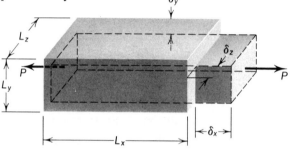

Figure 1–14

Knowing Poisson's ratio and the modulus of elasticity E of a material, the deformation of the material, and subsequently its volume change, can be calculated. To illustrate this, assume that a specimen is placed in tension, as shown in Figure 1–14. The axial strain ϵ_x is δ_x/L, the transverse strain $\epsilon_y = -\mu\epsilon_x$ (contraction, thus the minus). Similarly, $\epsilon_z = -\mu\epsilon_x$. Letting L_y and L_z represent the original cross-sectional dimensions of the specimen, we have

$$\epsilon_x = \frac{+\delta_x}{L_x}$$

$$\epsilon_y = \frac{-\delta_y}{L_y} = -\mu\epsilon_x$$

$$\epsilon_z = \frac{-\delta_z}{L_z} = -\mu\epsilon_x$$

The original volume is

$$V_0 = L_x \cdot L_y \cdot L_z$$

The new length is

$$L_x + \delta_x = L_x + L_x\epsilon_x = L_x(1 + \epsilon_x)$$

Similarly, the new cross-sectional dimension in the z direction is

$$L_z - \delta_z = L_z - L_z \cdot \mu \epsilon_x = L_z (1 - \mu \epsilon_x)$$

In the y direction, the new dimension is

$$L_y - \delta_y = L_y - L_y \mu \epsilon_x = L_y (1 - \mu \epsilon_x)$$

Thus, the final volume is

$$V_f = L_x (1 + \epsilon_x) \cdot L_y (1 - \mu \epsilon_x) \cdot L_z (1 - \mu \epsilon_x) = V_0 (1 + \epsilon_x)(1 - \mu \epsilon_x)^2$$

Expanding the right-hand factors and neglecting the higher powers of ϵ (because ϵ is a very small quantity in comparison to unity), the final volume is $V_f = V_0(1 + \epsilon_x - 2\mu \epsilon_x)$. The change in volume $\Delta V = V_f - V_0$, or

$$\Delta V = V_0 \epsilon_x (1 - 2\mu) \tag{1–10}$$

The *unit volume change* is

$$\frac{\Delta V}{V_0} = \epsilon_x (1 - 2\mu) \tag{1–10a}$$

Because it is unlikely that a material will undergo a diminishing change in volume, from Equation (1–10) we can say that $(1 - 2\mu) \geq 0$. Thus, it is most likely that $\mu \leq \frac{1}{2}$. The value of $\frac{1}{2}$ is approached by some materials, such as plastics and rubber. For steel, it is about 0.3, although for concrete, μ varies between 0.1 and 0.2.

For biaxial or triaxial stress (stresses in the x, y, and z directions) Poisson's ratio permits the development of useful expressions relating stresses and strains in the elastic region.

Assuming the material is isotropic (the material has the same properties in all direction), the unit deformations are as follows:

$$\epsilon_x = \frac{\sigma_x}{E} - \frac{\mu}{E}(\sigma_y + \sigma_z) \tag{1–11a}$$

$$\epsilon_y = \frac{\sigma_y}{E} - \frac{\mu}{E}(\sigma_x + \sigma_z) \tag{1–11b}$$

$$\epsilon_z = \frac{\sigma_z}{E} - \frac{\mu}{E}(\sigma_x + \sigma_y) \tag{1–11c}$$

The expressions for stress are

$$\sigma_x = \frac{E}{1 + \mu}\left[\epsilon_x + \frac{\mu}{1 - 2\mu}(\epsilon_x + \epsilon_y + \epsilon_z)\right] \tag{1–12a}$$

$$\sigma_y = \frac{E}{1 + \mu}\left[\epsilon_y + \frac{\mu}{1 - 2\mu}(\epsilon_x + \epsilon_y + \epsilon_z)\right] \tag{1–12b}$$

$$\sigma_z = \frac{E}{1 + \mu}\left[\epsilon_z + \frac{\mu}{1 - 2\mu}(\epsilon_x + \epsilon_y + \epsilon_z)\right] \tag{1–12c}$$

Experimentally, ϵ_x, ϵ_y, and ϵ_z can be measured directly by strain gages. Thus, knowing E and μ for a material, we are able to determine the stresses in that material from Equations (1–12a), (1–12b) and (1–12c).

Example 1–11:

Given: A nominal 4-in. (or 10.16 cm)-diam, solid steel piston fits into a steel ring. The inward thrust on the piston is 75,000 lb (or 334 kN), the outward pull is 50,000 lb (or 222.40 kN).

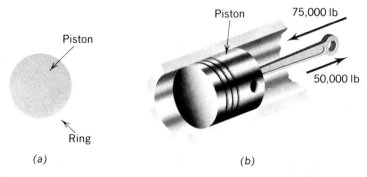

(a)   (b)

Find: The total radial deformation of the piston. Assume $\mu = 0.3$, $E = 30 \times 10^6$ psi.

Procedure: The inward thrust causes a radial expansion, the outward pull causes a radial contraction of the piston. The corresponding stresses causing these deformations are, for inward thrust

$$\sigma_x = \frac{P}{A} = \frac{75,000}{(\pi/4)(4)^2} = 6000 \text{ psi, } compression$$

for outward pull

$$\sigma_x = \frac{P}{A} = \frac{50,000}{(\pi/4)(4)^2} = 4000 \text{ psi, } tension$$

Let ϵ_x = axial strain and ϵ_r = radial or lateral strain. Then for inward thrust

$$\epsilon_r = -\mu\epsilon_x = -0.3\frac{\sigma_x}{E} = (-0.3)\left(\frac{-6000}{30 \times 10^6}\right)$$

$$\epsilon_r = +6 \times 10^{-5} \text{ in./in. (expansion)}$$

and for outward pull

$$\epsilon_r = (-0.3)\left(\frac{+4000}{30 \times 10^6}\right) = -4 \times 10^{-5} \text{ in./in. (contraction)}$$

Thus, the maximum unit deformation = $(4 + 6) \times 10^{-5}$ in./in. For the 4-in. diameter, the total deformation of the piston is

$$\delta_r = 4 \cdot (4 + 6) \times 10^{-5} = \mathbf{40 \times 10^{-5} \text{ in. (or } 1.01 \times 10^{-2} \text{ mm)}}$$

PROBLEMS

1-61 An aluminum block 2 in. × 2 in. × 15 in. long is subjected to a compressive axial load of 40,000 lb. Assuming $\mu = \frac{1}{3}$, $E = 10 \times 10^6$ psi, find (a) the new dimensions of the block, and (b) the unit change in volume.

1-62 A 2-ft-long bar has a cross section 2 in. × t in. Subjected to an axial stress, the bar elongates 12×10^{-3} in. The lateral deformation for the 2-in. side is 3×10^{-4} in., and for the other side, the deformation is 1.875×10^{-4} in. Determine the thickness t of the bar.

1-63 For steel the following are commonly assumed values: $E = 30 \times 10^6$ psi, $\mu = 0.3$, $\gamma = 0.284$ lb/in.³ Assuming the bar in Figure P 1-63 to be prismatic, show that the increase in volume is

$$\Delta V = 1.33 \times 10^{-8} \, Al(\sigma_x + 0.142l) = \text{in.}^3$$

where l = inches, A = inches squared, and $\sigma_x = P/A$, psi.

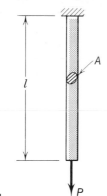

Figure P 1–63

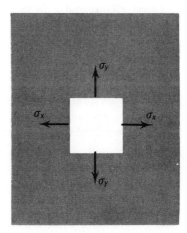

Figure P 1–64

1-64 Figure P 1-64 shows an element subjected to biaxial stresses σ_x and σ_y. Assuming the material is prismatic, show that the unit strain in the z direction is

$$\epsilon_z = -\mu/E \cdot (\sigma_x + \sigma_y)$$

1–65 Figure P 1–65 shows an element subjected to triaxial stresses σ_x, σ_y, σ_z. Show that the strains in the x, y, and z directions are

$$\epsilon_x = 1/E[\sigma_x - \mu(\sigma_y + \sigma_z)]$$
$$\epsilon_y = 1/E[\sigma_y - \mu(\sigma_x + \sigma_z)]$$
$$\epsilon_z = 1/E[\sigma_z - \mu(\sigma_x + \sigma_y)]$$

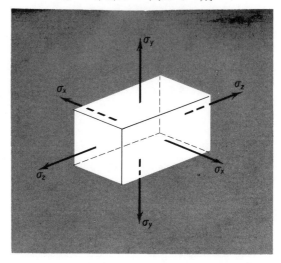

Figure P 1–65

1–66 The strains ϵ_x and ϵ_y in the x and y directions, respectively, in a steel plate subjected to biaxial stresses σ_x and σ_y were measured with *SR*–4 electric-resistance strain gages: $\epsilon_x = 1 \times 10^{-3}$ in./in. and $\epsilon_y = 0.5 \times 10^{-3}$ in./in. For $\mu = 0.3$, $E = 30 \times 10^6$ psi, determine σ_x and σ_y.

1–67 Derive an expression for the strain ϵ_z in the z direction in terms of ϵ_x and ϵ_y, and μ for a case of biaxial stresses σ_x and σ_y. [Hint: Refer to Problem 1–64 and Equations (1–12a) and (1–12b).]

1—9 Statically Indeterminate Axially Loaded Members

There are numerous combinations of structural members for which the equations of equilibrium alone are not sufficient to determine uniquely the internal forces in these members. For such structures, the number of unknown internal forces exceed the number of independent equations of equilibrium. These cases are *statically indeterminate* and we must supplement the equations of equilibrium by additional equations, which depend upon the *geometry* of the structure. In later chapters we consider indeterminate structures in torsion and bending; here we shall be concerned with indeterminate members subjected only to axial load.

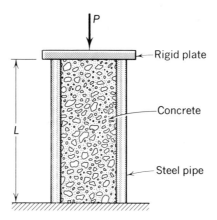

Figure 1–16

Consider the pipe column filled with concrete, subjected to an axial load P through the rigid plate, as shown in Figure 1–16. The load P is obviously distributed to both the concrete and the steel pipe, but equilibrium alone does not provide enough equations to determine the portion of P that the concrete takes and that which the pipe takes. From statics, the only applicable equations are

$$\Sigma F_y = 0; \qquad P_{Co} + P_{St} = P \qquad (a)$$

where P_{Co} = force taken by the concrete and P_{St} = force taken by the steel pipe.

So far we have only one equation and two unknowns, P_{Co} and P_{St}. To provide another independent equation containing these two unknowns, we make use of the geometry of deformation; that is, we note that because the plate is rigid, the deformation of the steel pipe is the same as that of the concrete:

$$\delta_{Co} = \delta_{St}$$

or,

$$\frac{P_{Co}L_{Co}}{A_{Co}E_{Co}} = \frac{P_{St}L_{Co}}{A_{St}E_{St}} \qquad (b)$$

Solving Equations (a) and (b) simultaneously, the values of P_{Co} and P_{St}, in terms of the total load P, may be found.

It is important to note that the deformation relationship, indicated by Equation (b), does not apply when the stress exceeds the proportional limit. For cases where the stresses exceed the proportional limit, the stress-strain relationship may be obtained directly from the stress-strain diagram. For example, the deformations are still equal so that

$$\delta_{Co} = \delta_{St}$$

from which

$$\epsilon_{Co}L_{Co} = \epsilon_{St}L_{St} \tag{c}$$

From the stress-strain diagram for the respective materials corresponding to the strains ϵ_{Co} and ϵ_{St}, we scale off the corresponding stresses σ_{Co} and σ_{St}, respectively. Equation (a) would still have to apply, that is,

$$\sigma_{Co}A_{Co} + \sigma_{St}A_{St} = P$$

The solution is a trial-and-error one.

The following examples should prove helpful in illustrating the procedure followed for the solutions of statically indeterminate problems. In this respect, it might be mentioned that a free-body diagram is frequently helpful.

Example 1–12:

Given: For the pipe column of Figure 1–16 the following data are given: $L = 2$ ft (or 609.60 mm); outside diameter of pipe, $d_o = 5.563$ in. (or 141.30 mm); inside diameter, $d_i = 5.047$ in. (or 128.20 mm), $E_{St} = 30 \times 10^6$ psi (or 200×10^6 kN/m²), $E_{Co} = 3 \times 10^6$ psi (or 20×10^6 kN/m²), $P = 50$ kips.

Find: (a) The stresses in concrete and pipe, and (b) the shortening of the column.

Solution: The relationship $\sum F_y = 0$ yields Equation (a).

$$P_{Co} + P_{St} = P \tag{a}$$

The deformation relationship yields Equation (b) if the deformation is assumed elastic.

$$\frac{P_{Co}L_{Co}}{A_{Co}E_{Co}} = \frac{P_{St}L_{St}}{A_{St}E_{St}} \tag{b}$$

$$A_{Co} = \frac{\pi}{4}(5.047)^2 = 20.1 \text{ in.}^2$$

$$A_{St} = \frac{\pi}{4}(5.563)^2 - A_{Co} = 24.4 - 20.1 = 4.3 \text{ in.}^2$$

Recognizing that $L_{Co} = L_{St}$, and substituting values in Equation (b), we obtain a relation between P_{Co} and P_{St}:

$$\frac{P_{Co}}{(20.1)\cdot(3 \times 10^6)} = \frac{P_{St}}{(4.3)\cdot(30 \times 10^6)}; \quad P_{St} = \textbf{2.14 } P_{Co}$$

Substituting this into Equation (a), we get

$$P_{Co} + 2.14\,P_{Co} = 50,000$$
$$P_{Co} = 15,900 \text{ lb}$$

Thus, $P_{St} = 50,000 - 15,900 = 34,100$ lb.

$$\sigma_{Co} = \frac{P_{Co}}{A_{Co}} = \frac{15,900}{20.1} = \textbf{780 psi (or 5378.10 kN/m}^2\textbf{)}$$

$$\sigma_{St} = \frac{P_{St}}{A_{St}} = \frac{34,100}{4.3} = \textbf{7900 psi (or 54,470.50 kN/m}^2\textbf{)}$$

The deformation $\delta = \delta_{Co} = \delta_{St} = \dfrac{PL}{AE} = \left(\dfrac{L}{E}\right)_{Co} \cdot \sigma_{Co} = \left(\dfrac{L}{E}\right)_{St} \cdot \sigma_{St}$

Therefore, $\delta_{St} = (7900) \cdot \dfrac{24}{30 \times 10^6} = \textbf{6.35} \times \textbf{10}^{-3}$ **in. (or 1.62 $\times$ 10^{-1} mm)**

The stresses for both steel and concrete are well within the elastic limit for both materials and therefore our assumption regarding elasticity is correct.

Example 1-13:

Given: The arrangement shown in Figure a. The modulus of elasticity of the brass and steel are $E_{Br} = 14 \times 10^6$ psi and $E_{St} = 30 \times 10^6$ psi, respectively.

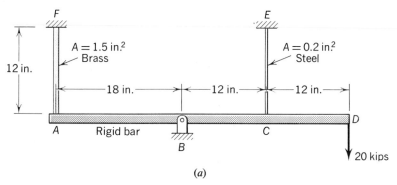

(a)

Find: (a) The vertical translation of point D, and (b) the forces in the brass and steel bars.

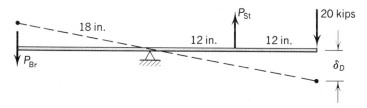

(b) P_{Br} = force in brass bar. P_{St} = force in steel bar.

Solution: It is advisable to draw a free-body diagram showing the forces acting on a body and the general geometry of deformation. This is shown in Figure *b*. The dashed line shows the exaggerated position of the bar after the load is applied. Assuming that the support *B* is unyielding, and the beam does not bend, the deflections of point *D* can be related to the deformations in bars *AF* and *EC*, permitting us to write a deformation equation.

From similar triangles:

$$\delta_{AF} : 18 \text{ in.} = \delta_D : 24 \text{ in.}$$

$$\delta_D = \frac{24}{18}\delta_{AF}; \quad \delta_D = \frac{4}{3}\delta_{AF} \tag{a}$$

Similarly,

$$\delta_{AF} = \frac{18}{12}\delta_{CE}; \quad \delta_{AF} = \frac{3}{2}\delta_{CE} \tag{b}$$

From $\Sigma M_B = 0$ (refer to Figure *b*)

$$P_{Br} \times 18 + P_{St} \times 12 = 24 \times 20 \tag{c}$$

In terms of *P*, Equation (*b*) becomes

$$\left(\frac{PL}{AE}\right)_{Br} = \frac{3}{2}\left(\frac{PL}{AE}\right)_{St} \tag{b'}$$

Substituting the values of various terms in Equation (*b'*)

$$\frac{P_{Br} \times 12}{1.5 \times 14 \times 10^6} = \frac{3}{2}\left(\frac{P_{St} \times 12}{0.2 \times 30 \times 10^6}\right)$$

Solving,

$$P_{Br} = \frac{21}{4}P_{St}$$

Substituting this into Equation (*c*) above,

$$P_{St}\left(\frac{21}{4} \times 18 + 12\right) = 24 \times 20$$

$$P_{St} = \frac{24 \times 20}{106.5} = \textbf{4.51 kips (or 20.06 kN) (tension)}$$

Therefore,

$$P_{Br} = \frac{21}{4}(4.51) = \textbf{23.7 kips (or 105.42 kN) (compression)}$$

The positive or plus values of P_{St} and P_{Co} indicate that the assumed sense or direction for these forces were correct. A negative value would have indicated a wrong assumption; then it would have been necessary to reverse the direction. The deformations in the bars are

$$\delta_{AF} = P_{Br}\left(\frac{L}{AE}\right)_{Br} = \frac{23{,}700 \times 12}{1.5 \times 14 \times 10^6} = 1.35 \times 10^{-2} \text{ in. (compression)}$$

$$\delta_{CE} = \frac{12}{18}\delta_{AF} = \frac{2}{3} \times 1.35 \times 10^{-2} = 0.9 \times 10^{-2} \text{ in. (elongation)}$$

From Equation (a) above,

$$\delta_D = \frac{24}{18}\delta_{AF} = \frac{24}{18} \times 1.35 \times 10^{-2} = \mathbf{1.80 \times 10^{-2}} \textbf{ in.}$$
$$\textbf{(or 4.57} \times \textbf{10}^{-1} \textbf{ mm) down}$$

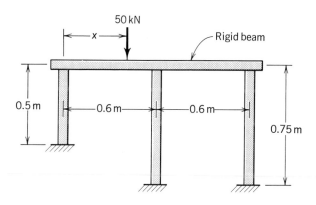

(a)

Example 1-14:

Given: Three posts of equal cross section and of the same material support the 50-kN load, as shown in Figure *a*.
Find: (a) The distance *x* (shown) such that the rigid beam remains horizontal, and (b) the force in each post in such a case.
Solution:

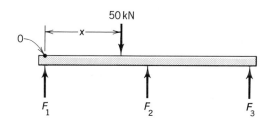

(b)

In Figure *b*, the forces in the posts are denoted by F_1, F_2, and F_3.

From

$$\sum F_y: \quad F_1 + F_2 + F_3 = 50 \text{ kN} \tag{1}$$

and from

$$\sum M_0: \quad 0.6F_2 + 1.2F_3 = (50)x \tag{2}$$

but, $\delta_1 = \delta_2 = \delta_3$, condition imposed via horizontal beam

$$\frac{0.50F_1}{AE} = \frac{0.75F_2}{AE} = \frac{0.75F_3}{AE} \tag{3}$$

or

$$F_2 = F_3 = \tfrac{2}{3}F_1 \tag{3'}$$

Substituting in Equation (1),

$$F_1 + \tfrac{2}{3}F_1 + \tfrac{2}{3}F_1 = 50 \text{ kN}$$

or

$$F_1 = \textbf{21.43 kN}$$

and

$$F_2 = F_3 = \tfrac{2}{3}F_1 = \textbf{14.29 kN}$$

From Equation (2),

$$x = \frac{1.8(14.29)}{50} = \textbf{0.514 m}$$

PROBLEMS

1–68 The reinforced concrete column in Figure P 1–68 is to support load *P*. Neglecting the weight of the column, determine the safe load *P* if the stress in the steel is not to exceed 18,000 psi, and the stress in the concrete is not to exceed 900 psi. Assume the moduli of elasticity of concrete and steel are $E_{Co} = 3 \times 10^6$ psi and $E_{St} = 30 \times 10^6$ psi, respectively.

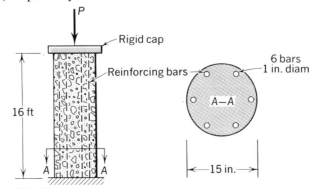

Figure P 1–68

1-69 Determine the axial deformation of the column of Figure P 1-68 when the load $P = 250$ kips. Assume the moduli of elasticity for the concrete and steel as $E_{Co} = 3 \times 10^6$ psi, and $E_{St} = 30 \times 10^6$ psi. What are the stresses in the steel reinforcement and in the concrete for this load?

1-70 Load P in Figure P 1-70 is placed on a square rigid plate such that the plate will remain horizontal after loading. Determine the distance x in terms of b. Assume elastic deformation.

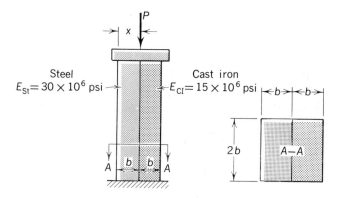

Figure P 1-70

1-71 In Figure P 1-70 assume the dimension $b = 3$ in. If the stresses in the steel and cast iron are not to exceed 20,000 psi and 10,000 psi, respectively, determine x and the maximum value that P can have.

1-72 In Figure P 1-72 determine the maximum weight W such that the stress in the middle column does not exceed 3000 psi.

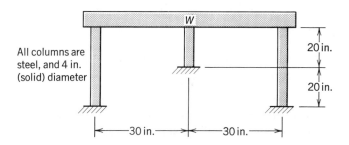

Figure P 1-72

1–73 Determine the axial deformation of the brass sleeve ($E_{Br} = 14 \times 10^6$ psi) and the stress in the steel bolt ($E_{St} = 30 \times 10^6$ psi) of Figure P 1–73 when the nut is turned ½ of a full turn if the thread pitch is ⅛ in., and the cross-sectional areas of the tube and bolt are: $A_{Br} = 1$ sq in., $A_{St} = 0.5$ sq in., respectively.

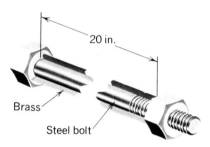

Figure P 1–73

1–74 Determine the number of turns the nut in Figure 1–73 may make beyond snug contact if the stresses in the steel bolt and brass sleeve are not to exceed 18,000 psi and 12,000 psi, respectively. $A_{Br} = 1$ sq in., $A_{St} = 0.5$ sq in., $E_{Br} = 15 \times 10^6$ psi, $E_{St} = 30 \times 10^6$ psi.

1–75 Determine the "wall" reactions R_1 and R_2 as shown in Figure P 1–75 if the wall is assumed to be unyielding.

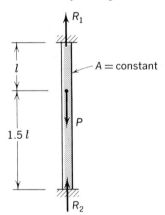

Figure P 1–75

1–76 If $P = 30$ kips and $l = 12$ in. (Figure P 1–75), what is the minimum cross-sectional area of the bar, if the stress is not to exceed 10,000 psi?

1–77 Assuming that the bar in Figure P 1–75 is made of steel ($E = 30 \times 10^6$ psi), 1 in. sq, and that $P = 20$ kips, $l = 12$ in., determine the reactions of R_1 and R_2 if the support R_1 yields 0.001 in. before assuming load.

1–78 Determine the stresses in the steel bar ($E_{St} = 30 \times 10^6$ psi) and the wood post ($E_{Wd} = 1.5 \times 10^6$ psi) in Figure P 1–78 when the load P is 5000 lb.

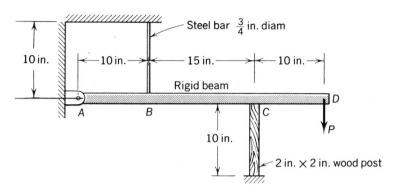

Figure P 1–78

1–79 Determine the maximum value P may have in Figure P 1–78 if the allowable stresses in the steel and wood are 15,000 psi and 1500 psi, respectively. ($E_{St} = 30 \times 10^6$ psi, $E_{Wd} = 1.5 \times 10^6$ psi.)

1–80 Determine the vertical deflection of point D in Problem 1–78.

1–81 Determine the location x if $P = 10$ kips (neglect the weight of the members) so that the bar remains horizontal if both the steel and copper bars in Figure P 1–81 have a diameter of 1 in.

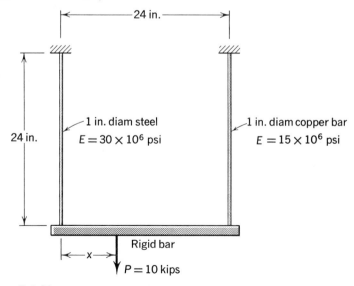

Figure P 1–81

1-82 Determine the location x and the maximum value P may have so that the rigid bar in Figure P 1–81 remains horizontal if the stresses in the steel and copper bars are not to exceed 20,000 psi and 15,000 psi, respectively.

1-83 The 40-kN load (neglect the weights of the bar) is supported by three 12-mm-diam steel bars, as shown in Figure P 1–83. Determine (a) the axial stress in each bar, and (b) the vertical travel of the 40-kN load.

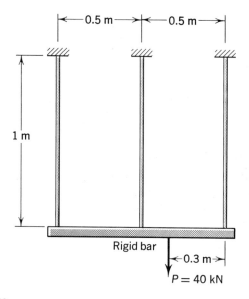

Figure P 1–83

1-84 Assuming the bars supporting the load in Figure P 1–83 are of steel ($E = 200 \times 10^6$ kN/m²) and are of the same diameter, determine the minimum diameter they may have if the maximum stress in any of the bars is not to exceed 140 MN/m²).

1-85 Three wood posts of equal cross-sectional areas support a 25-kip load, as shown in Figure P 1–85. Determine x such that the rigid beam remains horizontal. $E_{\mathrm{Wd}} = 1.5 \times 10^6$ psi.

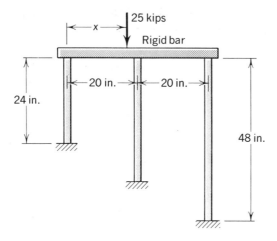

Figure P 1–85

1–86 If the wood posts ($E_{\text{wd}} = 1.5 \times 10^6$ psi) in Figure P 1–85 are 2 in. $\times$ 4 in., determine the stresses in each of the posts when $x = 12$ in.

1–87 All the truss members in Figure P 1–87 have a cross-sectional area of 3 sq in. Assuming hinged joints (typical assumption for a truss), determine the force in each member. [Hint: see Example 1–7.]

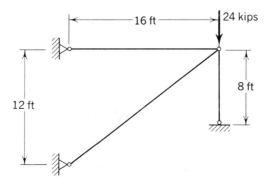

Figure P 1–87

1–88 Determine the vertical travel of the 10-kip load in Figure P 1–88. Neglect the weight of the members. $E_{St} = 30 \times 10^6$ psi.

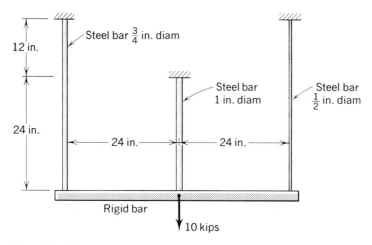

Figure P 1–88

1–89 Determine the stresses in the three bars of Figure P 1–88. $E_{St} = 30 \times 10^6$ psi.

1–90 Three steel bars of equal cross-sectional area support the 100-kN force, as shown in Figure P 1–90. Determine the value of the cross-sectional area of one bar if the stress is not to exceed 140 MN/m².

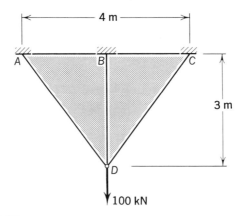

Figure P 1–90

1–91 Determine the vertical deflection of point D, Figure P 1–90, if the three bars are steel ($E_{St} = 30 \times 10^6$ psi), and ¾-in. diameter.

1–92 Determine the maximum value that P in Figure P 1–92, can be if the stresses in the steel bars and the wood post are not to exceed 25,000 psi and 1500 psi, respectively. $E_{Wd} = 1.5 \times 10^6$ psi, $E_{St} = 30 \times 10^6$ psi.

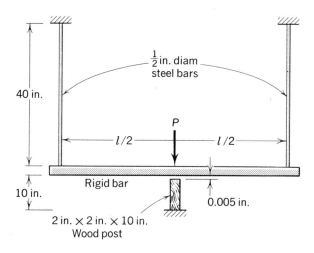

Figure P 1–92

1–93 Determine the stresses in the wood post and steel bars in Figure P 1–92 if $P = 12$ kips. $E_{Wd} = 1.5 \times 10^6$ psi, $E_{St} = 30 \times 10^6$ psi.

1—10 Thermal Stresses

When free to deform, most engineering materials will either expand or contract when subjected to changes in temperature; that is, they expand when heated and contract when cooled. The total unrestrained deformation δ_T is expressed by

$$\delta_T = \alpha \cdot \Delta t \cdot L \qquad (1\text{--}13)$$

where α is the *coefficient of thermal expansion*, a constant for the range of temperatures we would normally deal with, Δt is the *change in temperature*, and L is the *dimension* in the direction in which the thermal deformation is investigated. Perhaps the most *common* units are: $\alpha = $ in./ in./°F, $\Delta t = $ °F, and $L = $ inches; this results in $\delta_T = $ inches, the same units as L.

When a member, which is subjected to temperature changes, is restrained partially or totally from deforming, *thermal stresses* are induced in that member. As an illustration, assume that a rod is anchored in two unyielding supports, as shown in Figure 1–17a. If the rod was not anchored at

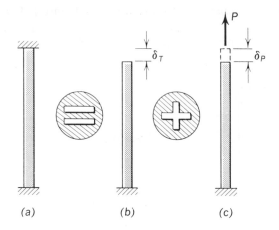

(a)   (b)   (c)

Figure 1–17

one end, a decrease in temperature would shorten the rod by an amount δ_T, as shown in Figure 1–17b. By applying an axial force P with such magnitude so as to make the deformation from this force equal to δ_T, as shown in Figure 1–17c, we obtain a net deformation consistent with the restriction derived from Figure 1–17a. In equation form, the deformation may be related as

$$-\alpha \cdot \Delta t \cdot L = \left(\frac{P}{A}\right)\frac{L}{E} = \sigma\left(\frac{L}{E}\right) \qquad (a)$$

from which

$$\sigma = -\alpha \cdot \Delta t \cdot E \qquad (b)$$

What we have determined is the net effect from the temperature change and the axial load P by adding algebraically the separate effects of Δt and P. In effect, we used the *principle of superposition*, which states that *if several causes (such as temperature and loads) act simultaneously on a system, and each effect (such as displacement and stresses at a point on a given plane) is directly proportional to its cause, the total effect is the sum of the individual effects when considered separately, provided Hooke's law is valid within the range of these effects, either individually or combined.* This principle, coupled with a free-body diagram, should prove a powerful tool in the visualization of the behavior of many structures under load, and in the simplification of the solution to those problems.

Example 1–15:

Given: A rigid bar is supported by two wires of cross-sectional area = 1.0 in.² (or 0.000645 m²) each, as shown in Figure *a*. When load *P* is applied, the temperature is T. $E_{St} = 30 \times 10^6$ psi (or 200×10^6 kN/m² = 200 GN/m²), $E_{Al} = 10 \times 10^6$ psi (or 70 GN/m²), $\alpha_{St} = 6.5 \times 10^{-6}$ in./in./°F [or $12(10^{-6})$/°C], $\alpha_{Al} = 12.8 \times 10^{-6}$ in./in./°F [or $23(10^{-6})$/°C]. *Find:* The Δt necessary for the rigid bar to assume a horizontal position when loaded.

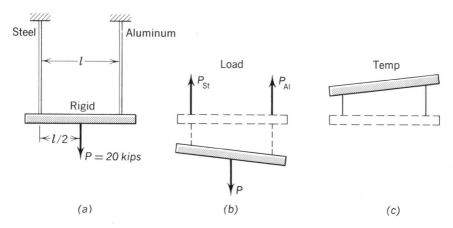

(a)   (b)   (c)

Procedure: The total effect of temperature and load can be determined by using the principle of superposition. To begin with, we guess that Δt shall be a decrease in temperature, and solve for Δt. If our assumption is incorrect, the result will be a *negative* Δt. The individual effects are shown in Figures *b* and *c*, where

$$(\delta_p - \delta_T)_{St} = (\delta_p - \delta_T)_{Al}$$

$$\left[\frac{(P/2)(l)}{AE} - \alpha \cdot \Delta t \cdot l \right]_{St} = \left[\frac{(P/2)(l)}{AE} - \alpha \cdot \Delta t \cdot l \right]_{Al}$$

Substituting the values for P, A, and E, we get

$$\frac{10,000}{1 \times 30 \times 10^6} - 6.5 \times 10^{-6} \, \Delta t = \frac{10,000}{1 \times 10 \times 10^6} - 12.8 \times 10^{-6} \, \Delta t$$

Multiplying both sides by 10×10^6,

$$\frac{10,000}{3} - 65 \, \Delta t = 10,000 - 128 \, \Delta t$$

from which

$$63 \, \Delta t = + \frac{20,000}{3}$$

and

$$\Delta t = \mathbf{106°F \ (or \ 41.11°C)}$$

Therefore, the plus indicates that our assumption of a *decrease* in temperature was correct.

PROBLEMS

For the following problems, assume the following constants:
steel: α_{St} 6.5 × 10⁻⁶ in./in./°F [or 12(10⁻⁶)/°C], E_{St} = 30 × 10⁶ psi (or 200 GN/m²);
aluminum: α_{Al} = 12.8 × 10⁻⁶ in./in./°F [or 23(10⁻⁶)/°C], E_{Al} = 10 × 10⁶ psi (or 70 GN/m²);
copper: α_{Cu} = 9.3 × 10⁻⁶ in./in./°F [or 17(10⁻⁶)/°C], E_{Cu} = 17 × 10⁶ psi (or 120 GN/m²);
brass: α_{Br} = 10.4 × 10⁻⁶ in./in./°F [or 19(10⁻⁶)/°C], E_{Br} = 14 × 10⁶ psi (or 100 GN/m²).

1-94 A brass bar shown in Figure P 1-94 is restrained from movement by two vertical walls. Derive a relation between stress and temperature change for this bar.

Figure P 1-94

1-95 Determine the range in temperature the bar in Figure P 1-94 may undergo from a reference temperature T at which the stress is zero, so that the compressive and tensile stresses do not exceed 10,000 psi and 15,000 psi, respectively.

1-96 The arrangement of Figure P 1-96 undergoes an increase in temperature of 100°F. Determine the stresses set up in the aluminum and steel bars if the stresses before the temperature rise were zero.

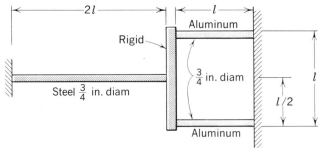

Figure P 1-96

1-97 Determine the range in temperature the members in Figure P 1-96 can undergo from a reference point T, if the permissible stresses in steel and aluminum are ± 20,000 psi and ± 15,000 psi, respectively. At T the stresses in the bars are zero.

1–98 Determine the change in stresses in the steel and aluminum bars of Figure P 1–98 if the temperature increases 100°F.

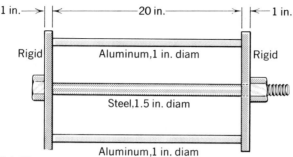

Figure P 1–98

1–99 If the steel rod in Figure P 1–98 is threaded with $\frac{1}{8}$-in. pitch threads, determine the number of turns the nut would have to make in order to result in equal stresses in the bars at a temperature increase of 80°F.

1–100 Figure P 1–100 shows a bar composed of three rods fixed between two unyielding walls. Determine the stresses in each rod if the temperature increases 75°C from a temperature at which the rods have no load in them.

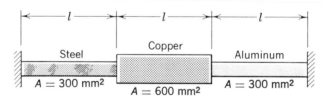

Figure P 1–100

1–101 Determine the range in temperature the rods in Figure P 1–100 may undergo simultaneously if the stresses in the rods are not to exceed ± 15,000 psi.

1–102 A steel band is to be "shrunk fit" over a solid steel shaft which has a diameter 0.003 in. larger than the inside diameter of the band. (a) At what temperature above normal must the band be raised to make this possible? (b) What is the average stress in the steel band when normal temperature is regained? Neglect the deformation in the shaft.

1–103 What is the maximum difference in the steel shaft diameter and inside diameter of the band in Figure P 1–102 that one may have if the stress in the steel band is not to exceed 140 MN/m²?

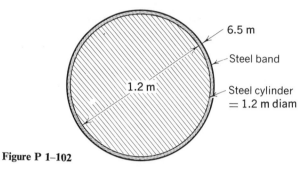

Figure P 1–102

1–104 Determine the stresses in the bars of Figure P 1–104 when the temperature increases 100°F.

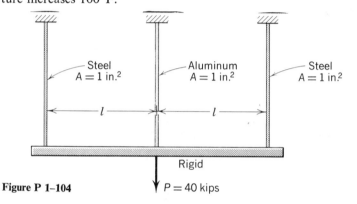

Figure P 1–104

1–105 Determine the permissible Δt if the stress in the steel bars of Figure P 1–104 is not to exceed 18,000 psi.

1–106 The members in Figure P 1–106 are subjected simultaneously to an increase of temperature of 50°F. If, at T, the stresses in the bars were zero, determine the stresses after the temperature increase.

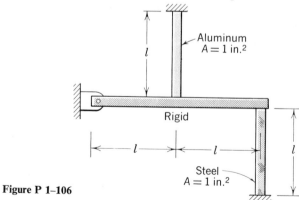

Figure P 1–106

1-107 Determine the range of temperatures the members of Figure P 1–106 may undergo if the stress in the aluminum is not to exceed ± 15,000 psi, and that in the steel, 20,000 psi.

1-108 Determine the stresses in the steel and aluminum wires in Figure P 1–108 if the members are subjected to a temperature increase of 100°F.

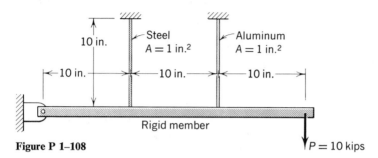

Figure P 1–108 $P = 10$ kips

1-109 Determine the permissible range in temperatures, if the stresses in the steel and aluminum wires in Figure P 1–108 are not to exceed + 20,000 psi and + 15,000 psi, respectively.

1-110 Determine the range of temperatures the members in Figure P 1–110 may undergo simultaneously, if the stresses are not to exceed ± 15,000 psi in any of the members.

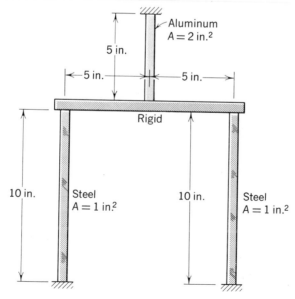

Figure P 1–110

1–111 In Figure P 1–111, the two wires, steel and aluminum, as shown, are subjected to a *decrease* in temperature of 100°C. Determine the stress in each wire.

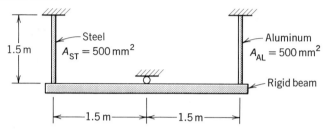

Figure P 1–111

1–112 In Figure P 1–112, determine the total vertical travel of the weight shown in the sketch below, $W = 40$ kN, if (*a*) there is constant temperature; (*b*) there is an increase in the temperature by 70°C (neglect the effect in the beams).

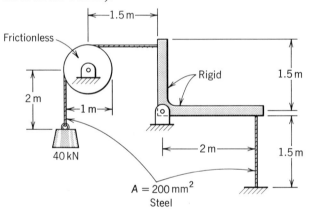

Figure P 1–112

1—11 Stresses on Inclined Planes

Now that we have discussed axial stresses, we shall move one step forward and discuss stresses on planes inclined with the axis of a uniaxially or biaxially loaded prism. As a start let us consider a prismatic member subjected to an axial load P, as shown in Figure 1–18a. From the length of this member, we cut a segment and set it in equilibrium, as shown in Figure 1–18b. On the inclined plane, the stress is

$$\sigma_o = \frac{P}{A/\cos \theta} = \left(\frac{P}{A}\right) \cos \theta = \sigma_x \cos \theta \qquad (a)$$

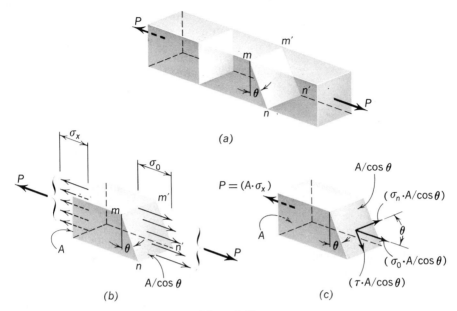

Figure 1–18

$(\sigma_0 A/\cos\theta)$ is then resolved into components, one $(\sigma_n \cdot A/\cos\theta)$, perpendicular to the plane and the other $(\tau \cdot A/\cos\theta)$, parallel to the plane as shown in Figure 1–18c. The corresponding stresses are

$$\sigma_n = \sigma_o \cos\theta = \sigma_x \cos\theta \cdot \cos\theta = \sigma_x \cos^2\theta \qquad (1\text{–}14a)$$

and

$$\tau = \sigma_o \sin\theta = \sigma_x \cos\theta \cdot \sin\theta \qquad (1\text{–}15a)$$

The above formulas might be more conveniently expressed in terms of "double angles"; that is, from the fundamental trigonometric identities,

$$\cos^2\theta = \frac{\cos 2\theta + 1}{2}; \quad \cos\theta \sin\theta = \frac{\sin 2\theta}{2}$$

we get

$$\sigma_n = \sigma_x\left(\frac{\cos 2\theta + 1}{2}\right) \qquad (1\text{–}14b)$$

and

$$\tau = \sigma_x\left(\frac{\sin 2\theta}{2}\right) \qquad (1\text{–}15b)$$

A graphical presentation of these two expressions is shown in Figure 1–19. It should prove helpful in visualizing more readily the variation in the normal and shearing stresses with respect to each other and with angle θ.

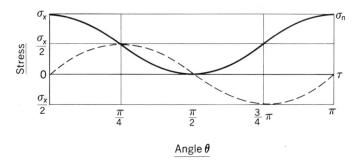

Angle θ

Figure 1–19

From the graph, the planes of maximum and minimum normal stresses are readily picked out at 0 and π, and at $\pi/2$ and $3\pi/2$, respectively. Note that at all points where the normal stresses are either minimum or maximum, the shear stress is always zero. We refer to these maximum and minimum normal stresses as *principal stresses*, and the planes on which they occur (planes of zero shear stress) as *principal planes*.

We observe further that the shear stress at θ has the same *magnitude* as the stresses at $(\pi/2 + \theta)$ or $(\pi + \theta)$, and so forth; that is, the shear stress on orthogonal planes has the same magnitude. Also, for the case we plotted, the maximum value of τ occurs on a plane making an angle $\theta = \pm(\pi/4)$ [or $(\frac{3}{4})\pi$, $(\frac{5}{4})\pi$, and so forth].

The maximum for both σ_n and τ can, of course, also be determined by the use of calculus by differentiating the stresses with respect to θ:

From

$$\frac{d\sigma_n}{d\theta} = -\sigma_x \sin 2\theta = 0, \quad \text{when } \theta = 0, \pi/2$$

and

$$\frac{d\tau}{d\theta} = +\sigma_x \cos 2\theta = 0, \quad \text{when } \theta = \pm\frac{\pi}{4}$$

we evaluate

$$\sigma_{n,\text{max}} = \sigma_x \text{ at } \theta = 0, \qquad \sigma_{n,\text{min}} = 0 \text{ at } \theta = \frac{\pi}{2}$$

and

$$|\tau_{\text{max}}| = |\tau_{\text{min}}| = \frac{\sigma_x}{2} \text{ for } \theta = \pm\frac{\pi}{4}$$

The fact that shear stresses on perpendicular planes of an element have the same magnitude can also be shown by cutting the body at these angles and placing the body in equilibrium, as was done in Figure 1–18, to obtain Equation (1–15a). Still another approach is to isolate a rectangular element from the body and place it in equilibrium, as shown in Figure 1–20. Because

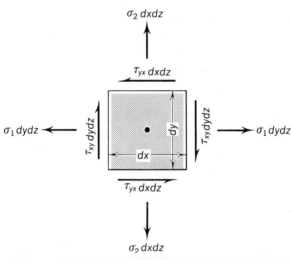

Figure 1–20

the normal stresses σ_1 and σ_2 acting on these faces form, respectively, collinear, oppositely directed forces of the same magnitude, their net translational and rotational effect is zero. Working only with the shear forces, therefore, and assuming the element to have a thickness dz, the shear forces on the respective faces are $\tau_{xy}\, dy\, dz$ and $\tau_{yx}\, dx\, dz$, as shown in Figure 1–20. The shear forces on the respective parallel faces are necessarily oriented opposite and are of the same magnitude in order to satisfy equilibrium regarding translation. Thus, we have two couples that necessarily oppose each other in order to satisfy equilibrium regarding rotation. The clockwise couple is $(\tau_{xy}\, dy\, dz)\, dx$; the counterclockwise couple is $(\tau_{yx}\, dx\, dz)\, dy$. Equating the two couples, we get

$$(\tau_{xy}\, dy\, dz)dx = (\tau_{yx}\, dx\, dz)dy$$

Hence,

$$|\tau_{xy}| = |\tau_{yx}|$$

Thus, we have established again, using equilibrium, that the shear stresses on all the perpendicular planes shown are equal. At this point, we shall adopt a sign convention for these stresses: *If the stress points in a direction that corresponds to a clockwise rotation, the stress is positive; a negative shear stress is associated with counterclockwise rotation.* As previously mentioned, the normal stresses are considered positive if there is a tensile load acting on that face; it is negative for a compressive load. Referring to Figure 1–20, stress τ_{xy} is, according to our convention, positive, τ_{yx} is negative, σ_1 and σ_2 are positive.

A similar procedure will yield expressions for the normal and shear stresses on inclined planes caused by normal and shear stresses oriented in both the x and y directions, as shown in Figure 1–21, and for negative as well as positive values.

Assume the element to be cut by a plane passing through $m\,n$, and place the left segment in equilibrium, as shown Figure 1–21b. If ΔA_n stands for the area on the inclined plane, the areas of vertical and horizontal faces of this segment are $\Delta A_n \cos\theta$ and $\Delta A_n \sin\theta$, respectively. The components of these forces, parallel and perpendicular to the mn plane, are shown by the

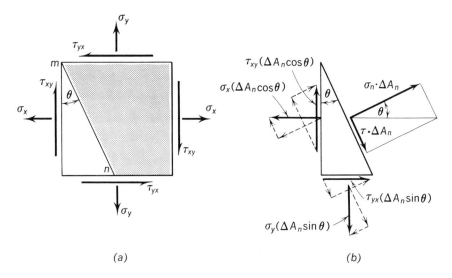

(a) (b)

Figure 1–21

broken lines. Summing forces in the direction normal to the plane, we have $(\Sigma F_n = 0)$

$$\sigma_n \Delta A_n = \sigma_x(\Delta A_n \cos\theta)\cos\theta - \tau_{xy}(\Delta A_n \cos\theta)\sin\theta + \sigma_y$$
$$(\Delta A_n \sin\theta)\sin\theta - \tau_{yx}(\Delta A_n \sin\theta)\cos\theta$$

which reduces to

$$\sigma_n = \sigma_x \cos^2 \theta + \sigma_y \sin^2 \theta - 2\tau_{xy} \sin \theta \cos \theta$$

and in terms of double angles,

$$\sigma_n = \left(\frac{\sigma_x + \sigma_y}{2}\right) + \left(\frac{\sigma_x - \sigma_y}{2}\right)\cos 2\theta - \tau_{xy} \sin 2\theta \qquad (1\text{-}16)$$

Summing forces in the direction parallel to the plane ($\Sigma F_\tau = 0$)

$$\tau \cdot \Delta A_n = \sigma_x(\Delta A_n \cos \theta)\sin \theta + \tau_{xy}(\Delta A_n \cos \theta)\cos \theta - \sigma_y(\Delta A_n \sin \theta)\cos \theta$$
$$- \tau_{yx}(\Delta A_n \sin \theta)\sin \theta$$

which reduces to

$$\tau = (\sigma_x - \sigma_y)\sin \theta \cos \theta + \tau_{xy}(\cos^2 \theta - \sin^2 \theta)$$

and, in terms of double angles,

$$\tau = \left(\frac{\sigma_x - \sigma_y}{2}\right)\sin 2\theta + \tau_{xy} \cos 2\theta \qquad (1\text{-}17)$$

The orientation of the principal plane is determined by setting the derivative $d\sigma_n/d\theta = 0$ from Equation (1–16), as follows:

$$-(\sigma_x - \sigma_y)\sin 2\theta - 2\tau_{xy} \cos 2\theta = 0$$

from which

$$\tan 2\theta = \frac{-\tau_{xy}}{(\sigma_x - \sigma_y)/2} \qquad (1\text{-}18)$$

To evaluate the maximum and minimum normal stresses, we evaluate Equation (1–16) at θ, as determined from Equation (1–18). At this point, we should note that there are values of θ, 90 degrees apart (2θ at 180 degrees apart) for which the tangent 2θ has the same sign. This may be more lucidly shown by Figure 1–22. Also,

For θ_1
$$\begin{cases} \cos 2\theta_1 = +\left(\frac{\sigma_x - \sigma_y}{2}\right)\Big/ \sqrt{\left(\frac{\sigma_x - \sigma_y}{2}\right)^2 + \tau_{xy}{}^2} \\[3mm] \sin 2\theta_1 = -\tau_{xy}\Big/ \sqrt{\left(\frac{\sigma_x - \sigma_y}{2}\right)^2 + \tau_{xy}{}^2} \end{cases}$$

For θ_2
$$\begin{cases} \cos 2\theta_2 = -\left(\frac{\sigma_x - \sigma_y}{2}\right)\Big/ \sqrt{\left(\frac{\sigma_x - \sigma_y}{2}\right)^2 + \tau_{xy}{}^2} \\[3mm] \sin 2\theta_2 = +\tau_{xy}\Big/ \sqrt{\left(\frac{\sigma_x - \sigma_y}{2}\right)^2 + \tau_{xy}{}^2} \end{cases}$$

where τ_{xy} and $(\sigma_x - \sigma_y)$ are positive quantities.

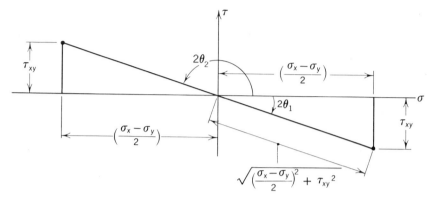

Figure 1–22

Keeping both values of the radical, that is, plus and minus, and substituting them in Equation (1–16), we obtain

$$\sigma_{max,min} = \left(\frac{\sigma_x + \sigma_y}{2}\right) + \left(\frac{\sigma_x - \sigma_y}{2}\right)\left[\pm \frac{(\sigma_x - \sigma_y)/2}{\sqrt{\left(\frac{\sigma_x - \sigma_y}{2}\right)^2 + \tau_{xy}^2}}\right]$$
$$- \tau_{xy}\left[\pm \frac{\tau_{xy}}{\sqrt{\left(\frac{\sigma_x - \sigma_y}{2}\right)^2 + \tau_{xy}^2}}\right]$$

which yields

$$\sigma_{max,min} = \left(\frac{\sigma_x + \sigma_y}{2}\right) \pm \sqrt{\left(\frac{\sigma_x - \sigma_y}{2}\right)^2 + \tau_{xy}^2} \qquad (1\text{–}19)$$

This applies equally well for negative σ_x or σ_y if the negative sign is inserted accordingly.

   In a similar manner, the orientation of the planes on which the maximum shear stress occurs (and the value of this stress) may be found. By differentiating from Equation (1–17) with respect to θ, we get

$$\frac{d\tau}{d\theta} = 0: \qquad (\sigma_x - \sigma_y)\cos 2\theta - 2\tau_{xy} \sin 2\theta = 0$$

Thus

$$\tan 2\theta' = \frac{(\sigma_x - \sigma_y)/2}{\tau_{xy}} \qquad (1\text{–}20)$$

which is a negative reciprocal of tan 2θ [Equation (1–18)]. Therefore, the two angles, 2θ and $2\theta'$, differ by 90 degrees and as a result, θ and θ' are 45 degrees apart. This shows that the *planes of maximum shear differ by 45 degrees from the principal planes.*

Substituting the value of θ' from Equation (1–20) into Equation (1–17), we obtain an expression for the maximum value of shear stress:

$$\tau_{max} = \pm\sqrt{\left(\frac{\sigma_x - \sigma_y}{2}\right)^2 + \tau_{xy}^2} \qquad (1–21)$$

When the principal stresses are known, the maximum shear stress may be found by substituting the τ_{max} from Equation (1–21) into Equation (1–19)

$$\sigma_{max} = \frac{\sigma_x + \sigma_y}{2} + \tau_{max}$$

$$\sigma_{min} = \frac{\sigma_x + \sigma_y}{2} - \tau_{max}$$

Subtracting, we get

$$\tau_{max} = \frac{\sigma_{max} - \sigma_{min}}{2} \qquad (1–22)$$

Example 1–16:

Given: A piece of wood, 2 in. × 2 in. × 6 in. (or 51 mm × 51 mm × 152 mm), is subjected to a compressive load *P* of 13,000 lb (or 57.82 kN), as shown in the figure.

Find: (*a*) The normal and shearing stresses on the inclined plane shown when $\theta = 40$ degrees, and (*b*) the shear and normal stress on a plane on which the shear stress is a maximum.

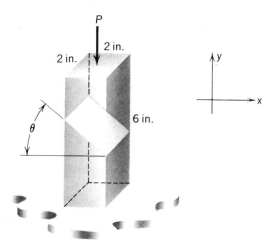

Solution: Slide rule accuracy shall be assumed sufficient.

$$\sigma_x = 0; \quad \sigma_y = \frac{P}{A} = \frac{13,000}{2 \times 2} = 3250 \text{ psi}$$

From Equations (1–14*b*) and (1–15*b*),

$$\sigma_n = 3250 \left(\frac{\cos 80 + 1}{2} \right) = \textbf{1910 psi (or 13.17 MN/m}^2\textbf{)}$$

$$\tau = 3250 \left(\frac{\sin 80}{2} \right) = \textbf{1620 psi (or 11.17 MN/m}^2\textbf{)}$$

(a)

The maximum shear stress occurs on a plane oriented 45 degrees with the vertical, as shown previously. Therefore, at $\theta = 45$ degrees,

$$\tau = \tau_{max} = 3250 \left(\frac{\sin 90}{2} \right) = \textbf{1625 psi (or 11.17 MN/m}^2\textbf{)}$$

$$\sigma_n = 3250 \left(\frac{\cos 90 + 1}{2} \right) = \textbf{1625 psi (or 11.17 MN/m}^2\textbf{)}$$

(b)

Example 1–17:

Given: An element that is stressed as shown in Figure *a*; $\sigma_x = 6000$ psi (tension), $\sigma_y = 6000$ psi (compression), $\tau_{xy} = 1000$ psi.
Find: (*a*) The principal stresses and the angle of the plane on which they occur, and (*b*) the maximum shear stress and the angle of the plane on which it occurs.

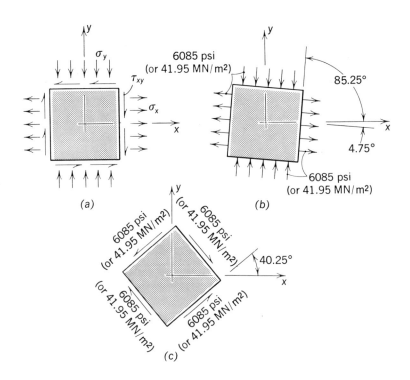

Procedure: Using Equation (1–19),

$$\sigma_{max,min} = \left(\frac{\sigma_x + \sigma_y}{2}\right) \pm \sqrt{\left(\frac{\sigma_x - \sigma_y}{2}\right)^2 + \tau_{xy}^2}$$

$$\sigma_{max,min} = \frac{6000 - 6000}{2} \pm \sqrt{\left(\frac{6000 + 6000}{2}\right)^2 + (1000)^2}$$

$$= \pm 6085 \text{ psi}$$

or,

$$\sigma_{max} = +6085 \text{ psi;} \qquad \sigma_{min} = -6085 \text{ psi}$$

$$(\text{or } +41.95 \text{ MN/m}^2) \qquad (\text{or } -41.95 \text{ MN/m}^2)$$

From Equation (1–18),

$$\tan 2\theta = \frac{-\tau_{xy}}{\dfrac{(\sigma_x - \sigma_y)}{2}} = \frac{-1000}{\dfrac{6000 - (-6000)}{2}} = \frac{-1}{6}$$

$$2\theta = -9.5° \quad \text{and} \quad 170.5°$$

Thus, $\theta = -4.75°$ and **85.25**, as shown in Figure *b*.

Substituting these angles into Equation (1–16), we find that for $\theta = -4.75$ degrees, $\sigma_n = +6085$ psi, and for $\theta = 85.25$ degrees, $\sigma_n = -6085$ psi. The orientation of the element and these stresses are shown in Figure *b*.

To determine the maximum shear stress, we may use either Equation (1–21) or Equation (1–22) because now we know σ_{max} and σ_{min}. Using Equation (1–22),

$$\tau_{max} = \frac{6085 - (-6085)}{2} = 6085 \text{ psi (or 41.95 MN/m}^2)$$

From Equation (1–20),

$$\tan 2\theta' = +6/1$$

where

$$2\theta' = 80.5° \text{ and } 260.5°$$

therefore,

$$\theta' = 40.25° \text{ and } 130.25°$$

From this we note that the maximum shear stress occurs on planes 45 degrees from the principal planes. Furthermore, at these angles, the normal stresses are zero, as may be shown using Equation (1–16). From Equation (1–17), we note that at $\theta' = 40.25$ degrees, the value of τ is positive (clockwise rotation of the element). We may now draw the element with the orientation and stresses as shown in Figure *c*.

PROBLEMS

1–113 Determine the normal and shear stresses on the plane oriented at $\theta = 30$ degrees in the element of Figure P 1–113 if $P = 2000$ lb.

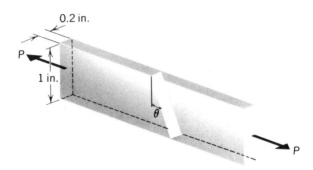

Figure P 1–113

1–114 Determine the normal and shear stresses on a plane of maximum shear stress for the element of Figure P 1–113 if $P = 3000$ lb.

1–115 Determine the normal and shear stresses on the plane oriented at $\theta = 55$ degrees in the element of Figure P 1–115 if $P = 20,000$ lb.

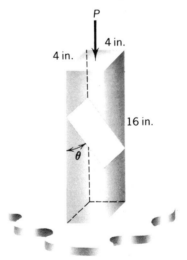

Figure P 1–115

1–116 For the block of Figure P 1–115, the allowable normal and shear stresses on a plane oriented at $\theta = 55$ degrees as shown are 1500 psi and 300 psi, respectively. Determine the allowable load P.

1–117 A concrete cylinder is subjected to a compressive load P, as shown in Figure P 1–117. Determine the maximum shear stress in the cylinder when $P = 56,600$ lb.

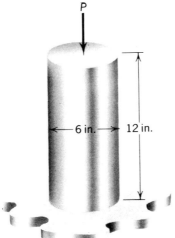

Figure P 1–117

1–118 Determine the normal and shear stresses on the inclined plane of Figure P 1–118, if $\sigma_x = 100$ MN/m².

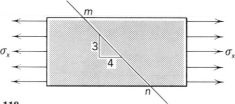

Figure P 1–118

1–119 The element in Figure P 1–119 was taken from a thin-walled cylindrical vessel subjected to internal pressure. For the stresses shown, determine the normal and shear stresses on the inclined plane shown if $\theta = 60$ degrees.

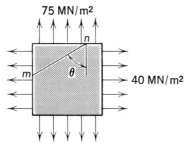

Figure P 1–119

1–120 For the element of Figure P 1–119, determine the maximum shear stress, and the normal stress on the plane on which the maximum shear stress occurs, and sketch these stresses on a properly oriented element.

1–121 For the element of Figure P 1–121, determine the shear and normal stresses on the inclined plane shown.

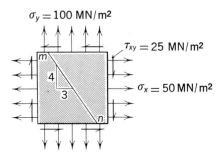

Figure P 1–121

1–122 Determine the principal stresses for the element of Figure P 1–121 and sketch them on a properly oriented element.

1–123 For the element of Figure P 1–121, determine the maximum shear stress, and the normal stress on the plane on which the maximum shear stress occurs, and sketch these stresses on a properly oriented element.

1–124 For the element of Figure P 1–124, determine the shear and normal stresses on a plane inclined at $\theta = 50$ degrees as shown.

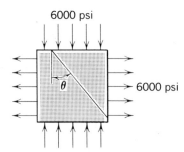

Figure P 1–124

1–125 For the element of Figure P 1–124, determine the maximum shear stress and the normal stress on the plane on which the maximum shear stress occurs, and sketch these on a properly oriented element.

1–126 For the element of Figure P 1–126, determine the shear and normal stresses on the inclined plane shown.

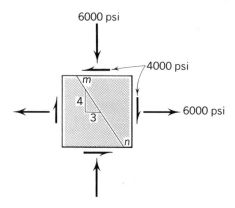

Figure P 1–126

1–127 Determine the principal stresses for the element of Figure P 1–126, and sketch these stresses on a properly oriented element.

1–128 For the element of Figure P 1–126, determine the maximum shear stress, and the normal stress on the plane on which the maximum shear stress occurs and sketch these on a properly oriented element.

1–129 Determine the principal stresses for the element of Figure P 1–129 and sketch these stresses on a properly oriented element.

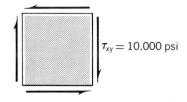

Figure P 1–129

–130 Determine the principal stresses for the element of Figure P 1–130, and sketch these stresses on a properly oriented element.

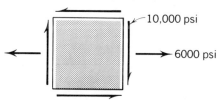

Figure P 1–130

1–131 For the element in Figure P 1–131, determine (a) the principal stresses and principal planes; (b) the maximum shearing stresses; (c) the normal and shearing stresses on the plane m–n shown.

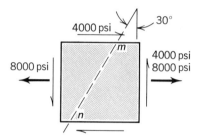

Figure P 1–131

1—12 Mohr's Circle for Stress

The normal and shear stresses given by Equations (1–16) and (1–17), respectively, may be represented graphically by an extremely useful device known as *Mohr's circle for stress*. It is named after the German engineer Otto Mohr (1835–1918) who devised it in 1882. The Mohr circle represents these equations in a manner that makes them more easily understood and re-membered, and brings out more lucidly their physical significance.

It is a rather simple matter to show that these are the equations of a circle in a σ_n–τ plane. For convenience, Equations (1–16) and (1–17) are rewritten below, Equation (1–16) being slightly rearranged.

$$\sigma_n - \frac{\sigma_x + \sigma_y}{2} = \frac{\sigma_x - \sigma_y}{2} \cos 2\theta - \tau_{xy} \sin 2\theta \qquad (a)$$

$$\tau = \frac{\sigma_x - \sigma_y}{2} \sin 2\theta + \tau_{xy} \cos 2\theta \qquad (b)$$

Squaring both sides of each equation, we get

$$\left(\sigma_n - \frac{\sigma_x + \sigma_y}{2}\right)^2 = \left(\frac{\sigma_x - \sigma_y}{2}\right)^2 (\cos 2\theta)^2 - (\sigma_x - \sigma_y)(\tau_{xy})(\cos 2\theta)(\sin 2\theta) + (\tau_{xy})^2(\sin 2\theta)^2$$

and

$$(\tau)^2 = \left(\frac{\sigma_x - \sigma_y}{2}\right)^2 (\sin 2\theta)^2 + (\sigma_x - \sigma_y)(\tau_{xy})(\cos 2\theta)(\sin 2\theta) + (\tau_{xy})^2(\cos 2\theta)^2$$

Adding the two equations, the result is

$$\left(\sigma_n - \frac{\sigma_x + \sigma_y}{2}\right)^2 + (\tau)^2 = \left(\frac{\sigma_x - \sigma_y}{2}\right)^2 [(\cos 2\theta)^2 + (\sin 2\theta)^2] + (\tau_{xy})^2[(\cos 2\theta)^2 + (\sin 2\theta)^2]$$

Hence,

$$\left(\sigma_n - \frac{\sigma_x + \sigma_y}{2}\right)^2 + (\tau)^2 = \left(\frac{\sigma_x - \sigma_y}{2}\right)^2 + \tau_{xy}{}^2 \qquad (1\text{–}23)$$

This is the equation of a circle of the form

$$(\sigma_n - a)^2 + \tau^2 = R^2$$

where the radius

$$R = \sqrt{\left(\frac{\sigma_x - \sigma_y}{2}\right)^2 + \tau_{xy}^2}$$

and $a = (\sigma_x + \sigma_y)/2$. The center of the circle lies at a point$[(\sigma_x + \sigma_y)/2, 0]$. Note that the center of the circle always lies on the σ_n axis.

The above discussion may be more meaningfully complemented and expanded by an actual construction of Mohr's circle. As a start, some basic steps might be outlined:

1. The *normal* stresses are plotted as horizontal coordinates. The tensile stresses are considered positive (plotted to the right of origin); the

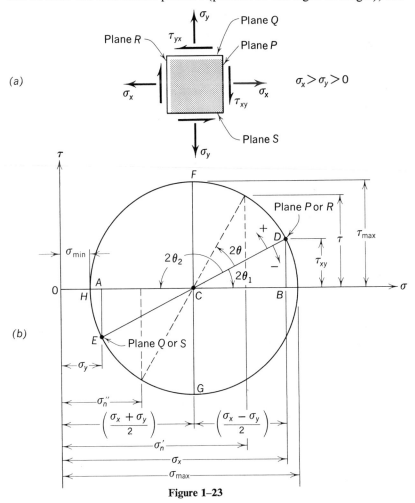

Figure 1–23

compressive stresses are considered negative (plotted to the left of origin).

2. The shearing stresses are plotted as vertical coordinates. Positive shear stresses (clockwise rotation of element □ ↓) are plotted above the origin; the negative stresses (□ ↑) are plotted below the origin.

3. *Positive* angles on the circle are obtained when measured in the *counterclockwise* sense; *negative* angles on the circle are obtained in the *clockwise* sense.[7] An angle of 2θ on the circle corresponds to an angle of θ on the element.

Figure 1-23b shows Mohr's circle for the assumed stresses given in Figure 1-23a at the stressed point, assuming that $\sigma_x > \sigma_y > 0$. The normal and shear stresses on vertical plane P give the coordinates of the one point on the circle, as indicated in Figure 1-23b by point D. Plane R, also vertical and 180 degrees from P, plots $2\theta = 360$ degrees on the circle from plane P, resulting in the *same* point. Similarly, the stresses on horizontal planes Q and S plot a different single point E. This shows that the stresses on parallel planes of an element oriented at any angle are correspondingly equal. The point of intersection of a line DE connecting these two points and the horizontal axis σ_n gives the location of the center of the circle. The diameter of the circle is the length of line DE.

The circle proves to be, indeed, a valuable visual aid. For example, the maximum shear stress is obviously equal to the radius of the circle. The orientation of this stress with respect to the original plane can either be measured or easily calculated with the help of trigonometry. The principal stresses σ_{max} and σ_{min} can, likewise, be easily spotted and determined, and orientation can easily be measured or calculated. Furthermore, on the principal planes (at σ_{max} and σ_{min}), the shear stresses are zero. The following example should prove helpful in further explaining the mechanics of constructing and interpreting a Mohr's circle.

Example 1-18:

Given: The stresses on an element representing a point in a loaded member. The magnitudes and orientations of the stresses are given in Figure *a*, on the following page.

Find: (a) The principal stresses and principal planes, (b) the maximum shearing stresses, and (c) the normal and shearing stresses on the plane *m–n* where $\theta = 30$ degrees. Construct Mohr's circle and draw a sketch of the elements.

Procedure: The Mohr's circle for our condition is shown in Figure *b*.

(a) The principal stresses occur on the σ axis. By inspection, they are (4000 ± 5000) psi, or $\sigma_{max} = +9000$ psi, $\sigma_{min} = -1000$ psi (compression). Relative to the P face on the element in Figure *a*, σ_{max} occurs on a

---

[7] This sign convention should not be confused with that for shear stresses.

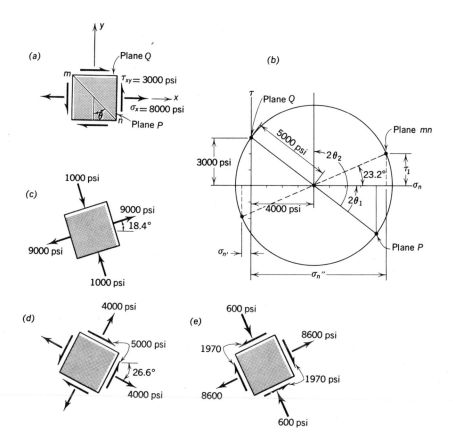

(a) Plane Q, $\tau_{xy} = 3000$ psi, $\sigma_x = 8000$ psi, Plane P, θ, m, n

(b) Plane Q, Plane mn, Plane P, 5000 psi, 3000 psi, 4000 psi, $2\theta_2$, $23.2°$, $2\theta_1$, τ_1, σ_n, $\sigma_{n'}$, σ_n''

(c) 1000 psi, 9000 psi, 18.4°, 9000 psi, 1000 psi

(d) 4000 psi, 5000 psi, 26.6°, 4000 psi

(e) 600 psi, 8600 psi, 1970, 1970 psi, 8600, 600 psi

plane which is rotated 2θ degrees counterclockwise. The other faces are rotated through the same angles, as shown in Figure c: $\theta_1 = \frac{1}{2}\tan^{-1} 3000/4000$; $\theta_1 = \frac{1}{2}(36.8°) = 18.4$ degrees. Thus, Figure c summarizes the answers to part (a).

(b) The maximum shear stress is $\tau_{max} = +5000$ psi. Relative to the vertical plane P, it occurs at $\theta_2 = (2\theta_1 + 90)/2$, or $\theta_2 = \theta_1 + 45$ degrees. The shearing and normal stresses are shown on a properly oriented element, Figure d.

(c) Corresponding to a "30-degree plane" on the element, we rotate 60 degrees, counterclockwise, on the circle. Referring to the circle, $\sigma_{n''}, \sigma_{n'}, = 4000 \pm 5000 \cos 23.2°$ and $\tau_{1,2} = \pm 5000 \sin 23.2°$. If the circle is constructed to scale, these values can be measured directly, eliminating the analytical work. The stresses on a properly oriented element are shown in Figure e.

PROBLEMS

1–132 Solve Problem 1–113 by use of Mohr's circle.
1–133 Solve Problem 1–115 by use of Mohr's circle.
1–134 Solve Problem 1–117 by use of Mohr's circle.
1–135 Solve Problem 1–118 by use of Mohr's circle.
1–136 Solve Problem 1–119 by use of Mohr's circle.
1–137 Solve Problem 1–121 by use of Mohr's circle.
1–138 Solve Problem 1–124 by use of Mohr's circle.
1–139 Solve Problem 1–125 by use of Mohr's circle.
1–140 Solve Problem 1–126 by use of Mohr's circle.
1–141 Solve Problem 1–129 by use of Mohr's circle.
1–142 Solve Problem 1–130 by use of Mohr's circle.

1—13 Thin-Walled Pressure Vessels

Pipes and tanks, hydraulic cylinders, or an inner tube from an automobile wheel, a basketball or a football are but a few items which may generally fall in the category of *pressure vessels*. Internal pressure produces tensile stresses in the walls of each of these "containers." Here we are concerned with the analysis of these vessels and the determination of the stresses in their walls.

The stresses in the walls of vessels subjected to internal pressure vary over the thickness of the wall, from a maximum at the inner surface to a minimum at the outer surface.[8] However, if the wall thickness is small in comparison with the principal radii of curvature, say a ratio of 1:10 or less, the stresses may be assumed uniformly distributed over the thickness of the wall. This "averaging" of the stresses results in but a small error for such thickness-to-radius ratios, perhaps less than 4 or 5 percent. However, the simplification of the stress calculation, as a result of the assumption, is appreciable. This will be the assumption we will use here to determine the stresses in the vessel walls.

Two types of curved surfaces will be considered here: (*a*) curvature in one direction only, and (*b*) curvature in two directions. Even though (*a*) is actually a special case of (*b*), it shall be considered separately. It is encountered frequently enough that it deserves this individual attention.

Curvature of Surface in One Direction. A thin-wall pipe or a cylindrical tank may be regarded as common examples of vessels whose walls have only one curvature. Figure 1–24*a* shows a closed circular cylindrical tank

---

[8] See, for example, S. Timoshenko and J.N. Goodier, *Theory of Elasticity*, 2nd ed., McGraw-Hill Book Company, Inc., New York, 1951.

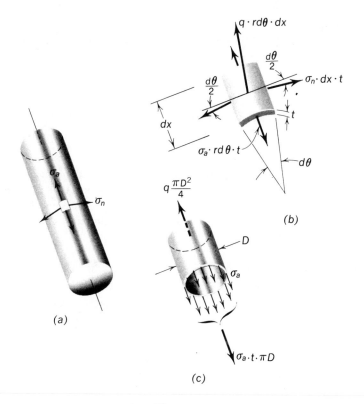

Figure 1–24

subjected to an internal pressure q, pounds per square inch. On an element from the wall, as shown in Figure 1–24a, there exists a *hoop* or *circumferential stress* σ_n and an *axial* or *longitudinal stress* σ_a, as shown.

The resultant pressure on the isolated element shown in Figure 1–24b is $q \cdot rd\theta \cdot dx$. Since σ_a has no component in this direction, the resultant pressure has to be resisted by the components of the hoop forces represented by $\sigma_n \cdot dx \cdot t$.

$$\Sigma F_n = 0: \qquad q \cdot rd\theta \cdot dx = 2\left(\sigma_n \cdot dx \cdot t \cdot \sin \frac{d\theta}{2}\right)$$

Because $\sin d\theta/2$ approaches $d\theta/2$ for small angles, the above expression may be written as

$$q \cdot rd\theta \cdot dx = 2\left(\sigma_n \cdot dx \cdot t \cdot \frac{d\theta}{2}\right)$$

Hence, the hoop stress is

$$\sigma_n = \frac{qr}{t} \qquad\qquad (1\text{–}24)$$

To determine the axial stress σ_a, let us use the isolated portion of the vessel shown in Figure 1–24c. The pressure resultant in the axial direction, $q[\pi D^2/4]$, has to be resisted by the forces in the wall of the cylinder, $\sigma_a \cdot t \cdot \pi D$; that is,

$$q\frac{(\pi D^2)}{4} = \sigma_a \cdot t \cdot \pi D$$

substituting $r = D/2$, the axial stress is

$$\sigma_a = \frac{qr}{2t} \tag{1-25}$$

Comparing Equations (1–24) and (1–25) we note that the *axial stress is only one-half the hoop stress.* Equation (1–25) is valid for circular cylinders only.

Curvature in Two Directions. Consider a vessel, perhaps shaped as the football mentioned above. The stresses in the two directions of curvature are denoted by σ_1 and σ_2, as shown in Figure 1–25a.

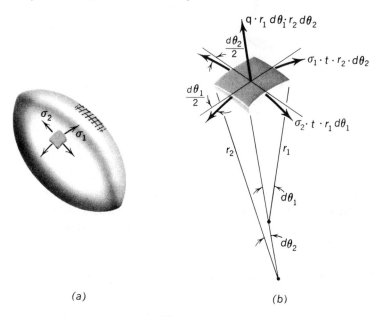

(a)   (b)

Figure 1–25

The lengths of the sides of the element are $r_1 d\theta_1$ and $r_2 d\theta_2$. Therefore, the resultant pressure $q \cdot r_1 d\theta_1 \cdot r_2 d\theta_2$, normal to the surface of the element, has to be resisted by the components of the forces parallel to the normal; that is,

$$q \cdot r_1 d\theta_1 \cdot r_2 d\theta_2 = 2\left(\sigma_1 \cdot t \cdot r_2 d\theta_2 \cdot \sin\frac{d\theta_1}{2}\right) + 2\left(\sigma_2 \cdot t \cdot r_1 d\theta_1 \cdot \sin\frac{d\theta_2}{2}\right)$$

and because $\sin d\theta/2 \approx d\theta/2$, for small angles,

$$q \cdot r_1 d\theta_1 \cdot r_2 d\theta_2 = \sigma_1 \cdot t \cdot r_2 d\theta_2 \cdot d\theta_1 + \sigma_2 \cdot t \cdot r_1 d\theta_1 \cdot d\theta_2$$

where $q \cdot r_1 r_2 = \sigma_1 t \cdot r_2 + \sigma_2 \cdot t \cdot r_1$. Dividing both sides by $r_1 \cdot r_2 \cdot t$, we get

$$\frac{\sigma_1}{r_1} + \frac{\sigma_2}{r_2} = \frac{q}{t} \tag{1-26}$$

Equation (1–26) becomes Equation (1–24) when the pressure vessel is cylindrical, and therefore one of the radii in Equation (1–26) is infinitely large.

Example 1–19:

Given: A spherical container of radius r, perhaps a basketball, which is subjected to an internal gas pressure q lb/sq in.
Find: An expression for the skin stress in terms of q, r, and wall thickness t.

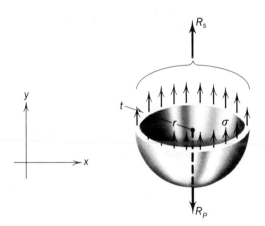

Solution: Assume that the sphere is cut into two equal sections and place one of the hemispheres in equlibrium, as shown in the figure. The total downward force due to pressure is equal to the pressure times the projected area in that direction; that is $R_P = q(\pi r^2)$. This must be resisted by the stresses in the walls σ times the wall's cross-sectional area $(2\pi r)t$; that is, $R_s = \sigma \cdot 2\pi r \cdot t$.

$$\Sigma F_y = 0: \quad q(\pi r^2) = \sigma \cdot 2\pi r \cdot t$$

Therefore,

$$\sigma = \frac{qr}{2t}$$

This may also have been obtained using Equation (1–26) if we recognized that $r_1 = r_2$ and $\sigma_1 = \sigma_2$.

Example 1–20:

Given: A 0.1-in. thick wall rubber inner tube from a truck tire has the dimensions shown in Figure *a*. The cross section is assumed to be circular.

Find: The stresses in the walls at points *A* and *C* shown in Figure *b* when the tube is inflated with a 15-lb/sq in. air pressure.

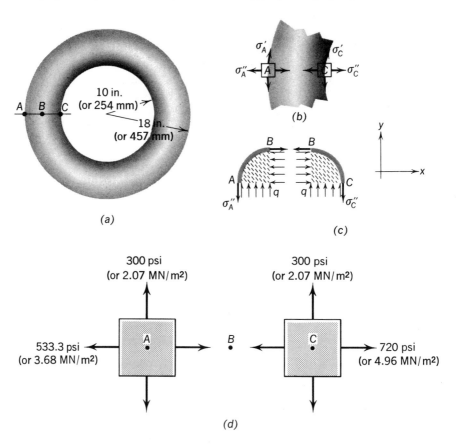

(a)

(b)

(c)

(d)

Solution: Equation (1–26) applies here with only a minor adjustment of signs to correct for a different "curvature" at *C*, as indicated in Equation (*b*) below.

For point *A*:
$$\frac{\sigma_A{}'}{18} + \frac{\sigma_A{}''}{4} = q/t \tag{a}$$

For point *C*:
$$-\frac{\sigma_C{}'}{10} + \frac{\sigma_C{}''}{4} = q/t \tag{b}$$

One of the unknowns from Equations (a) and (b) can be found by setting one-quarter of the cross section of the tube in equilibrium, as shown in Figure c. For convenience, the resultant pressure for the individual quarters is resolved into component in the y and x directions, acting on the projected planes. The upward resultant force for the outer segment is resisted by the forces in the tube wall along the outer edge, through A; that is,

$$\Sigma F_y = 0: \qquad \underbrace{\sigma_A''(t \cdot 2\pi \cdot 18)}_{\text{Wall area}} = \underbrace{q\pi(18^2 - 14^2)}_{\text{Projected area in } y \text{ direction.}}$$

(for outer quarter)

Therefore,

$$\sigma_A'' = q\frac{(18^2 - 14^2)}{t \cdot 2 \cdot 18} = \frac{15(128)}{(0.1)(2)(18)} = \textbf{533.3 psi (or 3.68 MN/m}^2\textbf{)}$$

$$\Sigma F_y = 0: \qquad \sigma_C''(t \cdot 2\pi \cdot 10) = q\pi(14^2 - 10^2)$$

(for inner quarter)

$$\sigma_C'' = \frac{q(14^2 - 10^2)}{t \cdot 2 \cdot 10} = \frac{15(96)}{2} = \textbf{720 psi (or 4.96 MN/m}^2\textbf{)}$$

Substituting in (a),

$$\frac{\sigma_A'}{18} + \frac{533.3}{4} = \frac{15}{0.1}$$

$$\sigma_A' = \left(\frac{15}{0.1} - \frac{533.4}{4}\right)18 = \textbf{300 psi (or 2.07 MN/m}^2\textbf{)}$$

Similarly, from Equation (b),

$$-\frac{\sigma_C'}{10} + \frac{720}{4} = \frac{15}{0.1}$$

$$\sigma_C' = \left(\frac{720}{4} - \frac{15}{0.1}\right)10 = \textbf{300 psi (or 2.07 MN/m}^2\textbf{)}$$

The stresses on the respective elements of walls are shown in Figure d.

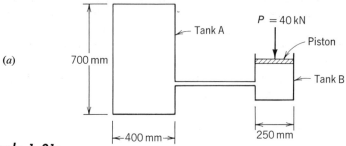

(a)

Example 1–21:

Given: In Figure *a*, an enclosed system supports a load *P* as shown. Assume no pressure *loss* between tank B and A.

Find: If an allowable stress of 70 MN/m² is given in the walls of tank A, determine the thickness of the wall of tank A.

Solution: The pressure

$$q = \frac{40 \text{ kN}}{(\pi/4)(0.25)^2} = 815 \text{ kN/m}^2$$

is transmitted from tank B to tank A. Therefore,

$$\sigma_{max} = \frac{qr}{t} = \frac{(815 \text{ kN/m}^2)\left(\frac{0.40}{2} \text{ m}\right)}{t} = 70 \text{ MN/m}^2$$

Solving,

$$t = 0.00233 \text{ m} = 2.33 \text{ mm}$$

PROBLEMS

1–143 A standard 12-in. diameter (nominal size) pipe, closed at its ends, of a wall thickness of 0.375 in., carries steam under 300-psi pressure. Determine (*a*) the average circumferential (or hoop) and axial (or longitudinal) stresses in the wall of the pipe, and (*b*) the principal stresses on an element taken from the wall of the pipe.

1–144 A cylindrical tank 6 ft in diameter, made of a ¾-in.-thick plate, is to store a certain gas under pressure. Determine the maximum pressure the tank can withstand if the allowable working stress in the tank is 20,000 psi, tension.

1–145 A spiral-welded penstock, 12 ft diameter and 48 ft high, made of ½- in.-thick plate, is filled with water. The pitch of the spiral is 15 ft. Determine the normal and shearing forces the weld carries per linear inch near the bottom. Assume the water weighs 62.4 lb/cu ft.

1–146 Determine the principal stresses in the walls near the bottom of the penstock of Problem 1–145.

1–147 Determine the maximum gas pressure that a spherical shell, 12 ft in diameter and ¼-in.-thick wall, can withstand if the tensile stresses are not to exceed 18,000 psi.

1–148 In Figure P 1–148, a boy pumps his basketball with a downward force of 25 lb on the pump, as shown. Determine (a) the internal pressure in the basketball, and (b) the maximum normal stresses in the skin of the ball.

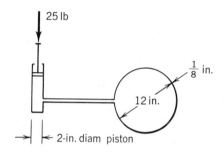

Figure P 1–148

1–149 Figure P 1–149 shows the external dimensions of a football. Determine the stresses on an element at A, if $t = $ ⅛ in., $q = 13$ psi. The larger radius at point A is 6.75 in.

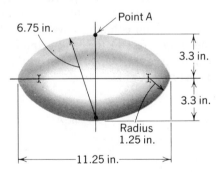

Figure P 1–149

1–150 Determine the maximum stress on an element from the skin of a basketball, $t = $ ⅛ in., when the ball is inflated to a pressure of $q = 13$ psi. Assume the basketball has a diameter of 12 in.

1–151 A 1-in.-diam copper tube is to be used as an ordinary home-water pipe. If the stress in the pipe is not to exceed 5000 psi, determine the minimum thickness the pipe must have for a water pressure of $q = 100$ psi. Neglect the axial stresses in the pipe.

1–152 The vessel shown in Figure P 1–152 is subjected to a gas pressure of 250 psi. Determine the stresses in the wall at point A if the thickness of the wall is ½ in.

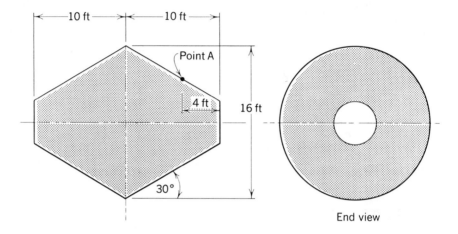

Figure P 1–152

1–153 A cylindrical tank 2 m long of 6-mm-thick plate is capped on each end by the spherical ends and is subjected to a hydrostatic pressure. Figure P 1–153 shows the longitudinal cross section of the tank. (a) What is the maximum column of water, H, the tank can safely withstand if the *shear* stress in the tank at point A is not to exceed 80 MN/m²? (b) What is the *normal* stress on the plane of maximum shear stress in part (a)?

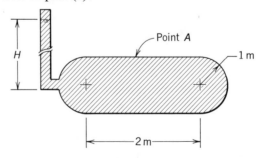

Figure P 1–153

1—14 Stress Concentrations

Up to now, in our discussion of uniform intensities of stress on members subjected to axial tension or compression, we have assumed the member to have a prismatic form. Also, we have implied that the stress at any one section anywhere along its length can be determined by merely dividing the load by the cross-sectional area at that point. Although this is frequently true, in many cases it is not. The stress distribution near the point where a concentrated load is applied, for example, is different than at some distance from this point, even for prismatic members. Furthermore, changing the cross section abruptly results in a changing *distribution* of stress, depending upon the magnitude and type of load (static, impact, repetitive, concentrated, uniform), upon the type of material (ductile or brittle), upon the type of change in the cross section (holes, fillets, notches, and so forth) and the relative size of the changed cross section to the others. At points of abrupt changes, the stresses are considerably greater than the average, making these points the most vulnerable spots for failures, especially under alternating loads, or in the case of brittle materials. We call this change in force intensity *stress concentration*, indeed an important factor in machine design.

Various experimental methods for determining the stress distribution around notches or holes are available. They are useful tools not only to check analytical results, but also to serve as substitutes when the analytical approach becomes unreasonably complex. The use of electrical resistance strain gages, bonded either directly to the prototype or to a model, and various "stress-coat" techniques have aided appreciably the experimental efforts in this regard. Flexible rubber models and photoelasticity depicts quite convincingly the concentration of stresses.

Much experimental and analytical work in this field has been done and the results are compiled in a way that is useful to a designer.[9] The peak stress is related to an average stress in the elastic range of the material by a term called the *stress concentration factor*, commonly represented by k. In equation form, $k = \sigma_{peak}/\sigma_{avg}$. As mentioned previously, the distribution and magnitude of stress, and thus the stress concentration factor, are functions of several variables, important among which are the relative size and shape compared to various dimensions of the stressed specimens, and the type of load.

Figures 1–26a to 1–26c are presented here to merely portray the type of stress rise that occurs in some commonly encountered shapes, and not to even attempt to summarize the much-more comprehensive tabulations presented and explained in other sources, for many other shapes and types

---

[9] See, for example, R. J. Roark, *Formulas for Stress and Strain*, McGraw-Hill Book Company, Inc., New York, 1954; or R. E. Peterson, *Stress Concentration Factors*, John Wiley & Sons, Inc., New York, 1953.

of load. Associated with each figure is the theoretically determined maximum k. For very small radii, such that when *the ratio of the radius to width approaches zero* (that is, when r/b and r/d approach zero, Figure 1–26) *the stress concentration factor approaches 3*, an interesting and useful fact to remember. The effect of a very small radius is particularly striking because the stress rises with the reduction of radius, and reduces as the ratios of radius to width increase, approaching 1 as the ratio approaches infinity. For this reason, the stress around cracks (small radius) in a machine part is often reduced by merely drilling a hole at that point.

These abrupt changes on the stress distribution are different after the stress passes the proportional limit; the stress becomes more uniform as the

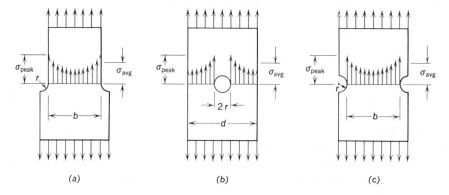

(a) (b) (c)

Figure 1–26 (*a*) Fillets. Theoretical $k = 3$, for $r/b = 0$; 1, for $r/b = \infty$.
(*b*) Circular hole. $k = 3$, for $r/d = 0$; $k = 1$, for $r/d = \infty$.
(*c*) Semicircular notches. $k = 3$, for $r/b = 0$; $k = 1$, for $r/b = \infty$.

material elongates. Also, ductile materials are less effected by abrupt changes of cross section than are brittle materials. And, finally, one should take particular note that the stress concentration is likely to have more disastrous effects when the object undergoes reversal of repeated loading than for static load.

1—15 Analysis of Plane Strain

The brief introduction of strain in Section 1–4 focused on the individual strains produced either by normal forces or by shear forces. Now we shall extend our discussion to a more general case of plane strain, and subsequently relate axial strains and shear strain in a manner analogous to that for plane stress which was introduced in Sections 1–11 and 1–12.

Let us consider an infinitesimal element subjected to a general case of plane stress, as shown in Figure 1–27a. The resulting deformation, shown

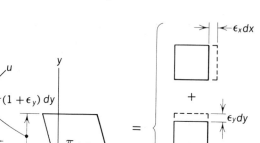

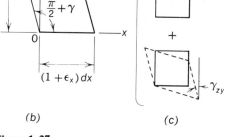

Figure 1–27

greatly exaggerated in Figure 1–27b, represents the superposition[10] of the small linear and angular deformations, illustrated in Figure 1–27c.

The normal strains (ϵ_x, ϵ_y) are considered positive if the element is elongated and negative if the element is compressed; the shear strain (γ_{xy}) is considered positive when the angle yOx is increased, as shown in Figure 1–27b. Hence, we proceed to relate the strains ϵ_x, ϵ_y, and γ_{xy} associated with the x–y coordinate system with strains of another coordinate system u–v which makes an angle θ with the x–y system, as shown in Figure 1–27a. For clarity, the three cases of strain in Figure 1–27c are magnified in Figures 1–28a, 1–28b, and 1–28c. If ds represents the diagonal length of the undeformed element, then it follows that

$$dx = ds \cdot \cos \theta \quad \text{and} \quad dy = ds \cdot \sin \theta$$

$$\frac{dx}{ds} = \cos \theta \quad \text{and} \quad \frac{dy}{ds} = \sin \theta$$

But strains ϵ_x, ϵ_y, and γ_{xy} cause a change in ds, as depicted in Figures 1–28a, 1–28b, and 1–28c, respectively. Hence, the corresponding strains in the μ and v directions, respectively, are from Figure 1–28.

$$\epsilon_u = \frac{\epsilon_x dx \cos \theta + \epsilon_y dy \sin \theta - \gamma_{xy} dx \sin \theta}{ds}$$

---

[10] For small strains, the order of superposition is irrelevant; for large strains where planes, straight lines, or parallel lines may become "deformed," care must be exercised in the superposing of strains. Here we shall assume only small deformations where the order of superposing is irrelevant; hence, powers higher than one will be ignored in any related equations.

Substituting for dx/ds and dy/ds, we obtain

$$\epsilon_u = \epsilon_x \cos^2 \theta + \epsilon_y \sin^2 \theta - \gamma_{xy} \cos \theta \sin \theta \qquad (a)$$

By merely substituting $(\theta + 90°)$ for θ in the above equation, we obtain ϵ_v:

$$\epsilon_v = \epsilon_x[\cos(\theta + 90)]^2 + \epsilon_y[\sin(\theta + 90)]^2 - \gamma_{xy} \cos(\theta + 90)\sin(\theta + 90)$$

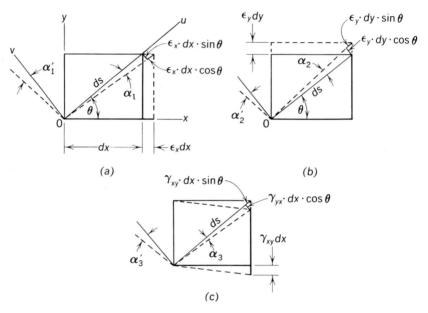

Figure 1–28

But $\sin(\theta + 90°) = +\cos\theta$ and $\cos(\theta + 90°) = -\sin\theta$. Hence, we get

$$\epsilon_v = \epsilon_x \sin^2 \theta + \epsilon_y \cos^2 \theta + \gamma_{xy} \sin \theta \cos \theta \qquad (b)$$

Equations (a) and (b) may be made less cumbersome if they are presented in terms of double angles. From the trigonometric identities, we have

$$\cos^2 \theta = \tfrac{1}{2}(1 + \cos 2\theta); \quad \sin^2 \theta = \tfrac{1}{2}(1 - \cos 2\theta)$$

which, when substituted into Equations (a) and (b), respectively, yield Equations (1–27a) and (1–27b).

$$\epsilon_u = \frac{\epsilon_x + \epsilon_y}{2} + \frac{\epsilon_x - \epsilon_y}{2} \cos 2\theta - \frac{\gamma_{xy} \sin 2\theta}{2} \qquad (1\text{–}27a)$$

and

$$\epsilon_v = \frac{\epsilon_x + \epsilon_y}{2} + \frac{\epsilon_x - \epsilon_y}{2} \cos 2\theta + \frac{\gamma_{xy} \sin 2\theta}{2} \qquad (1\text{–}27b)$$

The expression for the shearing strain γ_{uv} may be determined in a similar manner. We first determine the angular deviation (expressed by α in Figure

1–28) of axis u and that of v (expressed by α'). Now we observe that α and α' rotate in opposite directions, and thus the total change in the right angle γ_{uv} is the algebraic difference of α and α'. Hence, from Figure 1–28, the angular deviation is the increment of length perpendicular to axis u divided by ds; that is,

$$\alpha = \frac{\epsilon_x dx \sin \theta}{ds} - \frac{\epsilon_y dy \cos \theta}{ds} + \frac{\gamma_{xy} dx \cos \theta}{ds}$$

Substituting $\cos \theta$ for dx/ds, and $\sin \theta$ for dy/ds, we have

$$\alpha = \epsilon_x \cos \theta \sin \theta - \epsilon_y \sin \theta \cos \theta + \gamma_{xy} \cos^2 \theta$$

Again substituting $(\theta + 90°)$ for θ, and observing that $\sin (\theta + 90°) = +\cos \theta$ and $\cos (\theta + 90°) = -\sin \theta$, we obtain α'; that is,

$$\alpha' = -\epsilon_x \sin \theta \cos \theta + \epsilon_y \cos \theta \sin \theta + \gamma_{xy} \sin^2 \theta$$

The total angle change γ_{uv} is therefore

$$\gamma_{uv} = (\epsilon_x - \epsilon_y)\cdot 2 \sin \theta \cos \theta + \gamma_{xy}(\cos^2 \theta - \sin^2 \theta)$$

From the trigonometric identities of $\sin \theta \cos \theta = \frac{1}{2} \sin 2\theta$, $\cos 2\theta = \frac{1}{2}(1 + \cos 2\theta)$, and $\sin^2\theta = \frac{1}{2}(1 - \cos 2\theta)$, we have

$$\gamma_{uv} = (\epsilon_x - \epsilon_y)\sin 2\theta + \gamma_{xy} \cos 2\theta$$

or

$$\tfrac{1}{2}\gamma_{uv} = \left(\frac{\epsilon_x - \epsilon_y}{2}\right)\sin 2\theta + \frac{\gamma_{xy}}{2} \cos 2\theta \qquad (1\text{--}28)$$

Equations (1–27) and (1–28) for plane strain reveal their complete resemblence in form to Equations (1–16) and (1–17) for plane stress. The normal (linear) strains ϵ_x, ϵ_y, and ϵ_u (or ϵ_v) in Equations (1–27) and (1–28) correspond, respectively, to stresses σ_x, σ_y, and σ_n(or σ_n') in Equation (1–16); the half strains $\gamma_{xy}/2$ and $\gamma_u/2$ in Equation (1–28) correspond, respectively, to τ_{xy} and τ in Equation (1–17).

1—16 Mohr's Circle for Plane Strain

We may construct a Mohr's circle for normal and shear strains in a manner analogous to that for normal and shear stresses. Rearranging terms in Equations (1–27) and (1–28) and squaring both sides of the equations, we have

$$\left[\epsilon_u - \left(\frac{\epsilon_x + \epsilon_y}{2}\right)\right]^2 = \left[\left(\frac{\epsilon_x - \epsilon_y}{2}\right)\cos 2\theta - \frac{\gamma_{xy}}{2} \sin 2\theta\right]^2 \qquad (a)$$

and

$$\left(\frac{\gamma_{uv}}{2}\right)^2 = \left[\left(\frac{\epsilon_x - \epsilon_y}{2}\right)\sin 2\theta + \frac{\gamma_{xy}}{2} \cos 2\theta\right]^2 \qquad (b)$$

By first expanding Equations (a) and (b) then simplifying and adding, we get

$$\left(\epsilon_u - \frac{\epsilon_x + \epsilon_y}{2}\right)^2 + \left(\frac{\gamma_{uv}}{2}\right)^2 = \left(\frac{\epsilon_x - \epsilon_y}{2}\right)^2 + (\gamma_{xy})^2$$

which is the equation of a circle of the form

$$(\epsilon_u - a)^2 + \left(\frac{\gamma_{uv}}{2}\right)^2 = R^2$$

where

$$R = \sqrt{\left(\frac{\epsilon_x - \epsilon_y}{2}\right)^2 + \left(\frac{\gamma_{xy}}{2}\right)^2} \quad \text{and} \quad a = \left(\frac{\epsilon_x + \epsilon_y}{2}\right)$$

The center of the circle lies at a point $[(\epsilon_x + \epsilon_y)/2, 0]$, and always on the ϵ axis. Thus, ϵ_x and ϵ_y are plotted on the abscissa, and $(\gamma_{xy}/2)$ on the ordinate. As in the case of stress, the angle 2θ on Mohr's circle for strain corresponds to an angle of θ on the element. Also, the maximum and minimum axial strains, denoted by ϵ_1 and ϵ_2 and called *principal strains*, occur at the two points where the circle cuts the abscissa; hence, at these points the shear strain is zero. The axis on which the principal strains occur is known as the *principal axis*.

We shall now complement the above discussion with a sketch of a strained element and the Mohr circle for the strain, as shown in Figures 1–29a and 1–29b, respectively. In our discussion, we will assume that $\epsilon_x >$

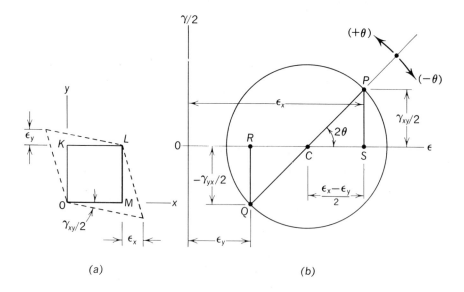

(a)　　　(b)

Figure 1–29

$\epsilon_y > 0$,[11] and that both are extensional strains; hence, both are positive and are plotted as positive abscissas. We note that $\gamma_{xy}/2$ is positive because line OM rotates clockwise relative to the x axis, increasing $\angle KOM$. Thus, on Mohr's circle we have point P, corresponding to the x direction; point Q corresponds to the y direction. Connecting points P and Q with a straight line (diameter of circle) we obtain the center of the circle, the diameter and, subsequently, the circle itself.

Several quantities can now be obtained directly (either measured if scale construction is carried out, or from geometric relationships), with the principal strains and principal axis being particularly important. From the geometry of Mohr's circle, we have

$$\text{Distance } OC = \frac{\epsilon_x + \epsilon_y}{2}$$

$$\text{Radius } CP = \sqrt{\left(\frac{\epsilon_x - \epsilon_y}{2}\right)^2 + \left(\frac{\gamma_{xy}}{2}\right)^2}$$

Hence,

$$\epsilon_{\text{max,min}} = \left(\frac{\epsilon_x + \epsilon_y}{2}\right) \pm \sqrt{\left(\frac{\epsilon_x - \epsilon_y}{2}\right)^2 + \left(\frac{\gamma_{xy}}{2}\right)^2} \qquad (1\text{--}29)$$

The direction of the principal axis with reference to line OM is given by

$$\tan 2\theta = -\frac{(\gamma_{xy}/2)}{\left(\frac{\epsilon_x - \epsilon_y}{2}\right)} = -\frac{\gamma_{xy}}{\epsilon_x - \epsilon_y} \qquad (1\text{--}30)$$

The maximum shear strain can also be easily determined from the circle:

$$\left(\frac{\gamma}{2}\right)_{\text{max}} = \text{radius} = \sqrt{\left(\frac{\epsilon_x - \epsilon_y}{2}\right)^2 + \left(\frac{\gamma_{xy}}{2}\right)^2} \qquad (1\text{--}31)$$

Again, we note the complete similarity between expressions for principal stresses [Equations (1–19) and (1–21)] and principal strains [Equations (1–29) and (1–31)].

---

[11] ϵ_x and ϵ_y are positve for elongation; $\gamma_{xy}/2$ is positive when the lower left-hand corner of the element is increased.

Example 1–22:

Given: The strain at a point in a body subjected to plane strain.
Find: The magnitude and direction of principal strains.

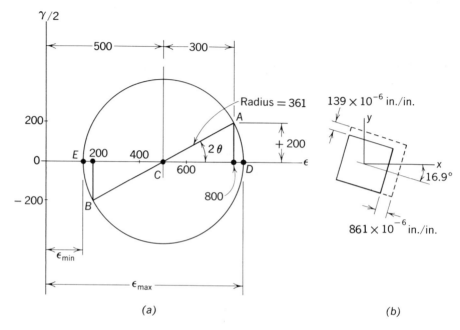

(a)   *(b)*

Procedure: For convenience, the factor (10^{-6}) will be omitted in the scaling of the circle, but will be reintroduced in the final result; that is, for constructing Mohr's circle we shall only scale $\epsilon_x = +800$, $\epsilon_y = + 200$, and $\gamma_{xy}/2 = +200$. Mohr's circle for these data is shown in Figure *a*. From the circle we observe the maximum and minimum strains (principal strains) to be distances $OD \times 10^{-6}$ and $OE \times 10^{-6}$, respectively; that is,

$$\epsilon_{\text{max}} = (OC + \text{radius}) \times 10^{-6} = (500 + 361) \times 10^{-6}$$
$$= 861 \times 10^{-6} \text{ in./in.}$$

$$\epsilon_{\text{min}} = (OC - \text{radius}) \times 10^{-6} = (500 - 361) \times 10^{-6}$$
$$= 139 \times 10^{-6} \text{ in./in.}$$

The angle between the maximum-strain axis and the x axis is equal to one-half the angle ACD, that is,

$$2\theta = \tan^{-1}\frac{200}{300} = 33.8°$$
$$\theta = 16.9° \text{ clockwise}$$

Figure *b* is an attempt to illustrate this state of strains (principal strains and orientation).

PROBLEMS

The state of plane strain at a point is defined by the data given in the following problems. For each problem (a) construct a Mohr's circle, (b) determine the principal strains and direction of the principal axis, and (c) draw a sketch of the element in the state of principal strains.

1–154 $\epsilon_x = +\ 900 \times 10^{-6}$; $\epsilon_y = -\ 100 \times 10^{-6}$; $\gamma_{xy} = +\ 300 \times 10^{-6}$.

1–155 $\epsilon_x = -\ 1100 \times 10^{-6}$; $\epsilon_y = +\ 500 \times 10^{-6}$; $\gamma_{xy} = -\ 200 \times 10^{-6}$.

1–156 $\epsilon_x = 0$; $\epsilon_y = 800 \times 10^{-6}$; $\gamma_{xy} = +\ 400 \times 10^{-6}$.

1–157 $\epsilon_x = 0$; $\epsilon_y = 0$; $\gamma_{xy} = 1200 \times 10^{-6}$.

1—17 Strain Measurement and Strain Rosette

It would be indeed convenient if we could obtain stresses, especially principal stresses, by direct measurement. This, however, is impossible. Hence, we resort to direct measurement of strain, which is possible, and subsequently compute stress in terms of strain from $\sigma = E\epsilon$.[12]

There are several methods of measuring strain, with (a) photoelasticity, (b) brittle lacquers, and (c) electrical (wire or foil) gages, the most prominant of the group. The latter, in particular, is an especially convenient method to measure linear strains on a surface. The most common of the electrical gages consists of a wire element (in the neighborhood of 0.001 in. in diameter) fixed to a tough paper base, which is cemented to the surface of the specimen. As the specimen is deformed, the paper and the wire also deform, causing a change in the diameter of the wire. The change in diameter, in turn, causes a change in the resistance of the wire and subsequently a change in the voltage passing through the gage. Calibrated instruments (the wheatstone bridge, for example) can then be used to transform these changes into useful strain data.

It would be a rather simple matter to determine the principal strains, as shown in the preceding section, if strains ϵ_x, ϵ_y, and γ_{xy} can be measured. In fact, γ_{xy} would not be needed if the direction of the principal strains were known; we would merely orient two gages in the direction of the principal strains and thus measure the principal strains directly. Unfortunately, however, the principal axes of strain are not usually known. Also, the shearing strain γ_{xy} is quite difficult to measure with satisfactory accuracy. Hence, we resort to other means.

If we measured *three* strains, ϵ_a, ϵ_b, and ϵ_c at a common point, and oriented in three arbitrary directions, θ_a, θ_b, and θ_c, as shown in Figure 1–30, then from Equation (1–27) we obtain the following three simultaneous

[12] Other approaches are emerging; see, for example, J. W. Dally and W. F. Riley, *Experimental Stress Analysis*, McGraw-Hill Book Company, Inc., New York, 1965, pp. 438–442.

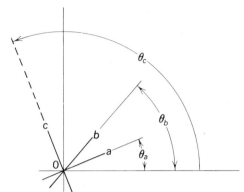

Figure 1–30

equations involving three unknowns ϵ_x, ϵ_y, and γ_{xy}:

$$\epsilon_a = \frac{\epsilon_x + \epsilon_y}{2} + \frac{\epsilon_x - \epsilon_y}{2} \cos 2\theta_a - \frac{\gamma_{xy}}{2} \sin 2\theta_a$$

$$\epsilon_b = \frac{\epsilon_x + \epsilon_y}{2} + \frac{\epsilon_x - \epsilon_y}{2} \cos 2\theta_b - \frac{\gamma_{xy}}{2} \sin 2\theta_b$$

$$\epsilon_c = \frac{\epsilon_x + \epsilon_y}{2} + \frac{\epsilon_x - \epsilon_y}{2} \cos 2\theta_c - \frac{\gamma_{xy}}{2} \sin 2\theta_c$$

The solution of these equation yields ϵ_x, ϵ_y, and γ_{xy}; hence, from Equation (1–29) or Mohr's circle we can determine the principal strains ϵ_1 and ϵ_2. Then the principal stresses σ_1 and σ_2 can be determined from

$$\sigma_1 = \frac{(\epsilon_1 + \mu\epsilon_2)E}{1 - \mu^2} \qquad \text{and} \qquad \sigma_2 = \frac{(\epsilon_2 + \mu\epsilon_1)E}{1 - \mu^2},$$

as given by Equation (1–12).

Electrical resistance strain gages are well adapted for measuring strain simultaneously in three or more directions near a point. (Actually, the strain is measured over a finite length; consequently, the strain measured is only an average over the length of the gage and, therefore, the principal strains only a good approximation of the actual strain.) The most common combination of gages consists of three gages whose axes are oriented either 45 degrees or 60 degrees apart. They are known as *strain rosettes*.

Example 1–23:

Given: A 60-degree rosette, attached to the surface of a strained specimen, yields the following strain reading near a point: $\epsilon_a = 1000 \times 10^{-6}$ in./in., $\epsilon_b = -500 \times 10^{-6}$ in./in., and $\epsilon_c = 400 \times 10^{-6}$ in./in. Assume $E = 30 \times 10^6$ psi, and $\mu = 0.30$.

Find: (a) The magnitudes and direction of the principal strains, and (b) the corresponding principal stresses.

Procedure: For convenience we may let the axis of gage *a* coincide with the *x* axis; that is, $\theta_a = 0$. Thus, it follows that $\theta_b = 60°$ and $\theta_c = 120°$. The three simultaneous equations would then become

$$\epsilon_a = \frac{\epsilon_x + \epsilon_y}{2} + \frac{\epsilon_x - \epsilon_y}{2}(1) - \frac{\gamma_{xy}}{2}(0) \tag{a}$$

$$\epsilon_b = \frac{\epsilon_x + \epsilon_y}{2} + \frac{\epsilon_x - \epsilon_y}{2}\left(-\frac{1}{2}\right) - \frac{\gamma_{xy}}{2}\left(\frac{\sqrt{3}}{2}\right) \tag{b}$$

$$\epsilon_c = \frac{\epsilon_x + \epsilon_y}{2} + \frac{\epsilon_x - \epsilon_y}{2}\left(-\frac{1}{2}\right) - \frac{\gamma_{xy}}{2}\left(\frac{-\sqrt{3}}{2}\right) \tag{c}$$

Solving, we obtain a useful generalization:

$$\epsilon_x = \epsilon_a$$

$$\epsilon_y = \frac{2(\epsilon_b + \epsilon_c) - \epsilon_a}{3}$$

$$\gamma_{xy} = \frac{2(\epsilon_c - \epsilon_b)}{\sqrt{3}}$$

Substituting for ϵ_a, ϵ_b, and ϵ_c we have

$$\epsilon_x = 1000 \times 10^{-6} \text{ in./in.}$$

$$\epsilon_y = \left[\frac{2(-500 + 400) - 1000}{3}\right] \times 10^{-6} = -400 \times 10^{-6} \text{ in./in.}$$

$$\gamma_{xy} = \frac{2(400 + 500) \times 10^{-6}}{\sqrt{3}} = 1030 \times 10^{-6} \text{ in./in.}$$

The corresponding Mohr's circle is shown in the figure below, from which we can either measure or calculate the principal strains and

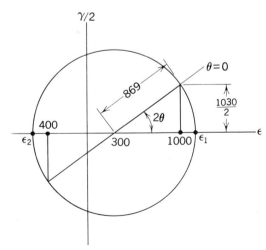

their directions. From the circle it is evident that:

$$\epsilon_1 = (300 + \text{radius}) \times 10^{-6} = (300 + 869) \times 10^{-6}$$
$$= \mathbf{1169 \times 10^{-6} \ in./in.}$$

and

$$\epsilon_2 = (300 - \text{radius}) \times 10^{-6} = -\mathbf{569 \times 10^{-6} \ in./in.}$$

ϵ_1 is obviously oriented $\theta°$, clockwise, with respect to the a axis; that is, from the figure, we have

$$\tan 2\theta = \frac{515}{700} = 0.734$$

or

$$\theta = \mathbf{18.10°} \text{ clockwise from } a \text{ axis, and so forth.}$$

The corresponding principal stresses are

$$\sigma_1 = \frac{(\epsilon_1 + \mu\epsilon_2)E}{1 - \mu^2} = \frac{(1169 - 0.3 \times 569)30}{1 - 0.09} = \mathbf{32,900 \ psi}$$

$$\text{(or } \mathbf{226.85 \ MN/m^2}) \text{ (tension)}$$

and

$$\sigma_2 = \frac{(\epsilon_2 + \mu\epsilon_1)E}{1 - \mu^2} = \frac{(-569 + 0.3 \times 1169)30}{1 - 0.09} = \mathbf{-7230 \ psi}$$

$$\text{(or } \mathbf{49.85 \ MN/m^2}) \text{ (compression)}$$

PROBLEMS

1–158 Derive the expression for ϵ_x, ϵ_y, and γ_{xy} in terms of ϵ_a, ϵ_b, and ϵ_c for a 45-degree rosette. (See Example 1–23.)

1–159 Show that for a 60-degree rosette, the expression for the principal strains is correct as

$$\epsilon_{\text{max,min}} = \frac{\epsilon_a + \epsilon_b + \epsilon_c}{3} \pm \frac{2\sqrt{\epsilon_a(\epsilon_a - \epsilon_b) + \epsilon_b(\epsilon_b - \epsilon_c) + \epsilon_c(\epsilon_c - \epsilon_a)}}{3}$$

1–160 Derive the expression for ϵ_{max}, in terms of ϵ_a, ϵ_b, and ϵ_c for a 45-degree rosette. (Problem 1–159 gives a corresponding result for a 60-degree rosette.)

1—18 Relationship between E, G, and μ

The relationship between E, G, and μ, expressed by Equation (1–6), may be obtained from a case of pure shear, as shown in Figure 1–31a. The corresponding Mohr's circle for this case is shown in Figure 1–31b. From

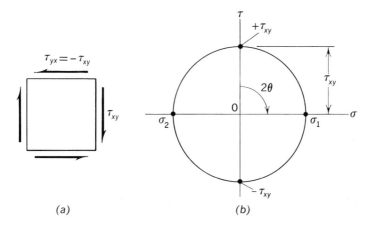

(a)   (b)

Figure 1–31

the circle we observe that $|\sigma_1| = |\sigma_2| = |\tau_{xy}| = |\tau_{yx}|$; also, $\sigma_1 = +\tau_{xy}$ and $\sigma_2 = -\tau_{xy}$. Now, from Equation (1–11), the principal strains may be expressed in terms of the shearing stress τ_{xy}; that is, from

$$\epsilon_1 = \frac{1}{E}(\sigma_1 - \mu\sigma_2) \quad \text{and} \quad \epsilon_2 = \frac{1}{E}(\sigma_2 - \mu\sigma_1)$$

we have

$$\epsilon_1 = \frac{\tau_{xy}}{E}(1 + \mu) \quad \text{and} \quad \epsilon_2 = \frac{-\tau_{xy}}{E}(1 + \mu) \qquad (a)$$

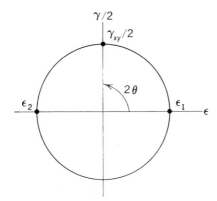

Figure 1–32

Plotting Mohr's circle for the principal strains, ϵ_1 and ϵ_2, as shown in Figure 1–32, we note that

$$\frac{\gamma_{xy}}{2} = \epsilon_1 = \frac{\tau_{xy}}{E}(1 + \mu)$$

or

$$\gamma_{xy} = \frac{2\tau_{xy}}{E}(1 + \mu) \qquad (b)$$

However, from Equation (1–5b), we have

$$\tau_{xy} = \gamma_{xy}G$$

which when substituted in (b), yields

$$\gamma_{xy} = \frac{2\gamma_{xy}G}{E}(1 + \mu)$$

or

$$G = \frac{E}{2(1 + \mu)} \qquad (1\text{–}6)$$

This equation was stated in Section 1–5.

Chapter 2 ◆ TORSION

2—1 Torsion of Circular Shafts

Torsion is a term used to denote a condition in which a body undergoes twisting. The causes of twisting are called *torsional moments* or *torques*. The stresses and deformations are *torsional stresses*, and *torsional* or *angular deformations*, respectively. In this chapter we are concerned with the problem of torsion as it relates to shafts and springs, with particular emphasis on circular cross sections.

In developing expressions for stress and deformation for round sections (either solid or tubular) due to a torsional moment, we make the following assumptions:

1. The material is homogeneous and isotropic.
2. The proportional limit of the material in shear is not exceeded, and Hooke's law applies.
3. Plane transverse sections remain plane transverse sections during and after twist.
4. The torque is in the planes perpendicular to the axis of the shaft.
5. The radial lines (radii or diameters) in a transverse plane remain straight during and after twist.

Before we proceed to develop various equations, it is perhaps appropriate to assure the student that these assumptions are common for most materials we encounter in normal use. Wood, for example, is not as strong in shear parallel to the grain as it is perpendicular to it, making our assumptions,[1] and therefore our results, less reliable than for steel. But wood is likely to be found in much less use as a torsional material than steel. As to the deformation of the shaft under torque, again we merely note that experimental results, using rubber models for visual demonstrations and strain

---

[1] Wood is an orthotropic material, and as such, violates assumption 1.

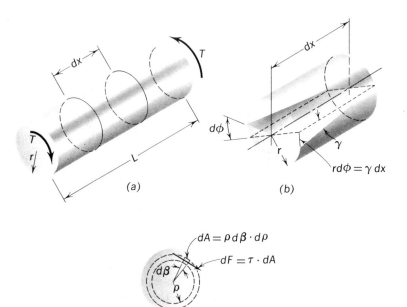

Figure 2–1

gages for a more accurate approach, agree well with those theoretically derived on these assumptions.

The last three assumptions can perhaps be represented pictorially. Figure 2–1a shows a circular shaft subjected to torque T. From assumption 3, the two transverse planes a distance dx apart, Figure 2–1b, remain plane and parallel to each other and to the "torque plane." For a small angle $d\phi$, the longitudinal stresses due to the twisting are assumed to be insignificant.[2] Furthermore, while a plane through the axis of the shaft would be distorted, as shown in Figure 2-1b, the axis remains straight, and any transverse line (diameter of the circle) also remains straight.

The shearing strain, indicated by the angle γ, may be closely approximated for small angles by $\gamma dx = r d\phi$, where $\tan \gamma = \gamma$. From this we get

$$\gamma = \frac{r d\phi}{dx} \qquad (a)$$

In the above expression $d\phi/dx$ represents the angle of twist per unit length of shaft, a constant within the length L. Thus, letting $d\phi/dx = \theta$,[3] the

---

[2] See S. Timoshenko, *Strength of Materials — Part II*, D. VanNostrand Company, Princeton, New Jersey, 1955, pp. 286–291.
[3] θ is measured in radians.

above expression becomes

$$\gamma = r\theta \qquad (b)$$

By virtue of assumption 5, the strain γ_ρ at some radius ρ can be approximated by $\gamma_\rho\, dx = \rho d\phi$. Thus, we have

$$\gamma_\rho = \rho\theta \qquad (c)$$

Although assumption 5 is difficult to substantiate directly by experimental results, the theoretical results based on this assumption (and of course on the other four also) are in close agreement with experimental results.

From Equation (1–5b), that is $G = \tau/\gamma$, the shearing stress at any point from the center of the shaft is

$$\tau = G\rho\theta \qquad (2\text{–}1)$$

The maximum shear stress occurs on the extreme fibers where $\rho = r$; that is, from Equation (2–1), we see that *the magnitude of the shearing stress is directly proportional to the distance from the axis of the shaft.* The

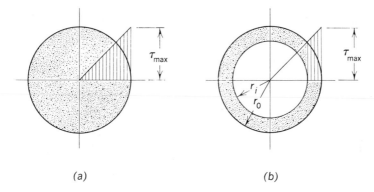

(a)   (b)

Figure 2–2

distribution of shearing stresses for both solid and tubular shafts are shown in Figure 2-2.

$$\tau_{\max} = Gr\theta \qquad (2\text{–}1a)$$

On a differential area dA, Figure 2–1c, the increment of shearing force τdA produces a resisting differential torque dT about the axis of the shaft and is equal to $(\tau \cdot dA)\rho$. The total resisting torque is therefore equal to the summation of these increments of torque. Hence, we have

$$T = \int_A (\tau \cdot dA)\rho \qquad (d)$$

From Equations (2–1) and (2–1a), we get

$$\frac{\tau}{\rho} = \frac{\tau_{max}}{r} = \text{constant}$$

or

$$\tau = \left(\frac{\tau_{max}}{r}\right)\rho \qquad\qquad (e)$$

and from this, Equation (d) may be written as

$$T = \int_A \left(\frac{\tau_{max}}{r}\, \rho \cdot dA\right)\rho = \frac{\tau_{max}}{r} \int_A \rho^2\, dA \qquad\qquad (f)$$

But

$$\int_A \rho^2 dA = J$$

where J represents the polar moment of inertia of the cross section.[4] Substituting from Equation (e) into (f) we get

$$\tau = \frac{T \cdot \rho}{J} \qquad\qquad (2\text{–}2)$$

Equation (2–2) is the *torsion formula for stress* at any point a distance ρ from the axis of the shaft. The maximum shear stress occurs, therefore, at $\rho = r$, that is, at the extreme fiber of the shaft.

We will now proceed to derive an expression for the total deformation of the shaft. By equating the expressions for stress, that is, Equations (2–1) and (2–2), we get

$$G\rho\theta = \frac{T \cdot \rho}{J}$$

from which we obtain the expression for the *unit angle of twist:*

$$\theta = \frac{T}{JG} \qquad\qquad (2\text{–}3)$$

For a shaft L units long, the *total angle* of *twist* ϕ becomes

$$\phi = \int_0^L \theta\, dx = \int_0^L \frac{T\, dx}{JG} \qquad\qquad (2\text{–}4)$$

If T, J, and G are constant over the length L, Equation (2–4) becomes

$$\phi = \frac{TL}{JG} \qquad\qquad (2\text{–}4a)$$

We can now note an interesting similarity between the expressions for axial and angular deformation:

axial deformation:

$$\delta = \frac{\rho L}{AE} \qquad\qquad (1\text{–}7a)$$

---

[4] $J = I_x + I_y$. $J = \pi d^4/32$ for a circular section, and $(\pi/32)(d_o{}^4 - d_i{}^4)$ for a tubular round section. See Appendix B, Table B–1.

angular deformation:

$$\phi = \frac{TL}{JG}$$

A little analysis will show that an element oriented parallel to the axis of a shaft subjected only to torsion is in a state of pure shear. Figure 2–3a shows a twisted shaft from which the element in Figure 2–3b is taken. On the faces A and B transverse to the axis, the magnitude of the shear stresses can be determined from Equation (2–2). The shear on the other two, faces C and D, must be the same in order to satisfy static equilibrium for rotation.[5] The normal forces on these faces are assumed to be insignificant

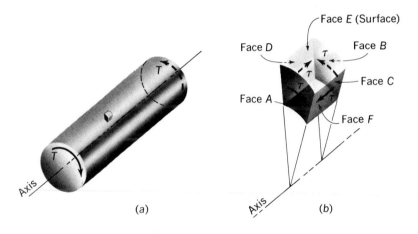

(a)   (b)

Figure 2–3

for small deformations. Finally, on faces E and F, the shear and normal stresses are zero, because there are no transverse loads applied to the shaft.

Example 2–1:

Given: A solid steel shaft is subjected to the torques shown in the figure. Assume $G = 12 \times 10^6$ psi (or 80×10^9 N/m²).

Find: (a) The maximum shearing stresses in the two sections, neglecting stress concentration, (b) the rotation of point ① relative to point ③, and (c) the horsepower developed when the shaft rotates at 126 rpm.

---

[5] See Section 1–11 for a discussion and proof.

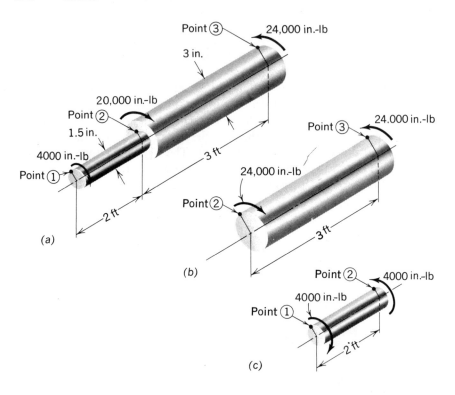

(a)

(b)

(c)

Solution: The free-body diagrams in Figures *b* and *c* show the moments that each section of the shaft must resist in order to satisfy equilibrium. Hence, one is able to easily visualize the type of deformation, and to determine the magnitudes of stresses and deformations. The horse-power each portion develops is proportional to the torque over that section:

$$\text{work/revolution} = T \cdot 2\pi, \text{ in-lb/rev}$$

$$\text{work/min at } n \text{ rpm} = T \cdot 2\pi \cdot n, \text{ in.-lb/min}$$

$$1 \text{ hp} = 33{,}000 \text{ ft-lb/min} = 396{,}000, \text{ in.-lb/min}$$

$$\text{Therefore, } T = \frac{396{,}000 \text{ hp}}{2\pi n} = \frac{63{,}000 \text{ hp}}{n}, \text{ in.-lb}$$

For the 1.5-in.-diam shaft,

$$J_{1,2} = \frac{\pi d^4}{32} = \frac{\pi (1.5)^4}{32} = 0.5 \text{ in.}^4$$

$$\tau_{1,2} = \frac{Tr}{J} = \frac{4000 \, (1.5/2)}{0.5} = 6000 \text{ psi}$$

The rotation of point ① with respect to point ②, $\phi_{1,2}$ is

$$\phi_{1,2} = \frac{T \cdot L}{JG} = \frac{4000 \times 24}{(0.5)(12 \times 10^6)} = 0.016 \text{ rad}$$

$$hp_{1,2} = \frac{T \cdot n}{63,000} = \frac{4000 \times 126}{63,000} = 8$$

For the 3-in.-diam shaft

$$J_{2,3} = \frac{\pi d^4}{32} = \frac{\pi(3)^4}{32} = 8 \text{ in.}^4$$

$$\tau_{2,3} = \frac{Tr}{J} = \frac{24,000 \, (\tfrac{3}{2})}{8} = 4500 \text{ psi}$$

$$\phi_{2,3} = \frac{TL}{JG} = \frac{24,000 \times 36}{8 \times 12 \times 10^6} = 0.009 \text{ rad}$$

$$hp_{2,3} = \frac{Tn}{63,000} = \frac{24,000 \times 126}{63,000} = 48$$

Hence, $\quad \tau_{1,2} = $ **6000 psi (or 41.37 MN/m²)**,

$\qquad \tau_{2,3} = $ **4500 psi (or 31.02 MN/m²)** $\hfill$ **(a)**

$\qquad \phi_{1,3} = \phi_{1,2} + \phi_{2,3} = 0.025 \text{ rad} = $ **1.43°** $\hfill$ **(b)**

$\qquad hp_{1,2} = $ **8**, $hp_{2,3} = $ **48** $\hfill$ **(c)**

PROBLEMS

For the following problems, assume $G = 12 \times 10^6$ psi, for steel.

2-1   Determine the maximum torque that a 2-in.-diam, 3-ft-long, solid steel shaft may transmit if the following limitations are imposed on it: (a) The shearing stress must not exceed 5000 psi, and (b) the rotation of one end relative to another must not be greater than ½ degrees.

2-2   A solid round steel shaft 6 ft long, is subjected to a torque T, which causes a stress of 6000 psi in its outer fibers, and a rotation of 1.146 degrees. Determine the diameter of the shaft.

2-3   A solid round steel shaft, 4 in. in diameter, 7.5 ft long, is used to transmit power in a machine system. When the maximum shearing stress is 8000 psi, determine (a) the horsepower transmitted at 100 rpm, and (b) the angle of twist of one end relative to the other.

2-4   The outside diameter d_o of a tubular shaft is 1.25 times the inside diameter d_i. (a) How much more torque can a solid shaft of diameter, $d_o = 5$ in. carry than the tubular one with the same outside diameter? (b) How much heavier is the solid shaft in part (a) than the tubular one?

2-5 Two shafts of 4-in. outside diameters, each 4 ft long, are each subjected to a torque of 48,000 in.-lb. If one of the shafts is tubular, with the cross-sectional area of one-half that of the solid shaft determine: (*a*) the ratio of maximum shearing stresses in the two shafts, and (*b*) the ratio of total angle of twist of the two shafts.

2-6 For the two shafts in Problem 2–5, determine the ratio of horsepower, solid to tubular, that the shafts may transmit at 252 rpm if the maximum shearing stress in each of the shafts is 10,000 psi.

2-7 How much weaker is a round tubular shaft, with an inside diameter one-half that of the outside, than a solid shaft of the same outside diameter, if both shafts are made of the same material?

2-8 A stepped solid steel shaft consists of two sections rigidly connected to each other, as shown in Figure P 2–8. Determine (*a*) the maximum shearing stress in each section, and (*b*) the angle of twist of one end relative to the other.

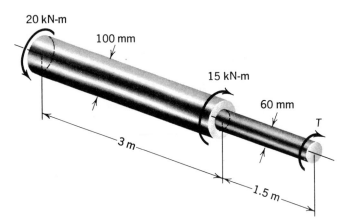

Figure P 2–8

2-9 Determine the maximum horsepower that may be transmitted across the shaft in Figure P 2–8 at 504 rpm if the shearing stress in either of the two sections is not to exceed 8000 psi. What total angle of twist will result from this torque?

2-10 For the shaft in Figure P 2–10, determine the magnitude of *T*, such that (assume solid steel shafts): (*a*) the shearing stress on either section does not exceed 70 MN/m² and (*b*) the angle of twist of the free end relative to the fixed end does not exceed 2 degrees.

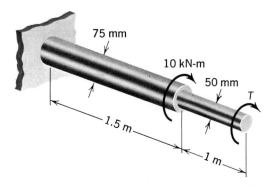

Figure P 2–10

2–11 For the shaft in Figure P 2–10, if the 1-m length was steel and the 1.5-m length was aluminum, determine the magnitude of T, such that: (a) the shearing stress at any section does not exceed 50 MN/m², and (b) the angle of twist of the free end relative to the fixed end does not exceed 2 degrees.

2–12 A system of gears transmit power via two solid steel shafts, as shown in Figure P 2–12. Determine the horsepower that can safely be transmitted from gear A to gear D if the shearing stress is not to exceed 6000 psi in any of the two shafts, when shaft AB rotates at 252 rpm. What is the stress in each of the shafts for this condition?

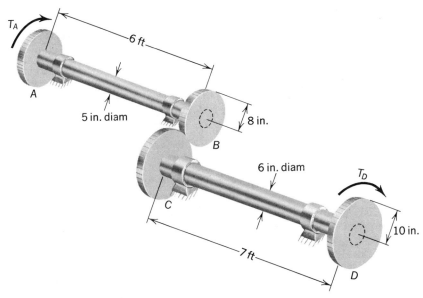

Figure P 2–12

2–13 Determine the relative rotation of gear A to gear D in Figure P 2–12 if $T_D = 200,000$ in.-lb. Assume that only the steel shafts deform; that is, neglect the deformation of the gears.

2–14 A step shaft made up of brass, aluminum, and steel resists the torque system shown in Figure P 2–14. Determine T_1, T_2, and T_3, oriented as shown, if the maximum shearing stress in each section is 10,000 psi.

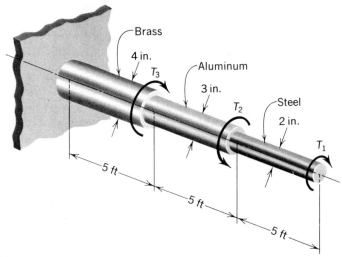

Figure P 2–14

2–15 In Figure P 2–14, determine the total angle of twist of the free end, relative to the fixed end, when $T_1 = 50,000$ in.-lb, $T_2 = 100,000$ in.-lb, and $T_3 = 200,000$ in.-lb. (Assume $G_{Al} = 4 \times 10^6$ psi, $G_{Br} = 6 \times 10^6$ psi.)

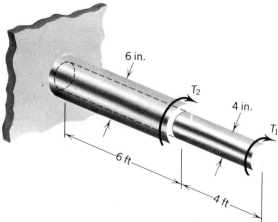

Figure P 2–16

2–16 A solid steel shaft is connected to a tubular steel shaft as shown in Figure P 2–16. Neglecting any deformation at the junction point of the two sections, determine T_1 and T_2, oriented in the sense indicated, if the maximum shear stress in each of the two sections is 7000 psi.

2–17 For the steel shaft of Figure P 2–16, determine the angle of rotation of the free end when $T_2 = 2T_1 = 150,000$ in.-lb.

2–18 The solid steel shaft in Figure P 2–18 is in equilibrium. Determine (a) the torque T_B, and (b) the maximum shearing stresses in the three sections.

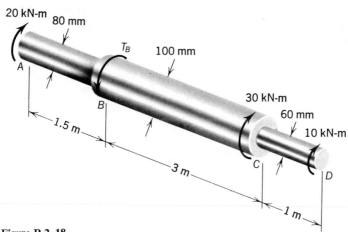

Figure P 2–18

2–19 Assuming the solid steel shaft shown in Figure P 2–18 to be in equilibrium, determine the relative rotation of end A with end D.

2–20 Determine the horsepower transmitted over the three sections of the steel shaft of Figure P 2–18 when it rotates 504 rpm.

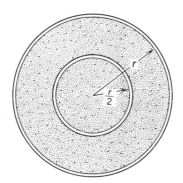

Figure P 2–21

2–21 Determine the ratio of the torque resisted by the outer "ring" to that by the inner ring at $\rho = r/2$, Figure P 2–21. Assume that the stress varies linearly from zero at the center of the shaft to a maximum at the outer fibers.

2–22 Divide a circular shaft into a number of rings of equal thickness, as shown in Figure P 2–21, and determine the relative torque resisted by each ring. Plot a graph showing, on the horizontal scale, the distance from the center of the shaft and, on the vertical scale, the percentage of the total torque each ring resists.

2–23 Figure P 2–23 is a *plan view* schematic of a torsion bar suspension system of an automobile. If the wheel load is 1500 lb, determine the maximum shear stress at the "support."

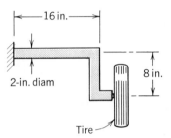

Figure P 2–23

2–24 A shaft supports the load shown in Figure P 2–24. (*a*) Determine the vertical distance the load will travel due to the torsional deformation of the solid steel shaft AB; (*b*) the maximum shear stress due to torsion.

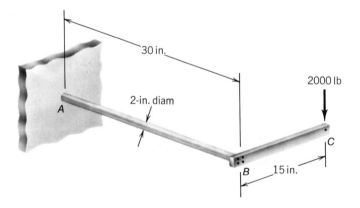

Figure P 2–24

2–25 A man pulls a weight (100 lb) as shown in Figure P 2–25. What torque is transmitted to the shaft when the man has to exert a 250-lb force for the impending upward motion shown?

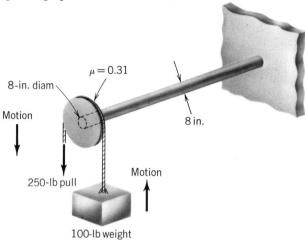

μ = 0.31

8-in. diam

Motion

8 in.

Motion

250-lb pull

100-lb weight

Figure P 2–25

2—2 Closely Coiled Helical Springs

A helical spring is formed by wrapping a wire or rod around a circular cylinder in such a manner that the rod forms a helix of uniformly spaced turns or coils. Here we are concerned only with springs whose coils are closely spaced and whose angle of pitch (of the coil) is quite small.[6] Furthermore, the wire or rod is assumed to be round, of radius r, and the helix is assumed to have a mean radius R. With these imposed conditions, we then apply the theory of torsion and the transverse shear effect to develop an expression for stress on the cross section of a spring coil.

A typical close-coiled spring is shown in Figure 2–4a. The externally applied load P must be resisted by the transverse shear force P and the counterclockwise couple PR acting on the cross section of the coil. Since the pitch of the coil is assumed to be small (that is, the coil is assumed to be nearly in a horizontal plane) the direct shear stress τ_p over the cross section of the rod is assumed to be uniform over the whole area, as indicated in Figure 2–4b, and may be approximated by

$$\tau_p = \frac{P}{A} = \frac{P}{\pi r^2} = \frac{4P}{\pi d^2} \qquad (a)$$

---

[6] For a rather comprehensive coverage of springs see A. M. Wahl, *Mechanical Springs*, McGraw-Hill Book Company, Inc., New York, 1963, 2nd edition.

The maximum shearing stress τ_T caused by the twist may be found by means of the torsion formula. From Figure 2–4c,

$$\tau_T = \frac{Tr}{J} = \frac{(PR)r}{J} = \frac{P \cdot R}{J/r} = \frac{16PR}{\pi d^3} \qquad (b)$$

If we superpose one stress on the other, we note that at point a (inside the coil) the torsional stresses and transverse-shear stress have the same direction; at point b, these stresses have opposite directions. We conclude, therefore, that the maximum shearing stress occurs at the inside

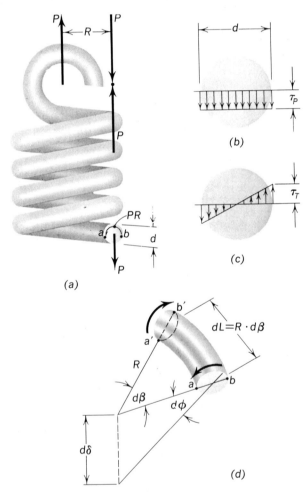

Figure 2–4

(point a) and has the magnitude:

$$\tau_{\max} = \tau_T + \tau_P = \frac{16PR}{\pi d^3} + \frac{4P}{\pi d^2}$$

or substituting $D = 2R$,

$$\tau_{max} = \frac{8PD}{\pi d^3}\left(1 + \frac{d}{2D}\right) \qquad (2\text{-}5)$$

An examination of the above expression shows that when the wire diameter d is small compared to mean coil diameter D, the maximum shearing stress is caused primarily by torsion. However, the influence of the transverse shearing stress becomes significant in heavy springs (for example, railroad-car springs) where the ratio of rod diameter d to coil diameter D is appreciable.

The torsional effect on the spring is not quite the same as assumed in deriving the expression (b) above because the rod or wire is not straight for the spring (the torsion formula $\tau = T\rho/J$ was derived for a straight bar). For the curved rod, as for the straight bar, the diameter of the rod is assumed to remain straight, thus making the displacement of point a relative to point a' the same as that of point b relative to point b', Figure 2–4d. $\gamma = rd\phi/aa' > \gamma = rd\phi/bb'$ because $bb' > aa'$. Thus, because the strain at point a is greater than the strain at point b, and because shearing stress is proportional to shearing strain ($\tau = \gamma G$), we conclude that the shearing stress is greater at the inside than at the outside points of the rod. Taking this into account, as well as the fact that the effect of the shearing force was only roughly approximated, we observe an inaccuracy in Equation (2–5). A more exact expression for maximum shearing stresses is given by[7]

$$\tau_{max} = \frac{16PR}{\pi d^3}\left(\frac{4m-1}{4m-4} + \frac{0.615}{m}\right) \qquad (2\text{-}6)$$

where $m = D/d$.

In calculating the deflection of a close-coiled spring, it is common practice to take into account only twisting of the coils (the effect of the transverse shearing strain is only a negligible percentage of that of the twist). Referring to Figure 2–4d, we note that the twist causes one end of the rod segment to rotate an angle $d\phi$ relative to the other. This corresponds to a deflection of $d\delta$ at the axis of the spring.

$$d\delta = R \cdot d\phi \qquad (c)$$

from Equation (2–4), $\phi = TL/JG$; hence,

$$d\phi = \frac{T \cdot dL}{JG} \qquad (d)$$

Thus,

$$d\delta = \frac{R \cdot T \cdot dL}{JG} = \frac{R \cdot (PR)R \cdot d\beta}{JG} = \frac{PR^3 d\beta}{JG} \qquad (e)$$

---

[7] See A. M. Wahl, *Mechanical Springs*, McGraw-Hill Book Company, Inc., New York, 1963, 2nd edition.

The total deflection of a spring, having n coils, is therefore

$$\delta = \int_0^{2\pi n} \frac{PR^3 d\beta}{JG} = \frac{64nPR^3}{d^4 G} \tag{2-7}$$

where the length of the coil was determined by neglecting the pitch angle.

A convenient form for expressing the relationship between the applied load and the resulting spring deflection (within the elastic limits) is given by the *spring constant k* as

$$k = \frac{P}{\delta} \tag{2-8}$$

or, in this case, $k = d^4 G/64nR^3$.

Example 2-2:

Given: The arrangement of Figure a.
Find: (a) The minimum diameter, d, of the solid steel shaft AB if the torsional stress is not to exceed 70 MN/m², and (b) the vertical travel of the 1-kN load due to the torsion in shaft AB when the stress is 70 MN/m².

(a)

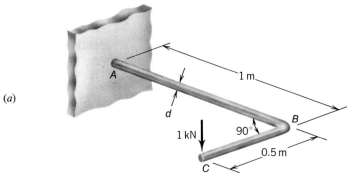

Solution:

$$T_{AB} = (1 \text{ kN})(0.5 \text{ m}) = 0.5 \text{ kN-m}$$

$$\tau = 70 \text{ MN/m}^2 = \frac{T(d/2)}{\frac{\pi}{32}(d)^4} = \frac{16T}{\pi d^3} = \frac{16}{\pi d^3}(0.5 \text{ kN-m})$$

or

$$d = 0.033 \text{ m} = 33 \text{ mm} \tag{a}$$

$$\delta_C = (\theta_B) \, 0.5 \text{ m} = \left(\frac{TL}{JG}\right) 0.50 = \frac{(0.5 \text{ kN-m})(1 \text{ m})(0.5 \text{ m})}{\frac{\pi}{32}(0.033)^4 (80 \times 10^6 \text{ kN/m}^2)}$$

$$\delta_C = 0.0268 \text{ m} = \textbf{26.8 mm}$$

Example 2–3:

Given: A closely coiled helical spring, $D = 2$ in. (or 51 mm), is made up of $\frac{1}{4}$-in. (or 6.35 mm)-diam wire and 15 turns. $G = 12 \times 10^6$ psi (or 80×10^9 N/m²).

Find: (a) The maximum axial load P that can be applied if the maximum shearing stress is not to exceed 60,000 psi, and (b) the deflection caused by the load in part (a).

Solution: From Equation (2–5)

$$60,000 = \frac{16 \cdot P \cdot 1}{\pi(\frac{1}{4})^3}\left(1 + \frac{\frac{1}{4}}{2 \times 2}\right)$$

$$P = \frac{60,000 \cdot \pi \cdot 1/64}{16 \cdot 1 \cdot (1 + 1/16)} = \textbf{173.5 lb (or 771.72 N)}$$

Note the negligible effect from transverse shear.

Using Equation (2–6) where $m = \dfrac{2}{\frac{1}{4}} = 8$

$$60,000 = \frac{16 \cdot P \cdot 1}{\pi(\frac{1}{4})^3}\left(\frac{4 \cdot 8 - 1}{4 \cdot 8 - 4} + \frac{0.615}{8}\right)$$

$$60,000 = 326\ P\ (1.187)$$

$$P = \textbf{155 lb (or 689.44 N)}$$

The latter P is the more exact of the two. However, for our purpose of illustrating the basic principle, Equation (2–5) will be used because it is simpler. Thus, assume that $P = 173.5$ is acceptable. From Equation (2–7)

$$\delta = \frac{64nPR^3}{d^4G} = \frac{64 \cdot 15 \cdot 173.5(1)^3}{(\frac{1}{4})^4 \cdot 12 \cdot 10^6} = \textbf{3.56 in. (or 90.4 mm)}$$

For this spring, the spring constant is

$$k = \frac{P}{\delta} = \frac{173.5}{3.56} = \textbf{48.8 lb/in.}$$

PROBLEMS

The following values of G are to be used for the following problems as needed: For steel, $G_{St} = 12 \times 10^6$ psi (or 80×10^9 N/m²); for phosphor bronze, $G_{Ph} = 6 \times 10^6$ psi (or 40×10^9 N/m²).

2–26 A $\frac{1}{2}$-in.-diam steel rod is closely coiled around a 4-in.-diam cylinder to form a spring of 12 complete turns. If a 250-lb compressive load is applied to the spring, determine (a) the maximum shear stress in the coil by both Equations (2–5) and (2–6) (Wahl's formula), and (b) the deflection of the spring.

2–27 For the spring of Problem 2–26, (a) determine the permissible load the spring can take if the maximum shear stress is not to exceed 75,000 psi, and (b) the deflection of spring caused by this load.

2–28 In Figure P 2–28, how much springing action (downward only) can a 200-lb man expect from only the spring when he is standing still at point C?

Spring = ½-in. diam   R = 3 in.,   n = 12 turns – steel

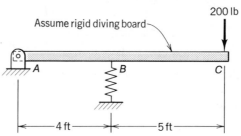

Figure P 2–28

2–29 A phosphor bronze spring consists of 10 turns of ¾-in. rod coiled on a 5¼-in.-diam cylinder. Compute (a) the maximum shearing stress in the spring when it is compressed 2 in., and (b) the spring constant of the spring.

2–30 Determine the maximum load the spring of Problem 2–29 can safely withstand if the maximum shearing stress is not to exceed 36,000 psi. How much will the spring deflect under this load?

2–31 In Figure P 2–31 the 300-lb load is transmitted to the spring via a 2-in.-diam solid steel shaft as shown. Determine (a) the distance the 300-lb load travels, and (b) the maximum shear stress in the shaft. The *steel* spring has the following dimensions: $D = 3$ in.; $d = ¼$ in.; $n = 10$.

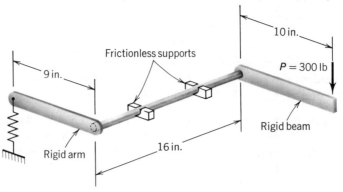

Figure P 2–31

2–32 Two closely coiled springs, one made of steel, are connected to a phosphor bronze spring, as shown in Figure P 2–32. If they each have a mean diameter of 2 in. and are made of ¼-in.-diam wire, determine (a) the ratio of number of turns of the steel to phosphor bronze if their deformations are to be the same, and (b) ratio of the spring constants of the spring.

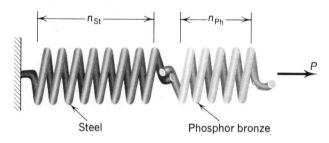

Figure P 2–32

2–33 If the shearing stresses in the steel and phosphor bronze springs in Problem 2–32 are not to exceed 60,000 psi and 40,000 psi, respectively, determine (a) the magnitude of force P allowable, and (b) the total "travel" of load P caused by the deformations of the springs when $n_2 = \frac{3}{4}n_1$, and (c) the ratio of the spring constants of the two springs.

2–34 In Figure P 2–34, two springs have identical characteristics except one: Spring 1 has $\frac{2}{3}$ as many coils as spring 2. The springs support a plank and a man. The plank and man have equal weight. If the plank was originally level before the man got on it, determine the location X at which the plank will again be level.

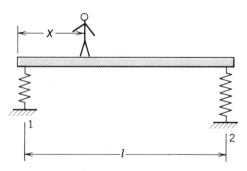

Figure P 2–34

2–35 Two springs, one made of steel, the other made of phosphor bronze and each made of $\frac{3}{8}$-in.-diam wire, fit one within the other, as shown in Figure P 2–35. They each have the same number of turns, $n = 18$. The steel is the outer one, of a mean diameter = 5 in.; the phosphor bronze spring (inside the steel spring) has a mean diameter of 4 in. Determine the total load the two springs can jointly support if the stresses in the steel and phosphor bronze are not to exceed 70,000 psi and 40,000 psi, respectively.

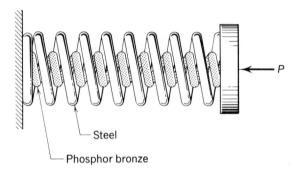

Steel

Phosphor bronze

Figure P 2–35

2–36 Determine the deflection of the two springs in Problem 2–35 when $P = 500$ lb. What are the stresses of the two springs at this load? What percentage of the load does each spring take?

2–37 Determine the spring constants of the two springs acting as one unit, in Problem 2–35.

2–38 In Figure P 2–38 find (a) the shear stress in the spring, and (b) vertical travel of point A.

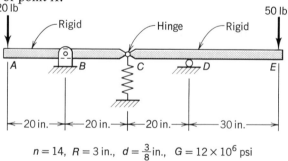

$n = 14, \ R = 3 \text{ in.}, \ d = \frac{3}{8} \text{in.}, \ G = 12 \times 10^6 \text{ psi}$

Figure P 2–38

2–39 Determine the relative spring constants of the two springs in Figure P 2–39 if they are to take an equal share of the load. The rigid bar is horizontal when $P = 0$.

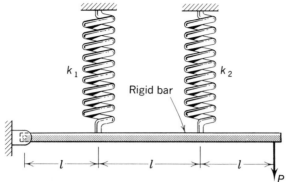

Figure P 2–39

2–40 Two identical springs are attached to rigid bars as shown in Figure P 2–40. Find the force carried in each spring.

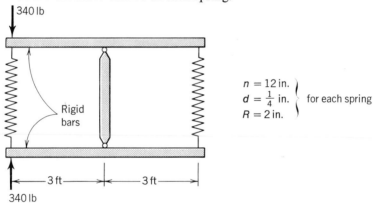

$n = 12$ in.
$d = \frac{1}{4}$ in. for each spring
$R = 2$ in.

Figure P 2–40

2–41 Two identical steel springs of $R = 4$ in. resist the two loads as shown in Figure P 2–41. (a) Determine the *minimum* size of the spring wire if the allowable shear stress in the spring is 70,000 psi. (b) What is the load in each spring?

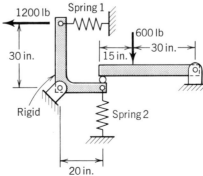

Figure P 2–41

2—3 Torsion of Statically Indeterminate Members

In Figure 2–5a, consider the shaft AB supported by two *fixed* supports at A and B, and subjected to the torque $P \cdot a$. The questions we may ask are: "What is the stress in each of the two sections of the shaft?"; and "how much will each section deform under the action of the applied torque?"

The equations of equilibrium are not sufficient to solve our problem: The use of $\Sigma F_x = 0$ and $\Sigma F_y = 0$ are meaningless in our case because the resultants of the applied forces in these directions are zero. The one equation that does apply is $\Sigma M = 0$. Thus, placing the shaft in equilibrium, as shown in Figure 2–5b, we may write

$$\Sigma M_0 = 0: \qquad\qquad P \cdot a = T_A + T_B \qquad\qquad (a)$$

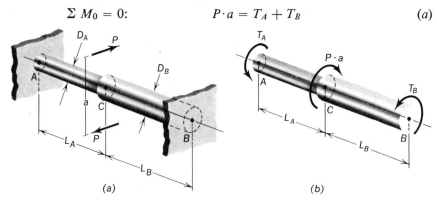

(a)   (b)

Figure 2–5

Because the above equation has two unknowns, T_A and T_B, we need an additional relationship to determine the unknowns. To solve this *statically indeterminate* case we make use of the deformation (rotation) relationship that exists between the two shafts. At point C, the two sections rotate the same amount, that is,

$$\phi_{AC} = \phi_{CB} \qquad\qquad (b)$$

and, in terms of the shaft mechanical properties, this is

$$\left(\frac{TL}{JG}\right)_{AC} = \left(\frac{TL}{JG}\right)_{CB} \qquad\qquad (c)$$

Thus solving Equations (a) and (c) simultaneously, the values of T_A and T_B may be determined; hence, stresses or angular deformations may then be determined. Example 2–3 will illustrate this.

Example 2–4:

Given: The shaft of Figure 2–5, with the following values: $a = 10$ in. (or 254 mm), $P = 5000$ lb (or 22.24 kN), $l_A = 3$ ft (or 914 mm), $l_B = 6$ ft (or 1828 mm), $D_A = 2$ in. (or 51 mm), $D_B = 3$ in. (or 76 mm), $G = 12 \times 10^6$ psi (or 80×10^9 N/m²).

Find: (a) The stresses in each section, and (b) the rotation of the shafts at point *C*.
Solution:

for section *AC*,   for section *BC*,

$$J = \frac{\pi d^4}{32} = \frac{\pi(2)^4}{32} = \frac{\pi}{2} \qquad J = \frac{\pi d^4}{32} = \frac{\pi(3)^4}{32} = \frac{81\pi}{32}$$

$$L = 3 \text{ ft} = 36 \text{ in.} \qquad L = 6 \text{ ft} = 72 \text{ in.}$$

$$P \cdot a = 50,000 \text{ in.-lb}$$

Therefore,

$$50,000 = T_A + T_B \qquad \text{[Equation (a)]}$$

and

$$\frac{T_A \cdot 36}{\pi/2 \cdot 12 \times 10^6} = \frac{T_B \cdot 72}{81\pi/32 \cdot 12 \times 10^6}$$

$$T_A = \frac{32 T_B}{81} \qquad \text{[Equation (b)]}$$

Solving (a) and (b) simultaneously,

$$50,000 = \frac{32}{81} T_B + T_B = \frac{113}{81} T_B$$

Therefore,

$$T_B = \textbf{35,800 in.-lb (or 4.04 kN-m)}$$

$$T_A = 50,000 - 35,800 = \textbf{14,200 in.-lb (or 1.60 kN-m)}$$

The maximum stresses in the corresponding sections are

$$\sigma_A = \left(\frac{Tr}{J}\right)_A = \frac{14,200 \cdot 1}{\pi/2} = \textbf{9040 psi (or 62.33 MN/m}^2\textbf{)}$$

$$\sigma_B = \left(\frac{Tr}{J}\right)_B = \frac{35,800 \cdot 1.5}{81\pi/32} = \textbf{6750 psi (or 46.54 MN/m}^2\textbf{)}$$

(a)

The rotation of transverse plane at *C* relative to those at *A* and *B* is

$$\phi_C = \left(\frac{TL}{JG}\right)_A = \frac{14,200 \cdot 36}{\frac{\pi}{2} \cdot 12 \times 10^6} = 0.027 \text{ rad} = \textbf{1.55}° \qquad \text{(b)}$$

PROBLEMS

For steel, $G_{St} = 12 \times 10^6$ psi (or 80×10^9 N/m²); for aluminum, $G_{Al} = 4 \times 10^6$ psi (or 28×10^9 N/m²); for brass, $G_{Br} = 5 \times 10^6$ psi (or 35×10^9 N/m²).

2–42 A tubular round aluminum shaft has a steel core as shown in Figure P 2–42. Determine (a) the maximum shearing stress in each material, and (b) the rotation of the free end relative to the fixed end.

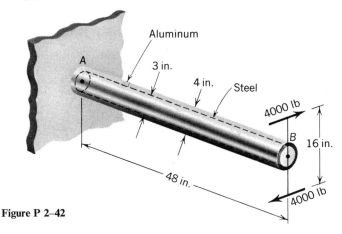

Figure P 2–42

2–43 Work Problem 2–42 if the steel core is replaced by a brass one of identical dimensions.

2–44 A tubular round aluminum shaft, 36 in. long, fits over a concentric solid steel shaft, 60 in. long, as shown in Figure P 2–44. Determine (a) the maximum shearing stress in each materal, and (b) the total rotation of the free end relative to the fixed end.

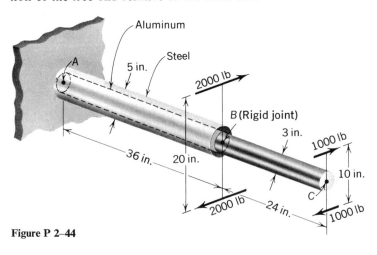

Figure P 2–44

2–45 Assume that for the composite shaft of Figure P 2–44 the torque at the free end remains as shown but the one at point B is changed to meet the following conditions: The maximum allowable working stress in the aluminum, $\sigma_{Al} = 10,000$ psi; the maximum stress in the steel, $\sigma_{St} = 12,000$ psi. Determine the magnitude of the torque at B (clockwise) to satisfy these conditions.

2–46 Determine the maximum shearing stresses in the composite shaft of Figure P 2–44 if the steel shaft is replaced by one of brass, of the same size.

2–47 In Figure P 2–47 the shaft is made up of a solid aluminum section AB and a tubular steel section BC rigidly joined at B. If $T_B = 15,000$ in.-lb, find (a) the maximum shear stress in the steel section, and (b) the maximum normal stress in the steel section (use Mohr's circle).

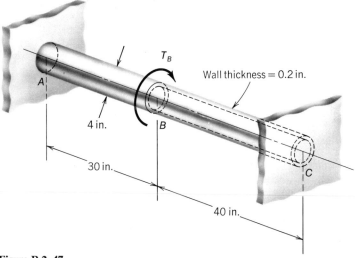

Figure P 2–47

2–48 Each of the two sections in Figure P 2–48 is solid, round and of steel. Determine (a) the stress in each section, and (b) the rotation at point C.

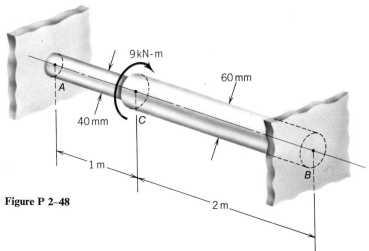

Figure P 2–48

2–49 Determine the vertical travel of the load shown in Figure P 2–49 due to torsion in bar *A-B*.

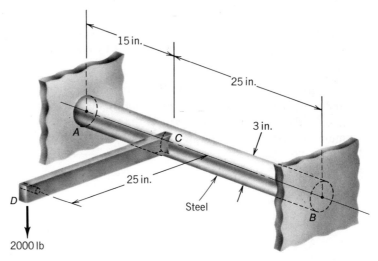

15 in.

25 in.

3 in.

A

C

25 in.

Steel

D

B

2000 lb

Figure P 2–49

2–50 In Figure P 2–50 two shafts of equal diameter and equal length support the load shown. The left shaft is aluminum, the right steel. Determine the percentage of *P* each supports.

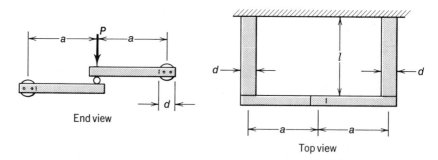

a

P

a

d

d

d

l

End view

a

a

Top view

Figure P 2–50

2–51 The shaft in Figure P 2–51 is made up of a solid steel section *AB* and a solid aluminum section *BC* of the same diameter, rigidly joined at point *B*. If $L_{AB} = L_{BC} = 40$ in. and $d = 4$ in., determine T_B allowable if the shearing stresses in the steel and aluminum are not to exceed 9000 psi and 6000 psi, respectively.

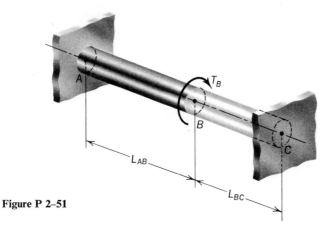

Figure P 2-51

2-52 Solve Problem 2–51 if section *BC*, instead of being aluminum, is brass.

2-53 Assume the following conditions apply to the combination shaft of Figure P 2-51: section *AB* = solid steel, L_{AB} = 48 in., d_{AB} = 5 in.; section *BC* = solid aluminum, d_{BC} = 5 in. Determine (*a*) the length of *BC* so that the stresses in both sections are the same, (*b*) the magnitude of the stresses for part (*a*), if T_B = 200,000 in.-lb, and (*c*) the rotation of a transverse plane at point *B* relative to a fixed end, when T_B = 200,000 in.-lb.

2-54 In Figure P 2-51 section *AB*, made of steel, is 60 in. long and has a 4 in. solid diameter. The section *BC*, connected rigidly at *C* is to be made out of brass, also of 4 in. solid diameter. If both sections reach a maximum shear stress of 7000 psi, simultaneously, determine (*a*) the length of *BC*, and (*b*) the value of T_B for this condition.

2-55 For the Figure P 2-55 assume the following data: L_{AB} = 6 ft, L_{BC} = 3 ft, d_{AB} = 5 in., d_{BC} = 3 in., material for *AB* is steel, for *BC* is aluminum. When T_B = 72,000 in.-lb, determine (*a*) the shearing stresses in each section of the shaft, and (*b*) the maximum rotation of each shaft section (degrees). The sections are rigidly joined at *B*.

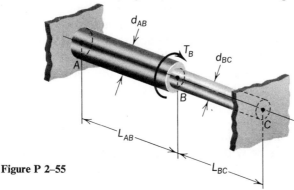

Figure P 2-55

2–56 If the maximum twist in the shaft of Figure P 2–55 is not to exceed 1.5 degrees, find (a) T_B, and (b) the shearing stresses in each section of the shaft.

2–57 Repeat Problem 2–55 if the aluminum section is replaced by a bronze shaft.

2–58 Refer to Figure P 2–58. Three steel springs of identical size support the 1000-lb load as shown. Determine (a) the load each spring supports, and (b) the total vertical travel of the 1000-lb load if $d = \frac{1}{2}$ in.; $D = 4$ in.; $n = 10$.

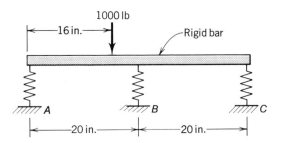

Figure P 2–58

2—4 Torsion of Thin-Walled Tubes

The elementary torsion theory presented up to this point was confined to members of circular cross sections, either solid or tubular. The results based on this theory would not be applicable with any acceptable accuracy to noncircular sections, such as the wing of an aircraft subjected to torsion. Here we are concerned with the analysis of thin-wall closed sections of noncircular form.

An exact analysis of such a problem is a rather involved procedure which takes into account equilibrium, consistency of deformation, and the relationship of stress to strain. Our approach, although admittedly *approximate*, is relatively easy to understand, and yet, a reliable and experimentally justified one. It is based on the assumptions that: (1) The wall thickness is small in comparison with other dimensions of the cross section; however, the thickness need not be uniform over the periphery of the shell. (2) There are no sudden changes in the thickness which might result in stress concentration. (3) The periphery is continuous; that is, no cuts in the section. (4) The torque lies in the transverse plane. (5) There is no buckling.

Figure 2–6a shows a thin-wall member which meets the above-stated assumptions. A free-body diagram of an element of the wall is shown in Figure 2–6b. Because there are no normal forces on either the inside or out-

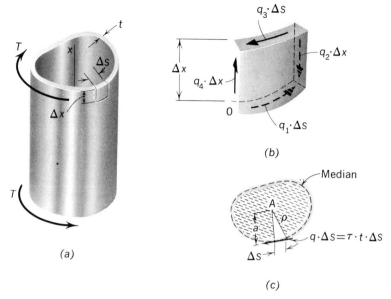

$q_3 \cdot \Delta s$

$q_2 \cdot \Delta x$

Δx

$q_4 \cdot \Delta x$

$q_1 \cdot \Delta s$

(b)

Median

A

ρ

$q \cdot \Delta s = \tau \cdot t \cdot \Delta s$

Δs

(a)

(c)

Figure 2–6

side surface of the wall (pure torsion), the element will be subjected only to the forces shown in Figure 2–6b. The quantities q_1, q_2, q_3, and q_4 represent a concept referred to as *shear flow*, commonly abreviated by q, and is defined as the *resultant shearing force per unit length* of the cross section $(q = \tau_{avg} \cdot t)$. The summation of forces in the horizontal and vertical direction, Figure 2–6b, yields

$$\sum F_s = 0: \quad q_1 \Delta s = q_3 \Delta s$$

and

$$\sum F_x = 0: \quad q_2 \Delta x = q_4 \Delta x$$

from which we see that $q_1 = q_3$ and $q_2 = q_4$. Then, summing moments about a point 0, we get

$$\sum M_0 = 0: (q_2 \cdot \Delta x) \cdot \Delta s = (q_3 \Delta s) \cdot \Delta x$$

Thus, $q_2 = q_3 = q_1 = q_4$, that is, *the shear flow is constant along any portion of a section regardless of the thickness of the shell.* From the definition, in terms of the *average shearing stress* τ, the shear flow is $q = \tau \cdot t \cdot 1$. From this relationship, we note that τ is independent of the radial distance ρ.

We shall now express the shear stress τ in terms of the externally applied torque T. In Figure 2–6c we assume that the increment of force ΔF acts through the median line in the direction shown. This force results in an increment of torque ΔT:

$$\Delta T = \Delta F \cdot a = (q \cdot \Delta s)a = (\tau \cdot t \cdot \Delta s)a$$

143

where a is the perpendicular distance from 0 to the force ΔF. The total torque is therefore

$$T = q \, \Sigma \, \Delta s \cdot a = \tau \, \Sigma \, t \cdot \Delta s \cdot a$$

We see that $\Delta s \cdot a = 2 \times$ area of small shaded triangle of base ds and height a.

Thus, $\sum \Delta s \cdot a = 2 \times A$, where A represents the area enclosed by the median line (see Figure 2–6c). Thus, for a special case of *constant thickness* t, we have

$$T = 2 \cdot \tau \cdot t \cdot A$$

where

$$\tau = \frac{T}{2 \cdot t \cdot A} \tag{2--9}$$

As the shell twists, the torque T does external work equal to $T\phi/2$. Equating this work with the stored energy in the shell we get[8]

$$\phi = \frac{TL}{4A^2G} \, \Sigma \, \frac{\Delta S}{t} \tag{2--10}$$

and if t *is a constant* over the periphery of the shell, we get

$$\phi = \frac{TLS}{4A^2Gt} \tag{2--10a}$$

---

[8] The relationship between angular deformation and applied torque is perhaps best illustrated by using the concept of energy; in this regard, it may indeed be advisable to read Sections 6–1 through 6–3. From Equation (6–9), the expression for the energy stored in a unit volume is

$$\mu = \frac{\tau^2}{2G}$$

Note the comparison with Equation (6–4a).

A strip from the shell in Figure 2–6, say of units ΔS wide, t thick, and L long, has a volume equal to

$$\Delta V = l \cdot t \cdot \Delta S$$

Therefore, the energy stored in such a strip is

$$\Delta U = u \Delta V = \frac{\tau^2}{2G} (L \cdot t \cdot \Delta S)$$

where U represents the *total* energy in the shell. Substituting from Equation (2–9), we get

$$\Delta U = \frac{T^2}{4t^2A^2} \frac{1}{2G} (L \cdot t \cdot \Delta S)$$

Thus, the total energy stored in all the strips (energy stored in the whole shell) is

$$U = \frac{T^2L}{8A^2G} \, \Sigma \, \frac{\Delta S}{t}$$

Hence, from

$$\frac{T\phi}{2} = \frac{T^2L}{8A^2G} \, \Sigma \, \frac{\Delta S}{t}$$

we get

$$\phi = \frac{TL}{4A^2G} \, \Sigma \, \frac{\Delta S}{t}$$

where ϕ = total deformation for the length L, and S represents the peripheral distance. T, A, G, and t represent torque, area (medial enclosure), modulus of rigidity, and wall thickness, respectively.

Example 2–5:

Given: A thin-wall, extruded aluminum section shown in the figure is subjected to a torque $T = 8000$ in.-lb (or 0.90 kN-m). Assume $G_{Al} = 4 \times 10^6$ psi (or 28×10^9 N/m²).
Find: (a) The maximum shearing stress using Equations (2–2) and (2–9) (that is, $\tau = Tp/J$ and $\tau = T/2 \cdot t \cdot A$), and (b) the angle of twist per foot of length using Equations (2–4) and (2–10) (that is, $\phi = TL/JG$ and $\phi = TSL/4A^2\,Gt$).

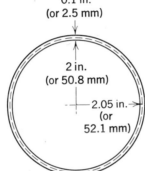

0.1 in.
(or 2.5 mm)

2 in.
(or 50.8 mm)

2.05 in.
(or 52.1 mm)

Solution:

$$S = 2\pi r = 2\pi(2.05) = 12.87 \text{ in.}$$

$$J = \frac{\pi}{32}(d_o{}^4 - d_i{}^4) = \frac{\pi}{32}(4.2^4 - 4^4) = 5.41 \text{ in.}^4$$

$$A = \pi r^2 = \pi(2.05)^2 = 13.19 \text{ in.}$$

Using Equation (2–2),

$$\tau = \frac{(8000)(2.1)}{5.41} = \textbf{3105 psi (or 21.40 MN/m}^2\textbf{)}$$

(a)

Using Equation (2–9),

$$\tau = \frac{8000}{2(0.1)(13.19)} = \textbf{3035 psi (or 20.93 MN/m}^2\textbf{)}$$

Using Equation (2–4),

$$\phi = \frac{8000 \times 12}{(5.41)(4 \times 10^6)} = \textbf{4.44} \times \textbf{10}^{-3} \textbf{ rad}$$

(b)

Using Equation (2–10),

$$\phi = \frac{8000 \times 12 \times 12.87}{4(13.19)^2(4 \times 10^6)(0.1)} = \textbf{4.44} \times \textbf{10}^{-3} \textbf{ rad}$$

This example provides a good basis for comparing the "approximate approach" in Section 2–4 with the more accurate approach covered in Section 2-1 for circular sections of thin walls.

Example 2–6:

Given: The thin-wall, extruded aluminum section shown in the figure is subjected to a torque $T = 8000$ in.-lb (or 0.90 kN-m) [assume $G_{Al} = 4 \times 10^6$ psi (or 28×10^9 N/m²)].
Find: (a) The shearing stress, and (b) the angle of twist per foot of length. (Neglect the stress concentration at the corners.)

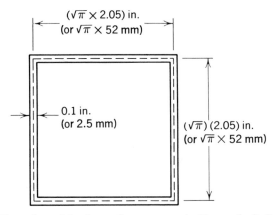

Solution: Note that A is about the same as in Example 2–5; that is,

$$A = (\sqrt{\pi})(2.05)(\sqrt{\pi})(2.05) = 13.19 \text{ in.}$$

$S = 4(\sqrt{\pi})(2.05) = 14.55$ in. (compared to 12.87 in., Example 2–5)

Using Equation (2–9),

$$\tau = \frac{8000}{2(0.1)(13.19)} = 3035 \text{ psi (or } 20.93 \text{ MN/m}^2) \tag{a}$$

Using Equation (2–10),

$$\phi = \frac{(8000)(14.55)(12)}{4(13.19)^2(4 \times 10^6)(0.1)} = 5.02 \times 10^{-3} \text{ rad}$$

(compared to 4.44×10^{-3}, Example 2–5) **(b)**

PROBLEMS

Note: For the following problems, neglect the possible stress concentration at any corners.

2–59 The section in Figure P 2–59 is of extruded aluminum, $r = 1.75$ in., $t = \frac{1}{8}$ in. Determine the stress and the angle of twist per foot of length when the tube is subjected to a torque of 7500 in.-lb, using both the corresponding equations of Sections 2–1 and 2–4 ($G_{Al} = 4 \times 10^6$ psi).

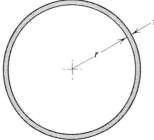

Figure P 2–59

2–60 The tubular section of Problem 2–59 is not to be stressed beyond 5000 psi nor to be twisted more than ⅙ degrees/ft of length. Using the dimensions given in Problem 2–59, determine the maximum torque that may safely be applied to this section.

2–61 A thin-wall, round tubular section as shown in Figure P 2–59 is to be designed to carry a torque of 10,000 in.-lb. The maximum shear stress in the section is not to exceed 4000 psi, and the angle of twist per foot of length is not to exceed 0.1 degrees. Assuming that $r/t = 15$, and the material to be aluminum ($G_{Al} = 4 \times 10^6$ psi), determine the smallest cross section that will suffice.

2–62 The section in Figure P 2–62 is subjected to torque of 100,000 in.-lb. For $t = 0.1$ in. and $G = 12 \times 10^6$ psi, determine (a) the shearing stress in the shell, and (b) the angle of twist per foot of length.

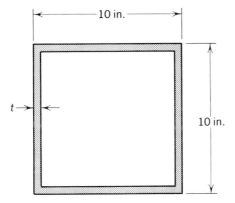

Figure P 2–62

2–63 If the section in Figure P 2–62 is to resist a torque of 150,000 in.-lb, and the shearing stress and angle of twist per foot of length are not to exceed 10,000 psi and 0.1 degrees, respectively, determine the minimum thickness t that will suffice. Assume $G = 12 \times 10^6$ psi.

2–64 The section in Figure P 2–64 is subjected to a torque of 100,000 in.-lb. For $t = 0.1$ in., and $G = 12 \times 10^6$ psi, determine (a) the shearing stress in the walls, and (b) the angle of twist per foot of length.

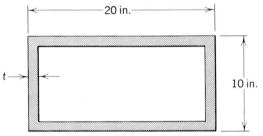

Figure P 2–64

2–65 If the section in Figure P 2–64 is to resist a torque of 150,000 in.-lb, and the shearing stress and angle of twist per foot of length are not to exceed 10,000 psi and 0.1 degrees, respectively, determine the minimum thickness t that will suffice. Assume $G = 12 \times 10^6$ psi.

2–66 The section in Figure P 2–66 is subjected to a torque of 100,000 in.-lb. For $t = 0.075$ in., and $G = 12 \times 10^6$ psi, determine (a) the shearing stress in the wall, and (b) the angle of twist per foot of length.

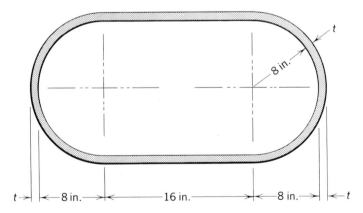

Figure P 2–66

2–67 If the section in Figure P 2–66 is to resist a torque of 200,000 in.-lb, and the shearing stress and angle of twist per foot of length is not to exceed 2000 psi and 0.5 degrees, respectively, determine the minimum thickness t that will suffice. Assume $G = 4 \times 10^6$ psi.

2–68 For the section in Figure P 2–68 determine which shape is stronger, a circular shape or a square shape of the same weight, with both shapes having the same wall thickness, t. (First determine the value of x; assume the shear stress in each section is τ.)

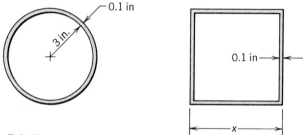

Figure P 2–68

2–69 The elliptical cross section in Figure P 2–69 is to resist a torque of 40,000 in.-lb. If $G = 4 \times 10^6$ psi $\{S = 2\pi[\sqrt{(a^2 + b^2)/2}]$; $A = \pi ab\}$ determine (a) the shearing stress in the wall, and (b) the angle of twist per foot of length.

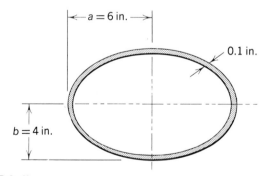

Figure P 2–69

2–70 Determine the maximum torque that the section of Figure P 2–69 may safely resist if the shearing stress in the shell and the angle of twist per foot of length is not to exceed 2000 psi and 0.5 degrees, respectively. Assume $G = 4 \times 10^6$ psi. (For S and A, see Problem 2–69.)

2–71 The section in Figure P 2–71 resists a torque of 100,000 in.-lb. For a uniform $t = 0.1$ and $G = 4 \times 10^6$ psi, determine (a) the shearing stress in the shell, and (b) the angle of twist per foot of length.

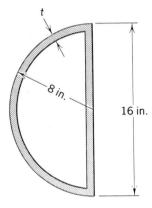

Figure P 2–71

2–72 Determine the maximum shear stress for a square cross section subjected to a torque of 10,000 in.-lb, if $a = b = 2.5$ in.

2–73 A rectangular section, $a = 2$ in., $b = 3$ in., is subjected to a torque T. Determine the maximum value T may have if the shearing stress is not to exceed 4000 psi and the maximum angle of twist per foot of length is not to exceed ⅙ degrees.

2–74 Compare the values of the maximum shearing stress of two sections, one rectangular, the other elliptical, subjected to the same torque, $T = 20,000$ in.-lb if (a) $a = 2$ in., $b = 3$ in., (b) $a = 4$ in., $b = 6$ in., (c) $a = 5$ in., $b = 7.5$ in.

2–75 Compare the angular deformations per foot of length of the respective sections of Problem 2–74 if (a) for both sections, $G = 30 \times 10^6$ psi, and (b) in the rectangular section, $G = 15 \times 10^6$ psi; in the elliptical, section, $G = 10 \times 10^6$ psi.

2–76 An equilateral triangle (steel) cross section, $a = 4$ in. is subjected to a torque of 30,000 in.-lb. A circular section, also steel, of 4 in. is to be subjected to the same torque. Find (a) the ratio of their maximum shear stresses, (b) the ratio of their unit angular deformation, and (c) ratio of their weight per unit length ($G_{St} = 30 \times 10^6$ psi).

2—5 Torsion of Noncircular Sections

The assumption that transverse sections remain plane during and after twist, as assumed for circular cross sections in Section 2–1, does not apply for solid, noncircular sections; neither does the approach used for thin-wall sections. Furthermore, the magnitudes of the shearing stresses for these sections are not proportional to the distance from the axis of the member, as was the case for circular sections. In fact, the stress at a *corner* of a rectangular section, which is the farthest away from the axis of the member, is actually *zero*.[9] The *maximum* stress occurs at the midpoint of the *long* side, the point on either of the two faces *nearest* to the axis of the member.

A rigorous analytical analysis of solid, noncircular sections subjected to torsion indicates that these sections *warp* when subjected to torque. It is beyond the scope of this text to present this complex mathematical approach here.[10] What is intended here is to merely make the student aware of the nature of the problem and to provide him with several solutions for stress

TABLE 2-1

Effect \ Shape	Rectangle *	Square	Ellipse	Equilateral Triangle
τ_{max}	$\dfrac{k_1 T}{ab^2}$	$\dfrac{4.81 T}{a^3}$	$\dfrac{16 T}{\pi ab^2}$	$\dfrac{20 T}{a^2}$
φ	$\dfrac{k_2 T}{ab^3 G}$	$\dfrac{7.10 T}{a^4 G}$	$\dfrac{(a^2 + b^2)T}{16 \pi a^3 b^3 G}$	$\dfrac{46 T}{a^4 G}$

*Values for k_1 and k_2 — Twist of Rectangular Sections

a/b	1.0	1.5	2.0	3.0	4.0	6.0	8.0	10.0	∞
k_1	4.81	4.33	4.07	3.75	3.55	3.34	3.26	3.19	3.0
k_2	7.10	5.10	4.37	3.84	3.56	3.34	3.26	3.19	3.0

---

[9] For further discussion of this see J. P. Den Hartog, *Advanced Strength of Materials*, McGraw-Hill Book Company, Inc., New York, 1952, pp.1–3.

[10] See, for example, S. Timoshenko and J.N. Goodier, *Theory of Elasticity*, McGraw-Hill Book Company, Inc., New York, 1951.

and angular deformations of some rather common sections. They are due *primarily* to a Frenchman named B. de St. Venant (1797–1886) who in 1853, developed a form of solution to some of these problems. Some of these solutions are further substantiated by mathematical analysis, and by a semi-experimental procedure known as membrane analogy, introduced by a German engineer named L. Prandtl (1875–1953) in 1903.

Prandtl observed that the basic differential equation for torsion has the same form as that governing the deflection of a thin membrane under a constant pressure (a soap film under a small gas pressure, for example). Briefly, the slope of the membrane at any point is proportional to the shearing stress at that point,[11] indeed a useful aid in estimating the intensities of stress at various points on the cross section.

Some of the useful expressions mentioned previously, which relate the stress and deformation to the properties of the members are cited in Table 2-1.

---

[11] A less important analogy states that the volume between the membrane and a plane over which the membrane spreads is an indication of the torque necessary for a desired angle of twist.

Chapter 3 ◆ STRESSES IN BEAMS

3—1 Shear Forces and Bending Moments

Beams are structural members whose primary purpose is to carry transverse loads. Most beams[1] carry some rather small loads in their axial direction (or perhaps torsional loads) along with transverse loads during their service, but these loads, and especially their effects on the beam, are usually small in comparison with the transverse loads. The beams we are mainly concerned with here are relatively slender members, assumed stable against lateral buckling. The loads considered are resolved into components parallel and perpendicular (transverse) to the longitudinal axis of the beam. Only the transverse components and their effects will be considered at this time.

The effects to be evaluated in this chapter are the *internal stress resultants*, and subsequently the *stresses* necessary to preserve equilibrium. The internal transverse force on any transverse plane of the beam necessary to maintain equilibrium in the transverse direction is called *shear force*, commonly represented by V. The corresponding shear stresses are commonly represented by τ. The internal resistance necessary to maintain rotational equilibrium in the plane parallel to the transverse loads is called a *bending moment*, frequently shortened to just *moment* and denoted by M. The corresponding stresses are called *bending* or *flexure stresses*, represented by σ.

It is perhaps advisable at this time to clearly classify the types of beams and the type of reactions (forces and moments) before we proceed to investigate the shear forces and moments. Figure 3–1 shows a number of beams supported by a variety of supports. These beams are separated into

---

[1] Shafts, airplane wings, and so forth, are often considered "beams." Hence, the term *beam* is intended to encompass a broad and generalized definition of transversely loaded members.

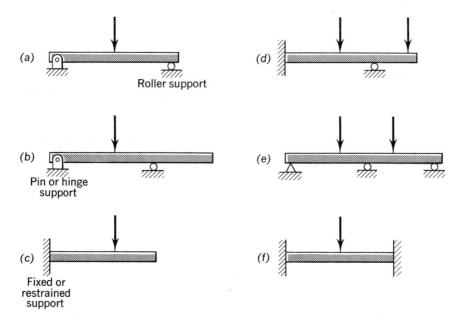

(a) Roller support

(b) Pin or hinge support

(c) Fixed or restrained support

Figure 3–1

(a)–(c) Statically determinate beams. (a) Simply supported beam. (b) Overhanging beam. (c) Cantilever beam. (d)–(f) Statically indeterminate beams. (d) Propped beam. (e) Continuous beam. (f) Fixed or restrained beam.

two groups, *statically determinate* and *statically indeterminate*. When the reactions can be obtained from the equations of equilibrium alone, the structure is *statically determinate*. When the equations of equilibrium are not sufficient to determine the unknown reactions, the structure is said to be *statically indeterminate*, a topic covered in Chapter 5.

There are three general types of supports: *pin* or *hinge*, which resists translation in all directions but cannot resist rotation (that is, $\Sigma M_{\text{pin}} = 0$); *roller*, which resists only translation (that is, $\Sigma M_{\text{roller}} = 0$, and $\Sigma F_{\text{axial}} = 0$); *fixed* or *restrained*, which resists translation in all directions and permits no rotation (fixed) or only a limited rotation (restrained). The schematic representations of these types are shown in Figures 3–1a to 3–1c.

Consider now the simply supported beam in Figure 3–2a which carries a uniform load w lb/lin ft and which is held in equilibrium by the reactions R_1 and R_2.[2] If we were to cut this beam by a transverse plane A, x distance from R_1, and were to place the two segments in equilibrium, as shown in Figures 3–2b and 3–2c, we see that the unbalanced *external* forces on either of the segments must be resisted by the *internal* forces on the plane A. To resist

---

[2] The weight of the beam is assumed to be contained in w.

the translational effect of R_1 and (wx), Figure 3–2b, we have the forces in the plane A which constitutes a magnitude V_x. Similarly, to resist the rotational effect of R_1 and (wx) we have an internal couple M_x. Using the equations of equilibrium, we have

$$\Sigma F_y = 0: \quad V_x = R_1 - wx = \frac{wl}{2} - wx$$

$$\Sigma M_x = 0: \quad M_x = R_1 x - (wx)\frac{x}{2} = \frac{wl}{2}x - \frac{wx^2}{2}$$

The quantity V_x is known as the resisting shear force (or just shear) and *is* simply *equal to the unbalanced, transverse external loads.* Its magnitude is found by merely summing the transverse forces on either side of the imaginary plane A; it also has a sense always opposite of the *resultant* external

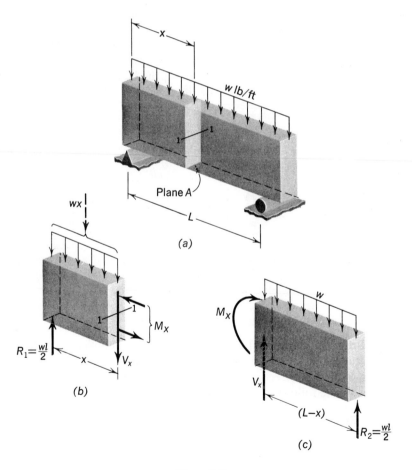

(a)

(b)

(c)

Figure 3–2

loads. The internal moment M_x is known as the *resisting moment,* and is determined by taking moments of the external forces on either side of plane A with respect to the horizontal axis in plane A, such as axis 1–1. Its sense always opposes the net rotational effect of the external loads. It may be well to repeat again that both V_x and M_x can be determined by analyzing *either* segment of the beam.

The sense of the shear force and moment may vary along the length of the beam. For example, from $V_x = wl/2 - wx$, the shear V_x is positive when $0 < x < l/2$, and *negative* when $l/2 < x < l$. It is therefore important that we adopt a sign convention to use when defining positive or negative shears and moments. The sign convention adopted is as follows:

Shear: The shear is positive when V_x results in a *clockwise rotation* with respect to a point *inside* the segment in equilibrium; V_x is negative when it creates counterclockwise rotation. Note that for the positive shear at a section, the left segment tends to move up with respect to the right segment; conversely for the negative shear.

Moment: The moment is *positive* when M_x tends to bend the segment *concave upwards.* M_x is negative when the reverse is true. A positive moment creates tension on the bottom fibers of the beam and compression in the top fibers; the opposite is true for the negative moment.

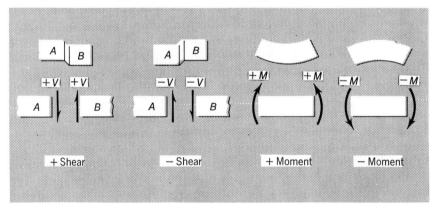

Figure 3–3

Figure 3–3 is a pictorial description of the sign convention relating to shear and moments. Further elaboration of the procedure for determining the magnitudes and direction of internal shears and moments is attempted by means of the following examples.

Example 3–1:

Given: The overhanging beam in Figure a is loaded with a uniform load, $w = 600$ lb/lin ft (or 8754 N/m), and a concentrated load, $P = 5$ kips (or 22.24 kN), at the center of the span.

Find: The shear and moments at a section at (*a*) 4 ft to the right of R_1 (at section *a–a*), (*b*) just to the left of *P* (section *b–b*), and (*c*) just to the right of *P* (section *c–c*).

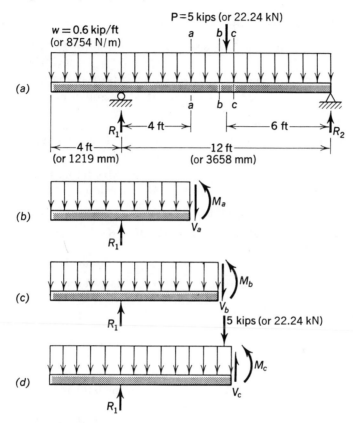

Solution: By summing moments about R_2, we find R_1. Then summing forces in the vertical direction yields R_2. The directions of *R*, *V*, and *M* are assumed. If the assumption is incorrect the result will be a negative quantity.[3]

$$\Sigma M_{R_2} = 0: \quad 12 \text{ ft} \times R_{1 \text{ kip}} = 5 \text{ kips} \times 6 \text{ ft} + (16 \text{ ft} \times 0.6 \text{ kip/ft})(8 \text{ ft})$$

$$R_1 = \textbf{8.9 kips (or 39.59 kN)}$$

$$\Sigma F_v = 0: \quad 16 \times 0.6 + 5 - 8.9 = R_2$$

$$R_2 = \textbf{5.7 kips (or 25.35 kN)}$$

The positive results indicate a correct assumption of the direction of

---

[3] An alternate approach is to always assume a positive shear and moment. Hence, a negative sign means a negative quantity for *V* or *M*.

R_1 and R_2, as shown in Figure a. To determine the shears at the desired points we put each segment in equilibrium, as shown in Figures b, c, and d. Again, the directions of the shears and moments are assumed, then corrected if wrongly assumed.

At section a–a (see Figure b),

$$V_a = R_1 - 8 \text{ ft} \times 0.6 \text{ kip/ft} = 8.9 - 4.8 = \textbf{4.1 kips (or 18.24 kN)}$$
$$\text{(positive } V \text{ — clockwise rotation)}$$

$$M_a = R_1 \times 4 \text{ ft} - (8 \times 0.6) \text{kips} \times 4 \text{ ft} = \textbf{16.4 kips-ft (or 22.24 kN-m)}$$
$$\text{(positive } M \text{ — tension at bottom of beam)}$$

At section b–b (see Figure c),

$$V_b = R_1 - 10 \text{ ft} \times 0.6 \text{ kip/ft} = 8.9 - 6.0 = \textbf{2.9 kips (or 12.90 kN)}$$
$$\text{(positive } V \text{ — clockwise rotation)}$$

$$M_b = R_1 \times 6 \text{ ft} - (10 \times 0.6) \text{kips} \times 5 \text{ ft} = \textbf{23.4 kips-ft (or 31.72 kN-m)}$$
$$\text{(positive } M \text{ — tension at bottom of beam)}$$

At section c–c (see Figure d)

$$V_c = R_1 - 10 \text{ ft} \times 0.6 \text{ kip/ft} - 5 \text{ kips} = 8.9 - 6 - 5 = \textbf{2.1 kips}$$
$$\textbf{(or 9.34 kN)}$$
$$\text{(negative } V \text{ — counterclockwise rotation)}$$

$$M_c = R_1 \times 6 \text{ ft} - (10 \times 0.6) \text{kips} \times 5 \text{ ft} = \textbf{23.4 kips-ft (or 31.72 kN-m)}$$
$$\text{(positive } M \text{ — tension of bottom of beam)}$$

Example 3–2:

Given: The beam as shown in Figure a.
Find: (a) The shear and moment close to the left support, and (b) the shear transmitted across the hinge.

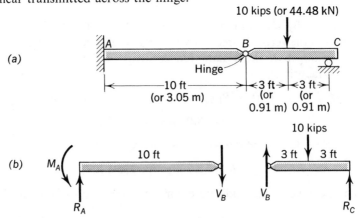

Solution: The hinge cannot transmit any moment, that is, $\Sigma M_{\text{hinge}} = 0$.

Thus,

$$\Sigma M_{hinge} = 0: \quad R_C = \frac{10 \text{ kips} \times 3 \text{ ft}}{6 \text{ ft}} = 5 \text{ kips}$$

$$V_B = 10 \text{ kips} - R_C = 10 \text{ kips} - 5 \text{ kips} = \textbf{5 kips (or 22.24 kN)}$$
$$\text{(positive } V \text{ — clockwise rotation)}$$

Note that on beam AB, Figure b, the shear V_B has a downward sense. (The beam must have this downward sense because Newton's third law of action and reaction applies.) Referring to Figure b, we note that, from

$$\Sigma F_y = 0: \quad R_A = V_B = 5 \text{ kips}$$
$$\Sigma M_A = 0: \quad M_A = V_B \times 10 \text{ ft} = 5 \text{ kips} \times 10 \text{ ft} = \textbf{-50 kips-ft}$$
$$\text{(or } \textbf{-67.79 kN-m)}$$
$$\text{(negative } M \text{ because tension is at the top of beam)}$$

$$R_A = V_B = \textbf{5 kips (or 22.24 kN)} \text{ (positive } V \text{ — clockwise rotation)}.$$

PROBLEMS

For the following problems neglect the weight of the beams (free-body diagrams should prove quite helpful).

3-1 For the cantilever beam in Figure P 3–1, determine the shear force V and bending moment M near the support and at a point $l/2$ from the support.

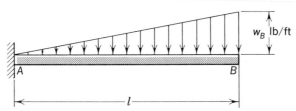

Figure P 3–1

3-2 For the cantilever beam in Figure P 3–2, determine the shear force V and bending moment M near the support and at a point $l/2$ from the support.

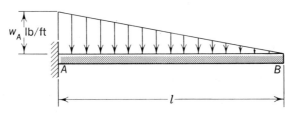

Figure P 3–2

3–3   For the cantilever beam in Figure P 3–3, determine the shear force *V* and bending moment *M* at points *A* and *B*.

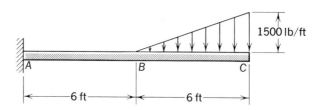

Figure P 3–3

3–4   For the cantilever beam in Figure P 3–4, determine the shear force *V* and bending moment *M* at point *A*, and just to the left of point *B*.

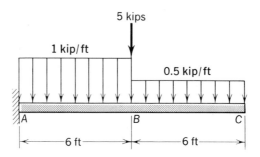

Figure P 3–4

3–5   For the cantilever beam in Figure P 3–5, determine the shear force *V* and bending *M* at the support.

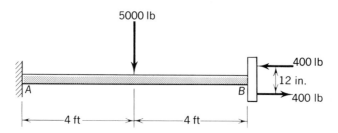

Figure P 3–5

3–6   For the simply supported beam in Figure P 3–6, determine the shear force *V* and bending moment *M* at any point along the span of the beam.

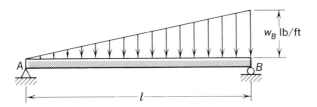

Figure P 3–6

3–7   For Problem P 3–6 determine the point (in terms of l) from the left support at which the moment is a maximum. What is the value of this moment? [Hint: Begin by writing an equation for the intensity of load at any point.]

3–8   For the simply supported beam in Figure P 3–8, determine the shear V and moment M at (a) the center of the span, and (b) just to the left of the 4.8-kip load.

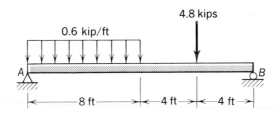

Figure P 3–8

3–9   For the simply supported beam in Figure P 3–9, determine the shear V and moment M at 5 ft and 10 ft from the left support.

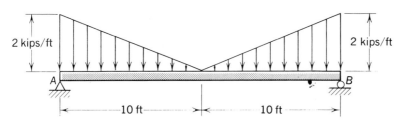

Figure P 3–9

3–10 For the simply supported beam with the overhanging ends in Figure P 3–10, determine the value of x (in terms of l) such that the moment at the center of the span is zero.

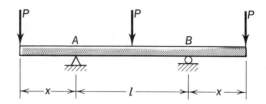

Figure P 3–10

3–11 For the overhanging beam in Figure P 3–11 determine the shear force V and bending moment M at points: (a) just to the left of B, and (b) just to the left of support C, when $P = 45$ N.

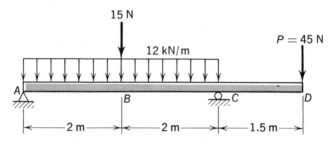

Figure P 3–11

3–12 For the beam in Figure P 3–11, determine the value P must have if the moment at B is to be zero. What will the shear be just to the right of B for this load?

3–13 For the symmetrically loaded beam in Figure P 3–13, determine the intensity of load w_0 at the free ends if the moment at the center of the beam is to be zero.

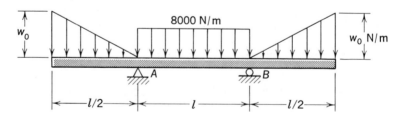

Figure P 3–13

3–14 For the beam in Figure P 3–14, determine (a) the shear force V and bending moment at 5 ft from point A, and (b) the shear force V transmitted across the hinge at C.

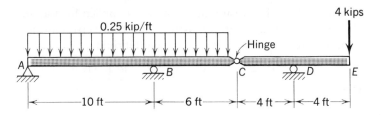

Figure P 3–14

3—2 Relations between Load, Shear Force, and Bending Moment

The load, shear force, and bending moment functions are most useful when presented graphically, that is, when the intensities of the load, shear force, and bending moment are plotted along the axis of the beam as shown in Section 3–3. Their usefulness in this form will be increasingly evident as they are made use of in one way or another in the following sections in this chapter, and in the following two chapters. For now we shall take the usefulness of the load, shear, and moment "diagrams" for granted, and shall proceed to show that these functions are interrelated by mathematical relationships which are extremely powerful as tools for the construction of these curves.

Consider an element of length Δx cut from a loaded beam placed in equilibrium, as shown in Figure 3–4, where, for convenience, w, V, and M

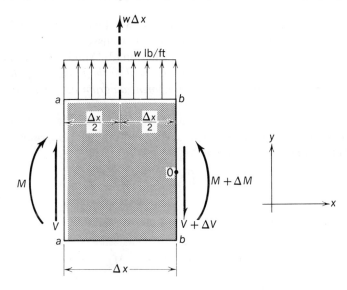

Figure 3–4

are assumed positive. The resultant of the total applied load acting on the element is $w \times \Delta x$, located at $\Delta x/2$ from the transverse planes $a\text{--}a$ and $b\text{--}b$.[4] Summing forces in the y direction, equilibrium gives us

$$\Sigma F_y = 0: \qquad V + w \cdot \Delta x = V + \Delta V$$

from which

$$\Delta V = w \cdot \Delta x \qquad \frac{\Delta V}{\Delta x} = w$$

As Δx approaches zero, the above becomes the first derivative of V with respect to x, that is,

$$\lim_{\Delta x \to 0} \frac{\Delta V}{\Delta x} = \frac{dV}{dx} = w \tag{3-1}$$

We note that dV/dx represents the slope of the shear "curve," and w represents the intensity of load. The sum of moments about point O, Figure 3-4, must also be zero; hence, we get

$$\Sigma M_O = 0: \quad M + \Delta M = V \cdot \Delta x + (w\Delta x)(\tfrac{1}{2}\Delta x) + M$$

or

$$\Delta M = V \cdot \Delta x + (\tfrac{1}{2}w)(\Delta x)^2$$

As Δx approaches zero, the above becomes the derivative of M with respect to x, that is,

$$\lim_{\Delta x \to 0} \frac{\Delta M}{\Delta x} = \frac{dM}{dx} = V \tag{3-2}$$

Here we note that dM/dx represents the slope of the moment curve at a point, and V represents the value of the transverse shear force at that point. A very useful interpretation of Equation (3-1) and (3-2) is as follows: *The slope of the shear curve at a point is equal to the intensity of the load at that point; similarly, the slope of the moment diagram at a point is equal to the intensity of the shear at that point.*

We may alter Equations (3-1) and (3-2) to get another, equally important relationship between the load, shear, and moment diagrams.

From Equation (3-1) we get

$$\int_{V_1}^{V_2} dV = \int_{x_1}^{x_2} w \, dx = V_2 - V_1 \tag{3-3}$$

and from Equation (3-2) we get

$$\int_{M_1}^{M_2} dM = \int_{x_1}^{x_2} V dx = M_2 - M_1 \tag{3-4}$$

---

[4] For small lengths Δx, the value of w may be assumed constant; that is, an average intensity w.

from which we conclude that: *The difference in the intensities (ordinates) of the shear at two points x_1 and x_2 along the beam is equal to the area under the load curve between these two points;* similarly, *the difference between the intensities of the moments at two points x_1 and x_2 along the beam is equal to the area under the shear curve between these two points.*

Both of the above relationships are very basic and, as previously emphasized, are extremely useful in constructing the related diagrams. Therefore, the student is urged to study and use these relationships so that he thoroughly understands them. Section 3–3 provides material for a suitable start in this direction.

3—3 Shear and Bending-Moment Diagrams

In the preceding two sections we have observed that, generally, both the shears and moments vary along a transversely loaded beam. We have also observed that if we integrate the load equation we get the shear equations; if we integrate the shear equations we get the moment equations, and so forth. In the next chapter we will see that these equations are useful in obtaining the deflections and slopes of deflected beams. However, in order to get to this, we must evaluate the constants of integration, not a readily evident task unless the shear and moment (and slope and deflections) are represented graphically. In addition, the maximum moments and shears, which result in maximum bending and shear stresses, are not nearly as evident from the equations as they are from the plotted functions. Finally, the graphical depiction of the moment function is a necessary part of some important procedures for determining deflections, and for analysis of statically indeterminante beams (the moment-area concept described and used in the following two chapters is such an example).

The shear and moment diagrams are graphs in which the abscissa indicates the position of the section along the beam, and the ordinates represent the corresponding value of shear or bending moment at that section. Hence, it is expedient and logical to place these diagrams directly under the loaded beam, as shown in Figure 3–5. This arrangement also greatly facilitates their construction as will become evident as we proceed.

To illustrate, let us consider the cantilever beam in Figure 3–5a. The first step is to determine the reactions resisting rotation or translation. These may be shown to act on the beam along with the loads, as shown in Figure 3–5b. The upward-acting loads are considered positive and thus are represented above the reference axis in Figure 3–5b. Similarly, the positive shears (Figure 3–5c) are placed above the reference axis; the negative shears and moments are placed below.

From $\Sigma F_y = 0$ and $\Sigma M_A = 0$, we get

$$R_A = P + \frac{wl}{2}$$

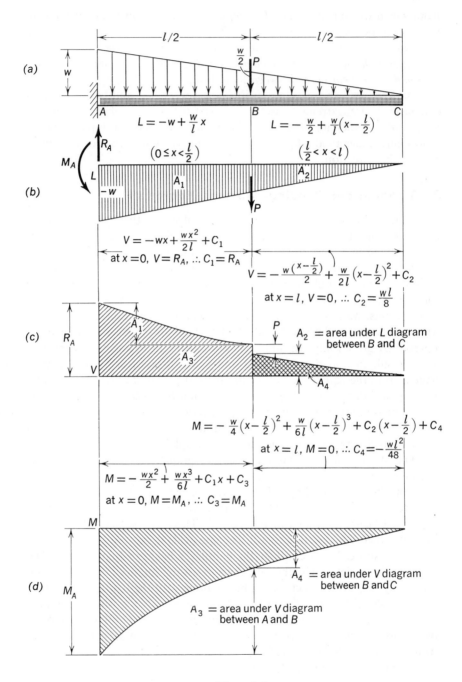

$$L = -w + \frac{w}{l}x \qquad L = -\frac{w}{2} + \frac{w}{l}\left(x - \frac{l}{2}\right)$$

$$\left(0 \leq x < \frac{l}{2}\right) \qquad \left(\frac{l}{2} < x < l\right)$$

$$V = -wx + \frac{wx^2}{2l} + C_1$$
at $x = 0$, $V = R_A$, $\therefore C_1 = R_A$

$$V = -\frac{w\left(x - \frac{l}{2}\right)}{2} + \frac{w}{2l}\left(x - \frac{l}{2}\right)^2 + C_2$$
at $x = l$, $V = 0$, $\therefore C_2 = \frac{wl}{8}$

$A_2 = $ area under L diagram between B and C

$$M = -\frac{w}{4}\left(x - \frac{l}{2}\right)^2 + \frac{w}{6l}\left(x - \frac{l}{2}\right)^3 + C_2\left(x - \frac{l}{2}\right) + C_4$$
at $x = l$, $M = 0$, $\therefore C_4 = -\frac{wl^2}{48}$

$$M = -\frac{wx^2}{2} + \frac{wx^3}{6l} + C_1 x + C_3$$
at $x = 0$, $M = M_A$, $\therefore C_3 = M_A$

$A_4 = $ area under V diagram between B and C

$A_3 = $ area under V diagram between A and B

Figure 3–5

and
$$M_A = P\left(\frac{l}{2}\right) + \left(\frac{wl}{2}\right)\frac{l}{3} = \frac{Pl}{2} + \frac{wl^2}{6}$$

The orientation of R_A and M_A is as shown in Figure 3–5b, that is, R_A is upward, M_A is negative.

The shear very close to point A is equal to R_A (from $\Sigma F_y = 0$); the moment very close to this point is negative ($\Sigma M_A = 0$). Thus, we have the ordinate of V_A and M_A on the respective diagrams. Using the relationship developed in Section 3–2, we now proceed to obtain the *shape* of the V and M diagrams. We note that the ordinate of the load at A is negative w, which is gradually diminishing with the distance x. Thus, the *slope* of the V diagram is negative w at A, varying with x. At point B we have the concentrated load P. Because we have two values for V_B (point of discontinuity), it is necessary to specify in our discussion the V just to the left or just to the right of B. Also, our equations for V and M will be valid only in the limits of $0 < x < l/2$ and $l/2 < x \leq l$. To the right of point B, the slope of the V diagram is equal to the intensity of the load at that point, namely, $-w/2$. The slope decreases to zero at point C, because the load is zero at C.

The moment curve has a positive slope at point A, corresponding to a positive V at that point; the slope becomes less positive with x, corresponding to a decrease in the positive V with x. At point B the moment curve has two slopes corresponding to two values of V, on the left and right of the point. At C the slope of the M-curve is zero, because the shear is zero.

Proceeding from a point where one ordinate is known, we may determine, step by step, the other ordinates of the graph by realizing that the *difference* of the intensities from point to point is equal to the area under the preceding curve between the same two points. This is explained in Figures 3–5b and 3–5c.

Equations for these curves will also give the above information for slope and ordinates when properly evaluated. For clarity, it is advisable to write the equations directly above the segment of the graph for which they apply. This procedure is shown is Figure 3–5. The constants of integration are obtained from known conditions, and this is one time where the shapes of the diagrams are helpful to us; for example, at $x = 0$, $V = R_A$, and $M = M_A$. This permits us to solve for C_1 and C_3. Similarly, from the diagrams, at $x = l$, $V = 0$, and $M = 0$; hence, we can solve for C_2, C_4 and so forth. The actual values of these constants in terms of load and length are given in Figure 3–5. The student should apply the above explanation to check the solution of Example 3–3.

Example 3—3:

Given: The overhanging beam in Figure a.
Find: (a) The shear and moment diagrams. Give "significant" values, and write the equations for V and M. (b) Determine the values of V

and M at points a, b, and c, and compare these values with those in Example 3–1.

Solution:

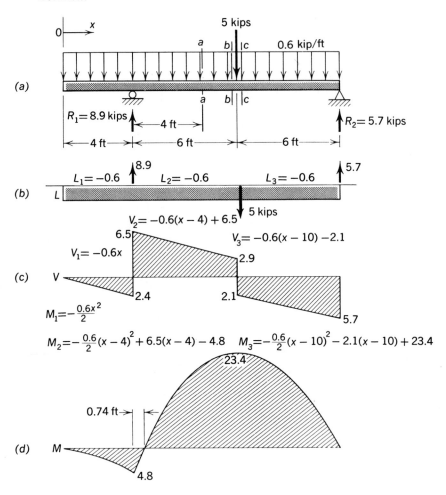

Example 3–4:

Given: The beam shown in Figure a.
Find: The shear and moment diagrams, and corresponding equations.
Solution:

$\sum M_3:$

$$R_4 = \frac{(20 \times 2) \times 3}{2} = 60 \text{ N}$$

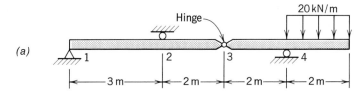

(a)

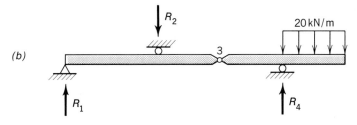

(b)

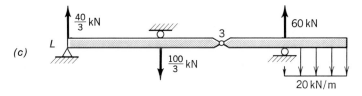

(c)

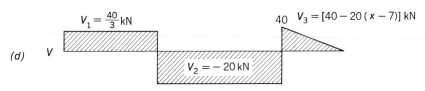

(d)

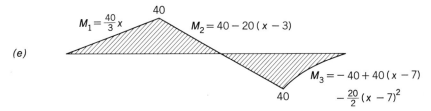

(e)

$$\Sigma M_1:$$

$$R_2 = \frac{(20 \times 2)8 - R_4 \times 7}{3}$$

$$= -\frac{100}{3} \text{ (downward)}$$

$$\Sigma F_y:$$

$$R_1 = \frac{40}{3} \text{ (upward)}$$

Example 3-5:

Given: The beam shown in Figure *a.*
Find: The shear and moment diagrams and write corresponding equations.

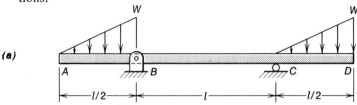

(a)

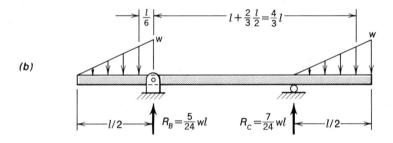

(b)

$$R_B = \frac{5}{24} wl \qquad R_C = \frac{7}{24} wl$$

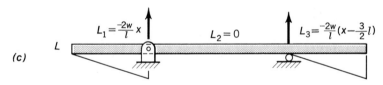

(c)

$$L_1 = \frac{-2w}{l} x \qquad L_2 = 0 \qquad L_3 = \frac{-2w}{l}\left(x - \frac{3}{2}l\right)$$

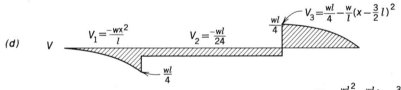

(d) V

$$V_1 = \frac{-wx^2}{l} \qquad V_2 = \frac{-wl}{24} \qquad V_3 = \frac{wl}{4} - \frac{w}{l}\left(x - \frac{3}{2}l\right)^2$$

(e) M

$$M_1 = \frac{-wx^3}{3l} \qquad M_2 = \frac{-wl^2}{24} - \frac{wl}{24}\left(x - \frac{l}{2}\right)$$

$$M_3 = \frac{wl^2}{12} - \frac{wl}{4}\left(x - \frac{3}{2}l\right) - \frac{w}{3l}\left(x - \frac{3}{2}l\right)^3$$

Solution:

$$\sum M_B:$$

$$R_c = \frac{\left(\dfrac{wl}{4}\right)\dfrac{4}{3}l - \left(\dfrac{wl}{4}\right)\dfrac{wl}{6}}{l} = \frac{7}{24}wl$$

$$\Sigma F_y:$$

$$R_B = \frac{wl}{2} - \frac{7}{24} wl = \frac{5}{24} wl$$

Example 3–6:

Given: The beam shown in Figure *a*.

Find: The shear and moment diagrams and write corresponding equations.

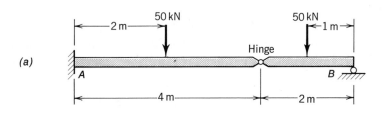

(a)

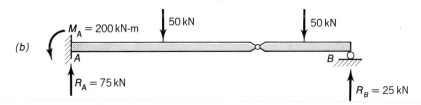

(b)

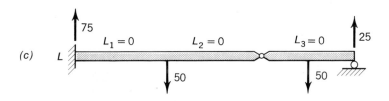

(c)

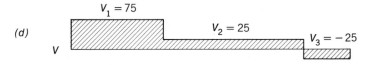

(d)

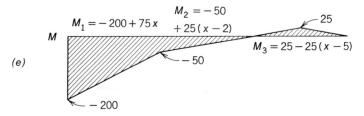

(e)

Solution:

$\sum M_{Hinge}$:

$R_B = 25$ kN

$\sum F_y$:

$\sum M_A$:

$R_A = 75$ kN

$M_A = 200$ kN-m

PROBLEMS

3–15 Construct to scale shear and moment diagrams for the beam in Figure P 3–15. Also, write equations and show significant values for these curves.

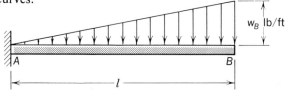

w_B lb/ft

A

B

l

Figure P 3–15

3–16 Construct to scale shear and moment diagrams for the beam in Figure P 3–16. Also, write equations and show significant values for these curves.

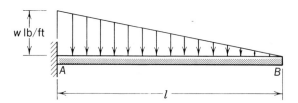

w lb/ft

A

B

l

Figure P 3–16

3–17 Construct to scale the shear and moment diagrams for the beam in Figure P 3–17. Also, write equations and show significant values for these curves.

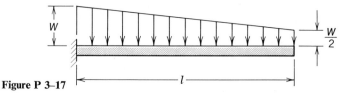

W

$\dfrac{W}{2}$

l

Figure P 3–17

3–18 Construct to scale shear and moment diagrams for the beam in Figure P 3–18. Also, write equations and show significant values for these curves.

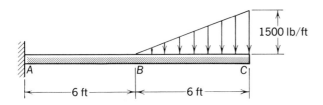

Figure P 3–18

3–19 Construct to scale shear and moment diagrams for the beam in Figure P 3–19. Also, write equations and show significant values for these curves.

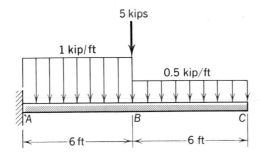

Figure P 3–19

3–20 Construct to scale shear and moment diagrams for the beam in Figure P 3–20. Also, write equations and show significant values for these curves.

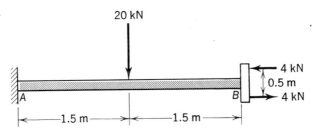

Figure P 3–20

3–21 Construct to scale shear and moment diagrams for the beam in Figure P 3–21. Also, write equations and show significant values for these curves.

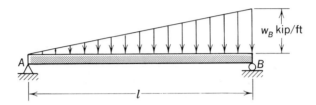

Figure P 3–21

3–22 Construct to scale shear and moment diagrams for the beam in Figure P 3–22. Also, write equations and show significant values for these curves.

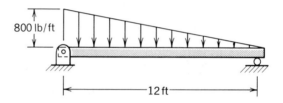

Figure P 3–22

3–23 Construct to scale shear and moment diagrams for the beam in Figure P 3–23. Also, write equations and show significant values for these curves.

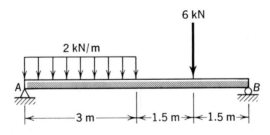

Figure P 3–23

3–24 Construct to scale shear and moment diagrams for the beam in Figure P 3–24. Also, write equations and show significant values for these curves.

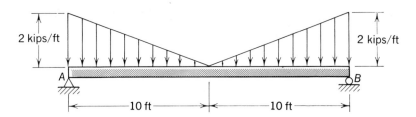

Figure P 3–24

3–25 Construct to scale shear and moment diagrams for the beam in Figure P 3–25. Also, write equations and show significant values for these curves.

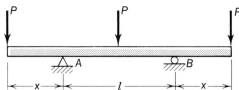

Figure P 3–25

3–26 Construct to scale shear and moment diagrams for the beam in Figure P 3–26. Also, write equations and show significant values for these curves.

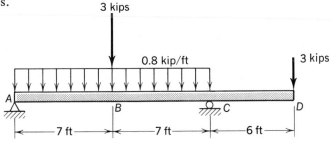

Figure P 3–26

3–27 Construct to scale the shear and moment diagrams for the beam in Figure P 3–27. Also, write equations and show significant values for these curves.

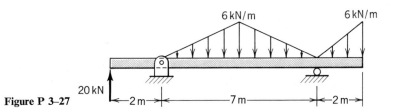

Figure P 3–27

3–28 Construct to scale shear and moment diagrams for the beam in Figure P 3–28. Also, write equations and show significant values for these curves.

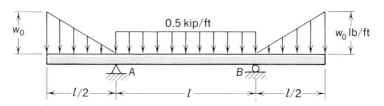

Figure P 3–28

3–29 Construct to scale shear and moment diagrams for the beam in Figure P 3–29. Also, write equations and show significant values for these curves.

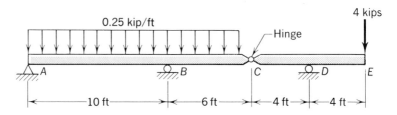

Figure P 3–29

3—4 Bending Stresses in Straight Beams

When a beam is subjected to a bending moment, the material of the beam deforms, resulting in a curvature of the beam. Perhaps we have all noted this at one time or another when bending an object such as a yard stick. We may have also observed that if the bending moment was increased enough, rupture resulted. The fibers of the yard stick on the outer side (convex side) broke in tension; however, those on the inner side (concave side) seemed to remain unaffected. Although the deformation and subsequent rupture of the yard stick may serve as evidence of a relationship between external moment and internal effects, they have to be analyzed in a more systematic manner if we are to relate cause and effect accurately and meaningfully.

We shall show the relationship existing between bending stress σ_x and bending moment M for a given cross section by beginning with the following assumptions:

1. Transverse planes before bending remain transverse planes after bending; that is, no warping takes place.

2. The material in the beam is homogeneous and isotropic and obeys Hooke's law. Here we assume that E is the same for tension as for compression.

3. The beam is straight and has a constant, prismatic, cross section.

4. The loads will cause neither twisting nor buckling of the beam. This condition is satisfied if the plane of loading contains an axis of symmetry of the cross section, and if the loads lie in this plane.

5. The load applied is a pure bending moment.

Later we shall see that we may deviate somewhat from these conditions without significant error in our results. However, at present it is convenient to maintain these conditions.

Let us begin by isolating a portion of the beam in Figure 3–6a and placing it in equilibrium, as shown in Figure 3–7a, assuming it to meet the

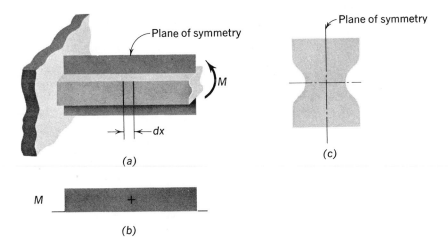

Figure 3–6 (c) Beam cross section.

above-stated conditions. It is evident that the moment is uniform (pure moment) over the length dx, and that it is the only external load on the beam. The segment will then take the form of a circular arc of a mean radius of curvature ρ, as shown in Figure 3–7a. The fibers on the concave side are compressed, and those on the convex side are elongated. Somewhere between the concave and convex side there is a layer of fibers which undergoes neither tensile nor compressive deformation. This curved surface is called the *neutral surface*, and the intersection of this surface with any transverse plane is called the *neutral axis*, shown in Figure 3–7b.

From assumption 1 above, the deformation varies linearly from zero at the neutral axis to a maximum at the extreme fibers. This may be verified experimentally by means of SR-4 strain gages. Thus, at some distance y

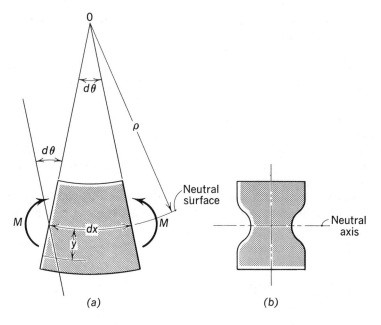

Figure 3–7

from the neutral axis, Figure 3–7a, the deformation is $y \cdot d\theta$, for a small angle. The strain at y is

$$\epsilon_x = \frac{y \cdot d\theta}{dx} \qquad (a)$$

From similar triangles, $y/\rho = yd\theta/dx$, which reduces to $d\theta/dx = 1/\rho$. Substituting this into Equation (a), we have

$$\epsilon_x = \frac{y}{\rho} \qquad (b)$$

The relationship $\sigma = \epsilon E$ may now be used here (assumption 2),

$$\sigma_x = \epsilon_x E = E\frac{y}{\rho} \qquad (c)$$

Equation (c) shows that the stress σ_x varies linearly with the distance from the neutral axis. This stress distribution is shown in Figure 3–8a. It is evident that the fibers above the neutral axis must be in compression, and those below in tension. Their action on the cross-sectional area forms a couple which must resist the moment M.

The increment of force acting on element of area dA in Figure 3–8b is

$$dF = \sigma_x dA = E\frac{y}{\rho}dA \qquad (d)$$

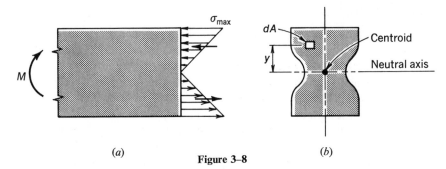

(a)

Figure 3–8

(b)

From assumption 5 above, the total compressive force must equal the total tensile force on the section, that is,[5]

$$\int_A \frac{Ey}{\rho} dA = \frac{E}{\rho} \int_A y \, dA = \frac{E}{\rho} A y_c = 0 \qquad (e)$$

Because $EA/\rho \neq 0$, $y_c = 0$; hence, we conclude that the *neutral axis of the cross section passes through the centroid of the section*.

The corresponding increment of moment is

$$dM = dF \cdot y = \left(\frac{E \cdot y^2}{\rho}\right) dA$$

The *total* moment resisted is

$$M = \int_A \frac{Ey^2}{\rho} dA = \frac{E}{\rho} \int_A y^2 \, dA \qquad (f)$$

Here we recognize that $\int_A y^2 \, dA = I$, the *second moment*, or more commonly, the *moment of inertia* of the cross-sectional area with respect to the neutral axis (n.a.). Thus, Equation (f) becomes

$$M = \frac{E}{\rho} I \qquad (3\text{-}5)$$

Also, from Equation (c), $\rho = E \cdot y / \sigma_x$, which, when substituted into Equation (3–5), gives *the flexure formula*:

$$\sigma_x = \frac{My}{I} \qquad (3\text{-}6)$$

It must be remembered that y represents the distance from the neutral axis to a particular fiber; thus the maximum stress occurs in the fibers furthest from the neutral axis.

---

[5] *Varignon's Theorem* (for parallel forces): The algebraic sum of the moments of two parallel forces with respect to any moment center or any axis is equal to the moment of their resultant with respect to the same center or axis.

Example 3–7:

Given: The overhanging beam in the figure. The material is cast iron for which the stress in tension and compression is not to exceed 10,000 psi and 20,000 psi, respectively.

Find: The maximum uniform load (neglecting the beam's weight) that can be safely carried by the beam.

Procedure: From the moment diagram we pick the maximum negative and maximum positive moments and, using Equation (3–6) we determine the tensile and compressive stresses at each point. Because the neutral axis of the section is not at mid-depth of the beam, we must first determine the centroidal distance, then, in Equation (3–6), substitute for y the values of $\bar{y}$ and $(8-\bar{y})$.

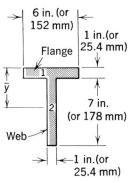

Beam cross section
(d)

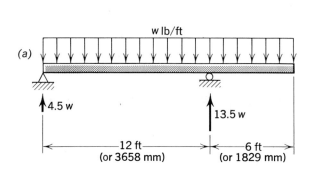

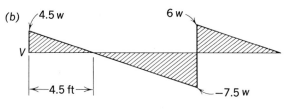

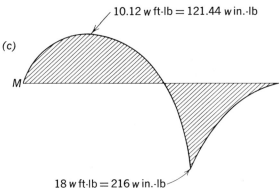

To find $\bar{y}$, we use composite areas. With respect to the top of the beam, we have

$$\bar{y} = \frac{A_1 y_1 + A_2 y_2}{A_1 + A_2}$$

$$\bar{y} = \frac{(6 \times 1)(0.5) + (7 \times 1)(4.5)}{(6 \times 1) + (7 \times 1)}$$

$$\bar{y} = \frac{34.5}{13} = 2.65 \text{ in. (or } 67.3 \text{ mm)}$$

The moment of inertia I with respect to the neutral axis is

$$I_{n.a.} = I_c + Ad^2$$

$$I_{n.a.} = \left[\frac{6 \times 1^3}{12} + (6 \times 1)(2.65 - 0.5)^2\right]_{\text{flange}}$$
$$+ \left[\frac{1 \times 7^3}{12} + (7 \times 1)(4.5 - 2.65)^2\right]_{\text{web}}$$

$$I_{n.a.} = (0.50 + 27.75) + (28.61 + 24) = 80.85 \text{ in.}^4$$

At the right support (maximum negative M),

$$\sigma_{\text{comp}} = 20,000 = \frac{M(8 - 2.65)}{80.85} = 0.0662 \times 216 \, w = 14.3 \, w_c$$

Therefore,

$$w_c = \frac{20,000}{14.3} = 1400 \text{ lb/ft (allowable as dictated by } \sigma_{\text{comp}})$$

$$\sigma_{\text{tension}} = 10,000 = \frac{(216w_t)(2.65)}{80.85} = 7.05 w_t$$

Therefore,

$$w_t = \frac{10,000}{7.05} = 1420 \text{ lb/ft (allowable as dictated by } \sigma_{\text{tension}})$$

At point of maximum positive M,

$$\sigma_{\text{comp}} = 20,000 = \frac{121.44 w_c(2.65)}{80.85} = 3.99 \, w_c$$

Therefore,

$$w_c = \frac{20,000}{3.99} = 5900 \text{ lb/ft} \qquad \text{(allowable as dictated by } \sigma_{\text{comp}})$$

$$\sigma_{\text{tension}} = 10,000 = \frac{121.44 w_c(5.35)}{80.85} = 8.04 \, w_t$$

Therefore,

$$w_t = \frac{10{,}000}{8.04} = 1240 \text{ lb/ft} \qquad \text{(allowable as dictated by } \sigma_{\text{tension}}\text{)}$$

Note that we have four values of w, at the two points where the peak moments occur; two values of w based on the assumption that tension controls, the other two values based on the assumption that compression controls. The *controlling* w is 1240 lb/ft (or 18.1 kN/m), the smallest of the four values.

Example 3–8:

(a)

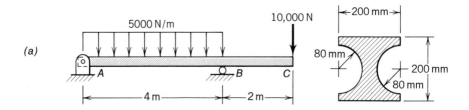

Given: The overhanging beam in Figure *a*.
Find: The maximum bending stress in the beam.
Solution:

The moment of inertia, via composite section, is

$$I = I_{\text{square}} - I_{\text{circle}}$$

$$= \frac{bd^3}{12} - \frac{\pi(2r)^4}{64} = \frac{(0.2)(0.2)^3}{12} - \frac{\pi}{64}(0.16)^4$$

$$= \frac{16}{12} \times 10^{-4} - 0.322 \times 10^{-4} = 1.01 \times 10^{-4} \text{ } 4$$

$$\sigma_{\text{max}} = \frac{(M_{\text{max}})(c)}{I}, \qquad \text{where } M_{\text{max}} = 20{,}000 \text{ N-m}$$

$$\text{and } c = 100 \text{ mm} = 0.1 \text{ m}$$

$$= \frac{(20{,}000 \text{ N-m})(0.1 \text{ m})}{1.01 \times 10^{-4} \text{ m}^4} = \textbf{19.80 MN/m}^2$$

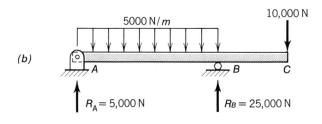

(b)

$R_A = 5,000$ N $R_B = 25,000$ N

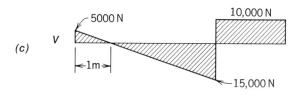

(c)

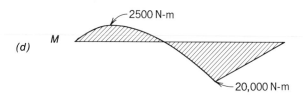

(d)

PROBLEMS

Although most of the following beams are subjected to other than a pure moment at the section where the bending stress is to be investigated, the calculations for bending stresses waiving the pure-moment condition give results which are quite acceptable; that is, $\sigma = My/I$ applies. Also, assume that the cross section of any one beam is constant for the whole length of the beam. See Appendix B, Table B–2, for the properties of some of the rolled sections used.

3–30 Construct to scale shear and moment diagrams for the beam in Figure P 3–30. Also, write equations and show significant values for these curves.

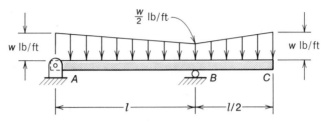

Figure P 3–30

3–31 Construct to scale shear and moment diagrams for the beam in Figure P 3–31. Also, write equations and show significant values for these curves.

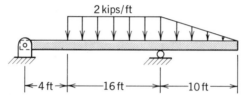

Figure P 3–31

3–32 The cantilever beam in Figure P 3–32 supports a load which varies from 0 at the support to w_B = 2000 lb/ft at point B. If the beam is a standard 10 in. I-25.4, determine the length l the beam may have if the allowable working stress in bending is 20,000 psi. (Neglect the weight of the beam.)

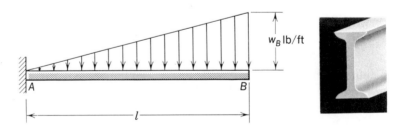

Figure P 3–32

3–33 Select the most economical section of a standard I beam to support the load shown in Figure P 3–33 for an allowable bending stress of 22,000 psi, if l = 14 ft.

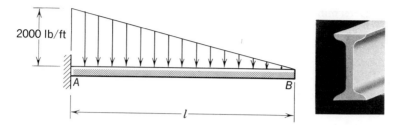

Figure P 3–33

3–34 Select the most economical wide-flange section (Appendix B) to support the load shown in Figure P 3–34 if the allowable bending stress is 24,000 psi.

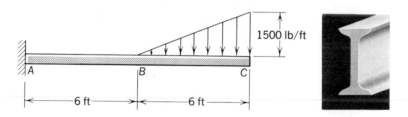

Figure P 3–34

3–35 Select the most economical wide-flange section (Appendix B) to carry the load shown in Figure P 3–35 for an allowable working stress in bending of 20,000 psi.

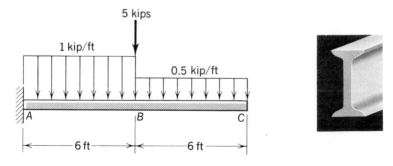

Figure P 3–35

3–36 Determine the maximum bending stress in the beam shown in Figure P 3–36.

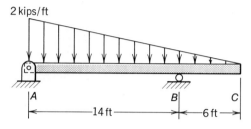

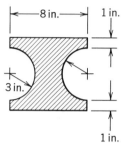

Figure P 3–36

3–37 The allowable bending stresses in the beam shown in Figure P 3–37 are 10,000 psi in tension and 15,000 psi in compression. Determine the maximum allowable w.

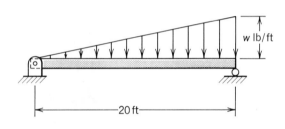

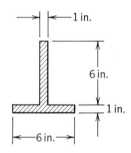

Figure P 3–37

3–38 For the cantilever beam in Figure P 3–38, (a) draw the shear and moment diagrams, and (b) determine the maximum compressive and tensile bending stresses.

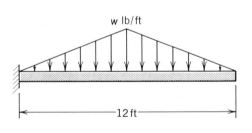

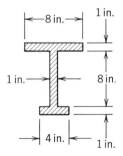

Figure P 3–38

3–39 The cantilever beam in Figure P 3–39 is an 8-in. WF, weighing 20 lb/ft. Determine the maximum bending stress in the beam.

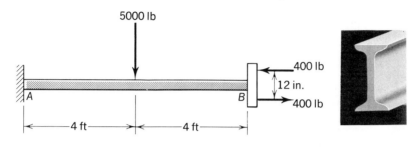

Figure P 3–39

3–40 For the beam in Figure P 3–40, (a) draw the shear and moment diagrams, and (b) determine the maximum bending stress.

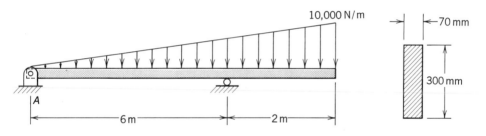

Figure P 3–40

3–41 Determine the load intensity w_B in Figure P 3–41 that the cast-iron section can take if the allowable bending stresses in tension and compression are 8000 psi and 12,000 psi, respectively. The span, $l = 10$ ft.

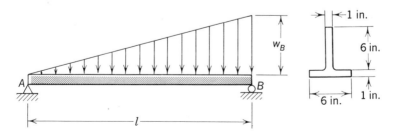

Figure P 3–41

3-42 In Figure P 3-42 we have a box beam fabricated by welding together four plates that are 4 in. wide and ¾ in. thick, as shown. Determine the maximum bending stress in the section for the given load. (Neglect the cross-sectional area of the weld when computing I.)

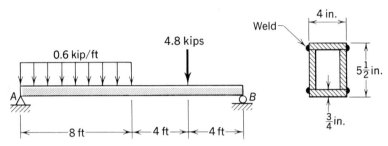

Figure P 3-42

3-43 Figure P 3-43 contains a box beam fabricated by welding together four plates 6 in. wide and 1 in. thick, as shown. Determine the maximum bending stress in the section for the given load. (Neglect the area of the weld when computing I.)

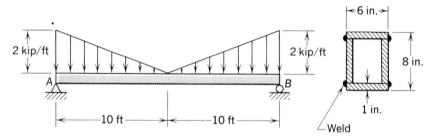

Figure P 3-43

3-44 The beam in Figure P 3-44 is made of a cast-iron cross section, as shown. If $l = 12$ ft and $x = 4$ ft, determine the maximum value P may have if the working stress of the cast iron in compression is 14,000 psi, and in tension is 8000 psi.

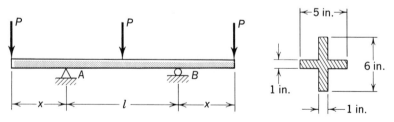

Figure P 3-44

3–45 The beam in Figure P 3–45 is an 8-in. WF 24. For a working stress in bending of 18,000 psi, determine (a) the maximum value that P may have, and (b) the maximum bending stress when P is zero.

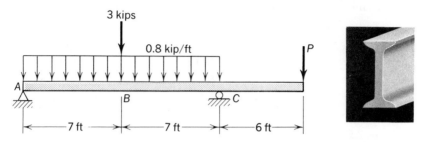

Figure P 3–45

3–46 Select an economical wide-flange beam to support the load of Figure P 3–46 where w_0 is 750 lb/ft, $l = 14$ ft, and the bending stress is not to exceed 20,000 psi.

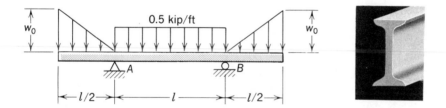

Figure P 3–46

3–47 Select an economical wide-flange beam to support the load of Figure 3–47 if the bending stress is not to exceed 22,000 psi.

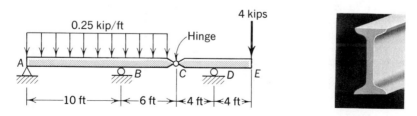

Figure P 3–47

3–48 The beam in Figure P 3–48 is to carry a 25-kN load, as shown. Points B and C are to be considered rigid (that is, they transmit both shear and moment). If the bending stress is not to exceed 125 MN/m², determine the minimum diameter beam AB may have.

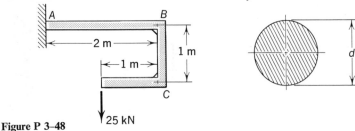

Figure P 3–48

3–49 The floor of a home is spanned by nominal 2-in. × 8-in. joists, spaced 16 in. center to center over a span of 14 ft, as shown in Figure P 3–49. If the bending stress in the joists is not to exceed 1000 psi, determine the allowable uniform load per square foot the floor may carry.

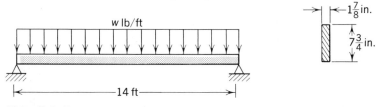

Figure P 3–49

3–50 Two beam sections are designed to take a given moment. The one in Figure P 3–50a is made up of 3 pieces, 2 in. × 4 in., completely glued to each other. In Figure P 3–50b, the 2 × 4's are assumed free to slide past each other; that is, each 2 × 4 is assumed to take $\frac{1}{3}$ of the total moment. Compare the moments that each section may safely carry at a maximum fiber stress of 1000 psi.

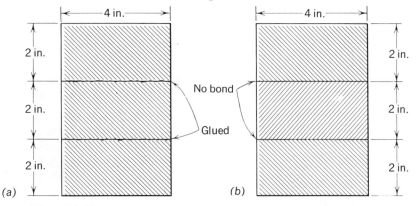

Figure P 3–50

3—5 Shearing Stress in Beams

In engineering we seldom encounter only a pure bending moment on a section of a member. The more common occurrence is transverse loads which cause both a bending moment and a shear force at a section. In Section 3–4 we have established the bending moment at a section to be the resultant of the bending stresses over the section, and proceeded to show the distribution of these stresses over the section. By induction, we may conclude that shear forces over the section would result in shear stresses over this section, and that these forces are indeed the resultants of these stresses. What the magnitude and distribution of the shearing stresses is at such a section, and how the force and stress are interrelated is the topic to be covered in this section.

Let us consider a short increment of length of a beam, as shown in Figure 3–9a. Because the bending moments on the two faces are different (by an amount dM), the bending stresses on the two faces will differ accordingly, as illustrated in Figure 3–9b. The forces acting on an element of area dA (Figure 3–9c) located y distance from the neutral plane are dF_a on the left and dF_b on the right. The resultants of these differential forces

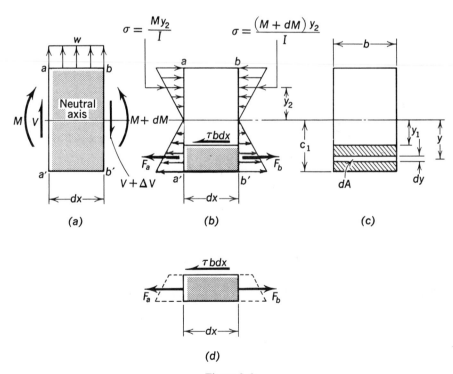

(a)

(b)

(c)

(d)

Figure 3–9

acting over the shaded area (Figure 3–9c) are F_a and F_b, respectively, acting as shown in Figure 3–9d. From $dF = (\sigma)dA = (My/I)dA$, we get

$$F_a = \int_{y_1}^{c_1} \frac{My}{I}\, dA = \frac{M}{I} \int_{y}^{c_1} y\, dA \qquad (a)$$

and

$$F_b = \int_{y_1}^{c_1} \frac{(M + dM)y}{I}\, dA = \frac{M + dM}{I} \int_{y_1}^{c_1} y\, dA \qquad (b)$$

The *difference* of these two forces must be resisted by a shear force composed of an *average* shearing stress τ times the area $b\,dx$ on which it acts (the upper face of the block), oriented as shown in Figure 3–9b. Hence, from $\Sigma F_x = 0$, we get

$$\tau b\, dx = \frac{M + dM}{I} \int_{y_1}^{c_1} y\, dA - \frac{M}{I} \int_{y_1}^{c_1} y\, dA$$

from which

$$\tau b\, dx = \frac{dM}{I} \int_{y_1}^{c_1} y\, dA$$

or

$$\tau = \frac{dM}{dx} \frac{1}{Ib} \int_{y_1}^{c_1} y\, dA \qquad (c)$$

In Section 3–2 we have shown that $dM/dx = V$ [Equation (3–2)]. Thus, expression (c) becomes

$$\tau = \frac{V}{Ib} \int_{y_1}^{c_1} y\, dA \qquad (3\text{–}7)$$

In Equation (3–7), $\int_{y_1}^{c_1} y\, dA$ is the *first moment*, a statical moment about the neutral axis, of that portion of the cross section below level y_1, at which the shearing stress τ is desired. In our case, the area in question corresponds to the shaded area. Hence, if we let $\int_{y_1}^{c_1} y\, dA = Q$, Equation (3–7) becomes

$$\tau = \frac{VQ}{Ib} \qquad (3\text{–}8)$$

We might note that Equation (3–8), as derived, expresses the average intensity of the shear stress in the *horizontal* direction. We know, however, that $|\tau_{\text{vert}}| = |\tau_{\text{horiz}}|$ (Section 1–11); hence, Equation (3–8) is also the expression for the transverse shear over any given cross section, oriented in the direction of the shear force V at that section.

We might also note that at any given point along the span of a beam the quantities V and I are constants; hence, $\tau = K(Q/b)$ (where $K = V/I$), obviously a variable if either Q or b (or both) varies. We may demonstrate this by selecting a *rectangular* section, assuming it to have the general dimensions shown in Figure 3–10a. Here we note that *b* is also a constant; thus, τ will vary directly with Q (Example 3–5 will illustrate a case where *b* also varies.)

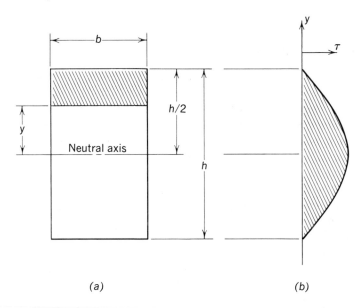

(a)   (b)

Figure 3–10

At any level *y*,

$$Q = \left(\frac{h}{2} - y\right)b \times \frac{(h/2 + y)}{2} = \frac{b}{2}\left(\frac{h^2}{4} - y^2\right)$$ (d)

where $(h/2 - y)b$ is the shaded area, and $(h/2 + y)/2$ is the moment arm. Here we note that at the extreme fiber the moment arm is a maximum but the area is zero, making $Q = 0$. Hence, at the extreme fiber $\tau = 0$. Plotting the right-hand side of Equation (d) will show this. Furthermore, it will also show that Q is a maximum where $y = 0$ (that is, at the neutral axis), and that the distribution of Q is parabolic over the depth *h*. Consequently, for our case, τ will have a parabolic distribution, varying from zero at the extreme fibers (top and bottom) to a maximum at the neutral axis, as shown in Figure 3–10b. In equation form, the average shearing stress for a *rectangular* cross section is

$$\tau = \frac{V}{Ib} \times \frac{b}{2}\left(\frac{h^2}{4} - y^2\right) = \frac{V}{2I}\left(\frac{h^2}{4} - y^2\right)$$ (3–9)

and the maximum, at the neutral axis, is for a *rectangular* section

$$I = \frac{bh^3}{12} \qquad y = 0$$

$$\tau_{max} = \frac{V}{2(bh^3/12)}\left(\frac{h^2}{4} - 0\right)$$

$$\tau_{max} = \frac{3V}{2bh} = \frac{3V}{2A} \tag{3-10}$$

A further look at this variation brings to light an interesting phenomenon related to a basic assumption made in analyzing beams in the elastic range: Plane sections before bending remain plane after bending. If we assume elastic deformation when we calculate the shearing stress, we conclude that if the shearing *stress* varies from top to bottom, the shearing *strain* must vary in a similar manner, because $\gamma = \tau/G$. At the top and bottom surfaces the strain is zero, varying to a maximum at the neutral axis. Thus, what were originally plane cross sections before bending become warped sections after bending, as shown in Figure 3–11. The warped plane at the top and bottom surfaces remains normal to these surfaces, and has the greatest slope with the vertical at the neutral axis. If the shear force V is constant over the length of the beam, the warping of any sections along this length is identical with any other in this length. Hence, the shear stresses do not have any effect on the normal (longitudinal) strains, and therefore, despite the warped section, the distributions of the normal strains (and consequently the normal stresses σ) remain as assumed for the case of pure bending; that is, proportional to the distance from the neutral axis. From the theory of elasticity, the above conclusion can be shown to be correct for a constant shear. For a varying V this is not the case; the warping of different planes differs along the beam, depending on the type and distribution of the loads. Thus, the normal strains are affected by the shear causing a degree of error in our assumptions of proportionality when using the flexure for-

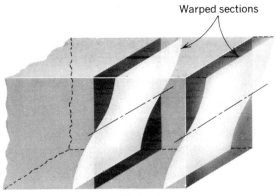

Warped sections

Figure 3–11

mula ($\sigma = My/I$) for beams not subjected to just pure bending. However for most cases this effect is quite small; it is actually negligible. Hence, the flexure formula is essentially correct even for cases of varying shear, a condition that was considered a proven fact in solving the problems of the last section.

Example 3–9:

Given: A standard 8-in. (or 203 mm) 18.4-lb per ft steel I beam supports a total load of 6800 lb (or 302 kN) uniformly distributed over a simple span, as shown in Figure *a*.
Find: (*a*) The maximum shear stress in the beam, (*b*) the shear stress distribution over the cross section, and (*c*) the percentage shear force (or 12.48 MN/m²) carried by the web.

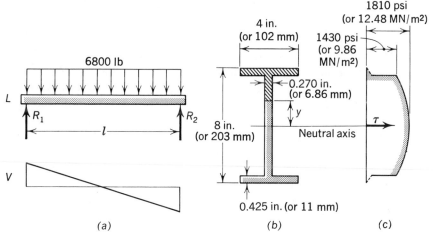

(*a*) Loading condition. (*b*) Beam's cross section. (*c*) τ distribution.

Solution: From Table B–2, Appendix B, we find the properties of the beam, some of which are shown in Figure *b*. Also from this table, $I = 56.9$ in.⁴ The shear force V is a maximum at one of the supports, as shown in Figure *b*, and is equal to one-half of the total load; that is, $V = 3400$. The shear stress is then

$$\tau = \left(\frac{V}{I}\right)\left(\frac{Q}{b}\right) = \frac{3400}{56.9}\left(\frac{Q}{b}\right) = 60\left(\frac{Q}{b}\right)$$

where Q and b are variables. The value of Q at a level y within the web (Figure *b*) is as follows:

$$Q = (4 \times 0.425)\left(4 - \frac{0.425}{2}\right) + 0.270\frac{(4 - 0.425)^2}{2} - 0.270\frac{y^2}{2}$$

$$Q = 6.44 + 1.72 - 0.135y^2 = 8.16 - 0.135y^2$$

Near the junction of the web and flange, $y = 3.575$ in. and thus $Q = 8.16 - 0.135 (3.575)^2 = 6.44$ in.^{3} At the neutral axis, $y = 0$ and $Q = 8.16$ in.^{3} Over the depth of the web, Q varies parabolically. The shear stress is therefore,

near flange:

$$\tau = \frac{60 \times 6.44}{0.270} = \textbf{1430 psi (or 9.86 MN/m}^2\textbf{)}$$

at the neutral axis, $\tau = \dfrac{60 \times 8.16}{0.270} = \textbf{1810 psi (or 12.48 MN/m}^2\textbf{)}$

We note that the maximum shear stress in the web occurs at the neutral axis, and that it is also the maximum shear stress over the section. The total shear force carried by the web may be determined as the product of the total area of the shear distribution times the thickness of the web, that is,

$$V_{\text{web}} = 0.270 \times 7.15 \, [1430 + \tfrac{2}{3}(380)] = \textbf{3250 lb (or 14.46 kN)}$$

This represents $(3250/3400) \, 100 = \textbf{95.5 percent}$ of total V.

From the above discussion we find that the portion of the shear force carried by the flange is negligible, a typical finding for I-beam sections. Furthermore, the distribution of the shear stress in the flange as determined by the VQ/Ib formula is rather misleading because one cannot assume a uniform stress distribution over the whole width of the flange. The stress is zero at the top and bottom of the flange where free surfaces exist, but is not zero at the junction of the web and flange, where a stress concentration exists. The dashed-line distribution shown in Figure c, is, at best, only a fabricated guess. Hence, the small effect of V that the flanges carry, combined with the uncertain stress distribution, and coupled with small stresses, makes the necessity of even considering the flanges in this regard indeed questionable.

Example 3–10:

Given: The overhanging beam shown in Figure a.

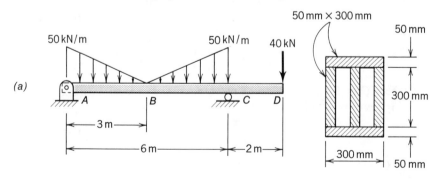

Find: (*a*) The maximum bending stress, (*b*) the maximum shear stress, and (*c*) the shear stress at the junction of flange and web.

Solution:

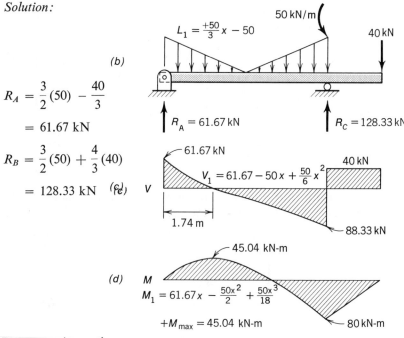

$$R_A = \frac{3}{2}(50) - \frac{40}{3}$$

$$= 61.67 \text{ kN}$$

$$R_B = \frac{3}{2}(50) + \frac{4}{3}(40)$$

$$= 128.33 \text{ kN} \quad \text{(e)}$$

(b)

$$L_1 = \frac{+50}{3}x - 50$$

50 kN/m

40 kN

$$R_A = 61.67 \text{ kN} \qquad R_C = 128.33 \text{ kN}$$

61.67 kN

40 kN

$$V_1 = 61.67 - 50x + \frac{50}{6}x^2$$

V

1.74 m

88.33 kN

45.04 kN-m

(d)   M

$$M_1 = 61.67x - \frac{50x^2}{2} + \frac{50x^3}{18}$$

$$+M_{max} = 45.04 \text{ kN-m} \qquad 80 \text{ kN-m}$$

By composite sections,

$$I = \frac{(0.3)(0.4)^3}{12} - \frac{(0.15)(0.3)^3}{12} = \frac{3 \times 10^{-4}}{12}[(4)^3 - (1.5)(3.0)^2]$$

$$= 12.62 \times 10^{-4} \text{ m}^4$$

$$\sigma_{max} = \frac{(80 \text{ kN-m})(0.2 \text{ m})}{(12.62 \times 10^{-4} \text{ m}^4)} = \textbf{12.67 MN/m}^2$$

$$\tau_{max} = \frac{VQ}{Ib} = \frac{(88.33 \text{ kN})Q_{max}}{(12.62 \times 10^{-4} \text{ m}^4)(0.15 \text{ m})}$$

$$Q_{max} \atop \text{(at n.a.)} = (0.2)(0.3)(0.1) - (0.15)(0.15)(0.075) = 4.4 \times 10^{-3} \text{ m}^3$$

Therefore,

$$\tau_{max} = \frac{(88.33 \text{ kN})(4.4 \times 10^{-3} \text{ m}^3)}{(12.62 \times 10^{-4} \text{ m}^4)(0.15 \text{ m})} = \textbf{2.05 MN/m}^2$$

At junction of "flange" and "webs,"

$$Q = (0.3)(0.05)(0.175) = 2.63 \times 10^{-3} \text{ m}^3$$

and

$$\tau = \textbf{1.22 MN/m}^2$$

PROBLEMS

3–51 Determine the horizontal shearing stress in the beam shown in Figure P 3–51 if $w_B = 1200$ lb/ft and $l = 12$ ft, at a point 2 ft from point A and at 3 in. below the top fiber of the beam.

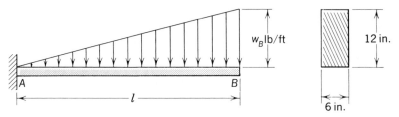

Figure P 3–51

3–52 Referring to Figure P 3–52, (a) draw the shear and moment diagram, and (b) find the maximum shear stress.

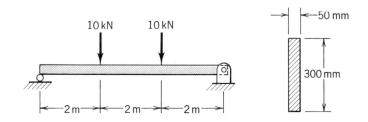

Figure P 3–52

3–53 In Figure P 3–53 assume the intensity of load at point B is w_B. If the allowable working stresses in bending and shear for the section shown are 1600 psi and 200 psi, respectively, determine the cantilever length l below which the shear stress governs and above which the stress governs.

3–54 The box section shown in Figure P 3–54 is fabricated by gluing together four 8-in. × 2-in. wood planks, as shown. If the allowable shearing stress in the glue is not to exceed 120 psi, determine (a) the length l that the beam may have, and (b) the maximum bending and shearing stresses in the beam.

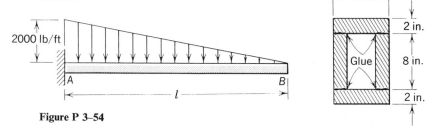

Figure P 3–54

3–55 For the beam in Figure P 3–55, (a) draw the shear and moment diagrams, (b) find the maximum shear stress, and (c) determine the maximum compressive and tensile stresses.

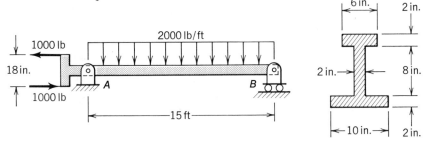

Figure P 3–55

3–56 The T beam in Figure P 3–56 carries the load as shown. If the allowable working stress in bending and shear are 20,000 psi and 10,000 psi, respectively, determine (a) the permissible magnitude of the concentrated load P at point B, and (b) the magnitudes of bending and shearing stresses for the load in part (a).

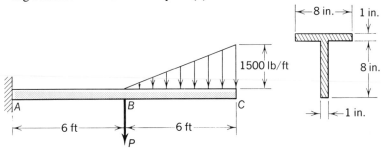

Figure P 3–56

3–57 For the beam in Figure P 3–57, determine (a) the maximum shear stress, and (b) the shear stress at the junction of the flange and webs.

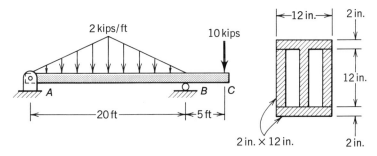

Figure P 3–57

3–58 Select the most economical wide-flange section to carry the load in Figure P 3–58 for an allowable working stress in bending and shear of 20,000 psi and 10,000 psi, respectively.

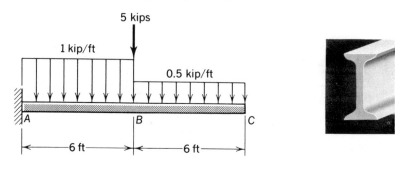

Figure P 3–58

3–59 The cantilever beam in Figure P 3–59 is an 8-in. WF, weighing 20 lb/ft. Sketch the shearing stress distribution and give the value of the maximum shearing stress.

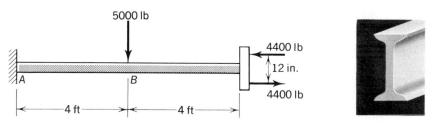

Figure P 3–59

3–60 Determine the maximum shear stress in the beam shown in Figure P 3–60.

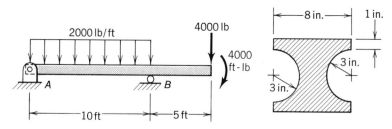

Figure P 3–60

3–61 The allowable working stresses for the beam shown in Figure P 3–61 are: tension, $\sigma_t = 8000$ psi; compression, $\sigma_c = 14,000$ psi; shear, $\tau = 7000$ psi. Determine the span l below which the shear stress governs and above which the bending stress governs.

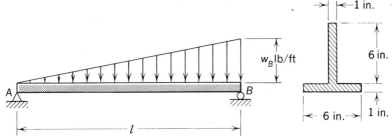

Figure P 3–61

3–62 Determine the load intensity w_B in Figure P 3–61 that the beam can carry (neglecting the weight of the beam) for a span of 10 ft if the allowable working stresses are: tension, $\sigma_t = 8000$ psi; compression, $\sigma_c = 14,000$ psi; shear, $\tau = 7000$ psi.

3–63 Determine the force/lineal inch that the weld in Figure P 3–63 must carry.

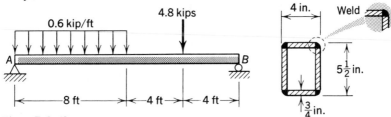

Figure P 3–63

3–64 Determine the force/lineal inch that the weld in Figure P 3–64 must carry.

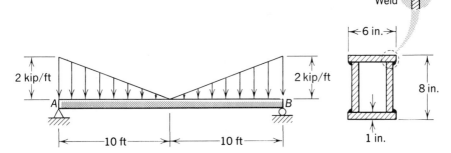

Figure P 3–64

3–65 For the beam shown in Figure P 3–65, determine (a) the maximum bending stress in each of the two beams (ABC and BCD), and (b) the maximum shear stress in the two beams.

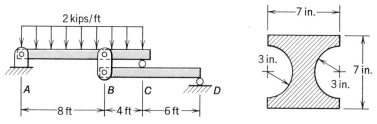

Figure P 3–65

3–66 The beam in Figure P 3–66 is of cast iron. If $l = 12$ ft and $x = 4$ ft, and the allowable working stress are: compression, $\sigma_c = 14,000$ psi; tension, $\sigma_t = 8000$ psi; shear, $\tau = 6000$ psi, determine (a) the maximum value P may have, and (b) the shearing stress distribution over the depth of the section. Give the value of the maximum shear stress.

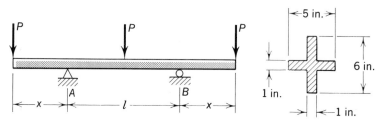

Figure P 3–66

3–67 A box beam is constructed by gluing 5 pieces of 2 in. × 12 in., as shown in Figure P 3–67. For a certain value of P the bending stress at point B is 1200 psi. What will be the maximum shear stress in the glue for this value of P? [Hint: The maximum shear force will be next to a support.]

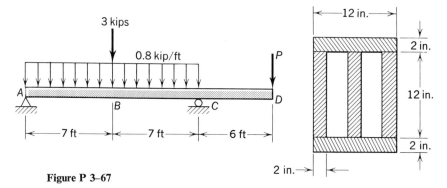

Figure P 3–67

3-68 In Figure P 3-68 the intensity of $w_0 = 750$ lb/ft as shown. If the shear stress in the glue of the channel section is not to exceed 150 psi, determine the permissible span l.

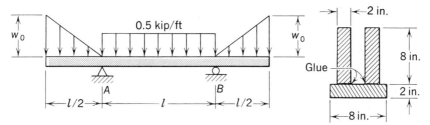

Figure P 3-68

3—6 Shear Center

Two of the assumptions made in the development of the bending-stress formula (Section 3–4), were that (a) the load applied was a pure bending moment (a couple), and (b) the loads were applied in a plane of symmetry, so that the member would not twist. Later we proceeded to apply the formula, with very satisfactory results, to cases where the pure-bending condition did not apply; that is, where we also encountered transverse shear forces. However, in all instances up to now we have either stated or implied that, regardless of the type of load, the loads were applied in a plane of symmetry. We have not covered the cases where either the member does not have a longitudinal plane of symmetry or the loads are not applied in a plane of symmetry, if one does exist. Under such circumstances the resultant of the shear stresses produced by the loads will be a force parallel to the plane of loading but not necessarily in that plane; consequently, the member will, in general, twist about its longitudinal axis. Bending without twisting is possible, however, if the loads are applied in the same plane as that in which the resultant of the shear stresses acts. More generally, bending without twisting is attained only when the resultant shear force passes through the *shear center*. This is defined as the point in the cross section of a member through which the resultant of the transverse shearing stresses must pass for any orientation of the transverse loads so that the member will bend without twisting.

Let us consider a thin, channel section, cantilever member loaded in a plane perpendicular to the plane of symmetry, as shown in Figure 3–12a. For convenience, the thickness of the flanges and web are considered small, although not necessarily the same, in comparison to the width and depth of the section. Such being the case, the percentage of the vertical shear carried by the flanges is negligible; we shall assume it is zero. Thus, on an element, shown as a free body in Figure 3–12b, the bending moments at the two sections 1–1 and 2–2 cause bending stresses, the resultants of which are

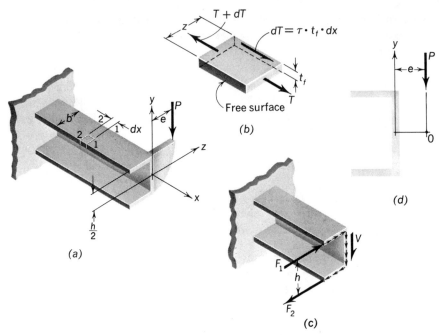

Figure 3-12

represented by T and $T + dT$, Figure 3-12b. Equilibrium dictates that the difference between the two forces must be resisted by a shear force acting on the inner face of the element parallel to the free surface (Figure 3-12b) because the free surface cannot carry any shear force. From the equations, $\sigma = My/I$ and $T = \sigma \cdot A$, we get

$$T + dT = \left(\frac{M_x + dM_x}{I_z}\right)\left(\frac{h}{2}\right)(z \cdot t) \qquad (a)$$

and

$$T = \left(\frac{M_x}{I_z}\right)\left(\frac{h}{2}\right)(z \cdot t) \qquad (b)$$

from which, we get by summing forces in the x direction,

$$dT = \left(\frac{dM_x}{I_z}\right)\left(\frac{h}{2}\right)(z \cdot t) \qquad (c)$$

From Figure 3-12b, $dT = \tau \cdot t \cdot dx$. We also recognize the $dM_x/dx = V$. These expressions alter Equation (c) to

$$\tau = \frac{1}{t \cdot dx}\left(\frac{dM_x}{I_z}\right)\left(\frac{h}{2}\right)(z \cdot t) = \left(\frac{dM_x}{dx}\right)\left(\frac{hz}{2I}\right) \qquad (d)$$

from which

$$\tau = \frac{V_x hz}{2I_z} \qquad (3\text{-}11)$$

Equation (3–11) shows that the shear stress in the flanges varies linearly from zero at the free surface where $z = 0$, to a maximum near the web where $z = b$.

From Section 1–11 covering complementary shear stresses, we conclude that the shearing stress distribution on the transverse plane, acting in the z direction, must be the same, as shown in Figure 3–12b. The direction is dictated by the orientation of the T forces; hence, for the bottom flange the direction would be the reverse of the top. The resultant forces F_1 and F_2 on these flanges are, consequently,

$$F_1 = (\tau_{avg})t_f\, b = \frac{1}{2}\left(\frac{V_x hb}{2I_z}\right)(t_f b) = F_2$$

acting as shown in Figure 3–12c. F_1 and F_2 form a couple which tends to twist the beam. In order to maintain rotational equilibrium,

$$V \cdot e = F_1 h = \left(\frac{Vb^2 ht_f}{4I_z}\right)h \qquad (e)$$

from which

$$e = \frac{b^2 h^2 t_f}{4I_z} \qquad (3\text{–}12)$$

Equation (3–12) specifies the distance from the center of the web to where the load is to be applied (P in Figure 3–12d, for example) so that there would be no twisting of the section. If the load were applied in the z direction, the plane of loading for zero twisting would have been the plane of symmetry, z-x. Thus, the shear center for our section is point O, Figure 3–12d. So long as the resultant shear force acts through point O, the member will bend without twisting, permitting us to treat the beam as for simple bending when determining stresses.

Example 3–11:

Given: An H section with unequal flanges as shown in the figure.
Find: The location of shear center.

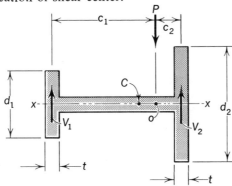

Procedure: We have shown previously where the flanges carried a negligible amount of the total shear when the load was applied parallel to the web. In the figure, the web in our case with the load parallel to the flanges may, therefore, be assumed to take only a negligible amount of the shear force—we will assume it to be zero. Also, we shall assume that the two flanges will have the same radius of curvature, that is, $\rho_1 = \rho_2$. Thus, taking moments about the shear center 0, we get

$$V_1 \cdot c_1 = V_2 \cdot c_2$$

or

$$V_1/V_2 = c_2/c_1 \qquad (a)$$

and if $\rho_1 = \rho_2$, from $1/\rho = M/EI$, we have

$$\frac{M_1}{EI_1} = \frac{M_2}{EI_2} \qquad (b)$$

where I_1 and I_2 represent the moments of inertia of the respective flanges about the x axis. At a section along the beam, say x units, $M_1 = x \cdot V_1$ and $M_2 = x \cdot V_2$. Thus, Equation (b) becomes

$$\frac{xV_1}{EI_1} = \frac{x \cdot V_2}{EI_2}$$

from which

$$\frac{V_1}{V_2} = \frac{I_1}{I_2} \qquad (c)$$

Combining Equations (a) and (c) we have the general location of point 0, the shear center.

$$\frac{c_2}{c_1} = \frac{I_1}{I_2}$$

Example 3-12:

Given: A semicircular section shown in Figure a.
Find: The shear center.
Procedure: At two sections 1-1 and 2-2, the bending moments are different by an amount dM. This causes a difference of force dT on the two "transverse" faces of the element in Figure b, which is resisted by a shear force, $\tau \cdot t \cdot dx$. Thus, $dT = \tau \cdot t \cdot dx$ and from $\sigma = Mc/I$,

$$dT = \left[\frac{dM}{I} \int_0^\theta (r \cos \theta) r t d\theta \right] = \frac{dM r^2 t}{I} \sin \theta$$

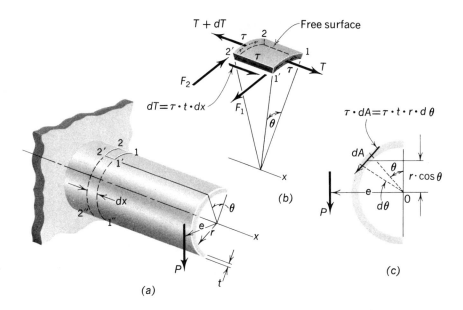

(a) (b) (c)

Thus, equating the two expressions for dT, we have

$$\tau \cdot t \cdot dx = \frac{dM r^2 t}{I} \sin \theta$$

from which

$$\tau = \frac{dM}{dx} \cdot \frac{r^2}{I} \sin \theta = V \frac{r^2}{I} \sin \theta$$

As previously mentioned, the average stress on the transverse faces is equal to that on the longitudinal face of the element, varying from zero at the free surface to a maximum at $\pi/2$, where $\sin \theta$ is a maximum. The increment of force on the area dA, Figure c, created by the shear stress is $\tau \cdot dA = \tau \cdot t \cdot rd\theta$. This creates an increment of moment about point O of $(\tau \cdot t \cdot rd\theta)r = \tau \cdot t \cdot r^2 d\theta$. The total moment created by the internal shear forces must be equal to the moment created by the external force P, that is,

$$P \cdot e = \int_0^\pi \tau \cdot t \cdot r^2 \cdot d\theta = 2 \frac{V r^4 t}{I} \int_0^{\pi/2} \sin \theta \, d\theta.$$

But, from $\Sigma F_y = 0$, $P = V$. Also, $I_x = \pi r^3 t/2$; hence,

$$e = \frac{2 r^4 t}{I} = \frac{4r}{\pi}$$

PROBLEMS

3–69 Determine the shear center for the channel section of Figure P 3–69.

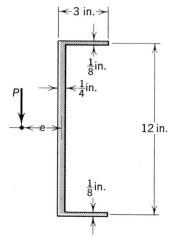

Figure P 3–69

3–70 Determine the shear center for the H section of Figure P 3–70.

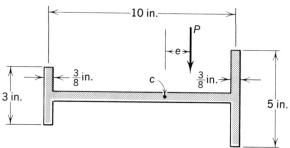

Figure P 3–70

3–71 Determine the shear center for the semicircular section shown in Figure P 3–71.

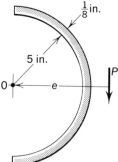

Figure P 3–71

3–72 Show that the shear center for the Z section shown in Figure P 3–72 lies at the centroid of the section.

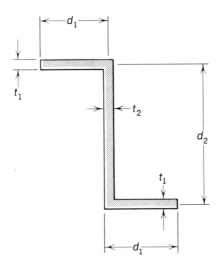

Figure P 3–72

3–73 Determine the shear center for the unbalanced H section of Figure P 3–73.

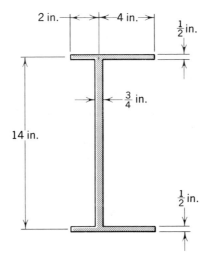

Figure P 3–73

3—7 Inelastic Bending of Beams

The preceding sections of this chapter discussed only members for which we assumed a linear distribution of the bending stresses over the depth of the member, varying from a maximum at the extreme fibers to zero at the neutral axis. Using such stresses as the criteria for selecting the size of a member is a convenient but rather inefficient method for material utilization, especially for rectangular sections because: (a) the maximum permissible stress is taken at the proportional limit although the ultimate stress is much higher (for most materials), and (b) especially for rectangular sections, only a small portion of the cross section is stressed even up to this level (hence, one reason for the development of I sections becomes readily evident).

We shall now consider transversely loaded members stressed beyond the proportional and yield points, and into the plastic region. This means that permanent deformation of the material occurs, and that we can no longer assume the proportionality between stress and strain. Thus, we would like to develop a method of analysis for such a condition, and as a result be able to select loads large enough to produce plastic deformations but not so large that they produce collapse of the structure. Proponents of·this method of analysis claim it to be an economical method of design. This, however, should not overshadow another advantage: it greatly simplifies the analysis of indeterminate structures.

An ideally ductile material is an important assumption we shall make for this analysis.[6] Structural steel is a material particularly suited for our purpose because it is ductile. For such a material we may idealize the stress-strain relationship as shown in Figure 3–13a with reasonable accuracy. This diagram assumes a straight-line relationship between stress and strain up to the yield point, and that the yield points and moduli of elasticity of·the material are the same for both tension and compression. Past the yield point the stress remains constant. We also assume that plane sections before bending remain plane in the plastic range, and, consequently, strains are proportional to the distance from the neutral axis. Finally, we shall assume a cross section that is symmetrical about an axis in the plane of loading.

Now let us examine the behavior of the rectangular cross section shown in Figure 3–13b subjected to a gradually increasing bending moment. As the moment is increased until the strain reaches ϵ_y, the relationship between stress and strain is a linear one, yielding a linear stress distribution as shown in Figure 3–13c. Increasing the moment further, the strains will also increase, the result being that some of the outer fibers will strain an amount ϵ_m. The

---

[6] The discussion is not applicable to brittle materials such as concrete or cast iron, or to materials subjected to any appreciable strain hardening.

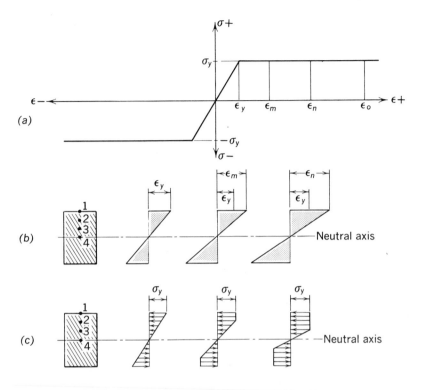

Figure 3–13

(a) Assumed stress-strain relationship. (b) Strain over section at various stages of load. (c) Stress over section at various stages of load.

corresponding stress for these fibers may be determined from the stress-strain relationship in Figure 3–13a. We note that it is still only σ_y for those fibers, decreasing to zero at the neutral axis for the intermediate fibers which have not strained past ϵ_y. Proceeding to increase the moment in this manner results in plastic strain for almost all fibers, and accordingly, the stress distribution becomes almost rectangular in shape as shown by the last sketch on the right in Figure 3–13c. The corresponding moment is called the *plastic moment*, usually denoted M_p, which represents the *limiting* strength of the beam.

To evaluate the plastic moment M_p, let us consider a general cross section shown in Figure 3–14a, stressed as shown in Figure 3–14b. For pure bending, the internal forces F_c and F_t must be parallel, equal in magnitude and opposite in sense, forming a couple which must resist the bending moment M_p. Thus, from $\Sigma F_x = 0$, we get

$$\sigma_y A_c = \sigma_y A_t$$

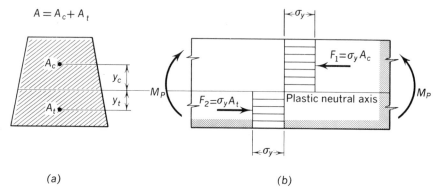

Figure 3-14 (*a*) Cross section. (*b*) Small length of beam.

from which

$$A_c = A_t = \tfrac{1}{2} A \tag{3-13}$$

which shows that the neutral axis of a plastically loaded section (pure moment) divides the cross section into two equal parts. For a general shape, this is a different neutral axis from that for elastic bending. For cross sections having symmetry about both the x–x and y–y axes (rectangular or circular section, for example), the neutral axis is at the centroid of the section for both elastic and plastic loading.

From $\Sigma M = 0$, we get

$$M_p = \sigma_y A_c \cdot \bar{y}_c + \sigma_y A_t \cdot \bar{y}_t = \sigma_y (A_c \bar{y}_c + A_t \bar{y}_t) \tag{b}$$

and because $A_c = A_t = \tfrac{1}{2} A$, Equation (*b*) becomes

$$M_p = \sigma_y [\tfrac{1}{2} A (\bar{y}_c + \bar{y}_t)] \tag{3-14}$$

From Equation (3-14) we may proceed to determine the plastic moment for various shapes and compare this moment with the *elastic moment* M_e, based on the elastic theory. For example, for the rectangular cross section,

$$A = b \times h$$

$$y_t = y_c = \frac{h}{4}$$

$$I = \frac{bh^3}{12}, \; c = \frac{h}{2}$$

$$M_P = \sigma_y \frac{bh}{2}\left(\frac{h}{4} + \frac{h}{4}\right) = \sigma_y \frac{bh^2}{4}$$

$$M_e = \sigma_y (I/c) = \sigma_y \left(\frac{bh^3}{12} \Big/ \frac{h}{2}\right) = \sigma_y \frac{bh^2}{6}$$

$$\frac{M_P}{M_e} = \frac{6}{4} = 1.5$$

For the circular cross section

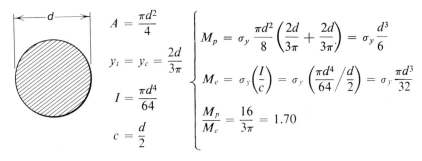

$$A = \frac{\pi d^2}{4}$$

$$y_t = y_c = \frac{2d}{3\pi}$$

$$I = \frac{\pi d^4}{64}$$

$$c = \frac{d}{2}$$

$$M_p = \sigma_y \frac{\pi d^2}{8}\left(\frac{2d}{3\pi} + \frac{2d}{3\pi}\right) = \sigma_y \frac{d^3}{6}$$

$$M_e = \sigma_y\left(\frac{I}{c}\right) = \sigma_y\left(\frac{\pi d^4}{64}\bigg/\frac{d}{2}\right) = \sigma_y \frac{\pi d^3}{32}$$

$$\frac{M_p}{M_e} = \frac{16}{3\pi} = 1.70$$

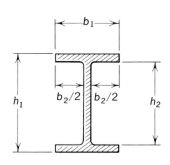

For the I beam section, by using composite areas, the strength of the I section may be determined by evaluating the section as a rectangular section, $b_1 \times h_1$ minus a rectangular section $b_2 \times h_2$. Thus, using the expression for M_p for the rectangular section derived above, we get for the I section,

$$M_p = \frac{\sigma_y}{4}(b_1 h_1{}^2 - b_2 h_2{}^2)$$

The moment of inertia for the I section is $(1/12)(b_1 h_1{}^3 - b_2 h_2{}^3)$. The elastic moment M_e is therefore

$$M_e = \sigma_y\left(\frac{I}{c}\right) = \frac{\sigma_y}{6h_1}(b_1 h_1{}^3 - b_2 h_2{}^3)$$

from which

$$\frac{M_p}{M_e} = \frac{3}{2}h_1\left(\frac{b_1 h_1{}^2 - b_2 h_2{}^2}{b_1 h_1{}^3 - b_2 h_2{}^3}\right)$$

Substituting specific values for b and h into the above expressions will show that the ratio M_p/M_e for an I section is not so great as for rectangular or circular sections of 1.50 and 1.70, respectively. (See Example 3–13.)

Example 3–13:

Given: The wide-flange section (18WF50) shown in the figure.
Find: (a) M_p and M_e for a yield stress $\sigma_y = 40,000$ psi (or 275 MN/m²), and (b) M_p/M_e ratio.

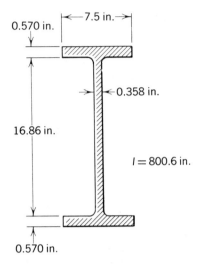

Procedure: Reference is made to the formula for M_e derived above. The dimensions shown in the figure are obtained from Table B–2, Appendix B.

$$h_1 = 18 \text{ in. (or 457 mm)}, \qquad b_1 = 7.5 \text{ in. (or 191 mm)}$$
$$h_2 = 16.86 \text{ in. (or 425 mm)}, \qquad b_2 = 7.142 \text{ in. (or 181 mm)}$$

$$M_p = \frac{\sigma_y}{4}(b_1 h_1{}^2 - b_2 h_2{}^2)$$

$$M_p = \frac{40,000}{4}(7.5 \times 18^2 - 7.142 \times 16.86^2) = 4 \times 10^6 \text{ in.-lb}$$

and

$$M_e = \frac{\sigma_y}{6h_1}(b_1 h_1{}^3 - b_2 h_2{}^3) = 3.526 \times 10^6 \text{ in.-lb}$$

Using $M_e = (\sigma_y)(I/c)$ gives an M_e slightly different because the I obtained from Table B–2 is an exact one accounting for fillets between web and flanges, rounded edges, and so forth, which we did not take into account. Thus,

$$\frac{M_p}{M_e} = \frac{4 \times 10^6}{3.526 \times 10^6} = 1.13$$

Thus, the plastic moment is only 13 percent higher than the maximum elastic moment, a typical finding for I-beam sections in this range of values.

PROBLEMS

3–74 Determine the ratio M_p/M_e for each cross section shown in Figures P 3–74a and b.

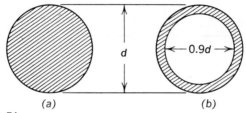

(a)   (b)

Figure P 3–74

3–75 For each section in Figure P 3–75, determine the M_p/M_e ratios.

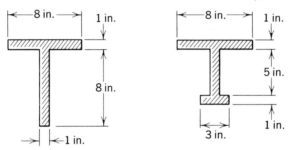

Figure P 3–75

3–76 For each section in Figure P 3–76, determine (a) the ultimate load moment M_p for $\sigma_y = 40,000$ psi, and (b) the M_p/M_e ratio.

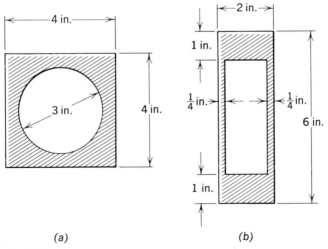

(a)   (b)

Figure P 3–76

3–77 In Figure P 3–77 determine the ultimate load for the beam shown if the yield stress of the material is 1800 psi.

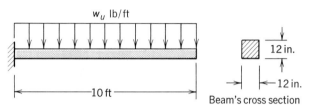

w_u lb/ft

10 ft

12 in.

12 in.

Beam's cross section

Figure P 3–77

3–78 Determine the ultimate load P the member in Figure P 3–78 can carry for $\sigma_y = 35,000$ psi. Neglect the weight of the beam.

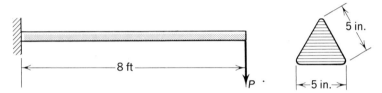

8 ft

P

5 in.

5 in.

Figure P 3–78

3–79 Determine the ultimate load $2P$ the member in Figure P 3–79 can carry for $\sigma_y = 40,000$ psi. Neglect the weight of the beam.

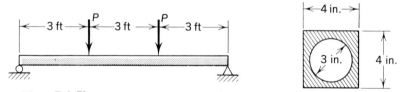

3 ft — P — 3 ft — P — 3 ft

4 in.

3 in.

4 in.

Figure P 3–79

3–80 Determine the ultimate uniform load that the member in Figure P 3–80 can carry for $\sigma_y = 40,000$ psi. Neglect the weight of the beam.

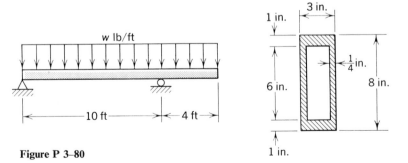

w lb/ft

10 ft — 4 ft

3 in.

1 in.

6 in.

$\frac{1}{4}$ in.

8 in.

1 in.

Figure P 3–80

3—8 Beams of Two Materials

Several prominent building materials possess compressive and tensile strengths that are appreciably different. Among these materials are wood, concrete, cast iron, and some special metals and plastics. For example, the tensile strength of concrete is only about 10 percent of its compressive strength; the tensile strength of cast iron is less than half its compressive strength. Hence, a beam's cross section, which is symmetrical about the neutral axis made of these materials, yields an inefficient use of the material. One solution to this is to fabricate a section that is unsymmetrical with respect to the neutral axis, such that the neutral axis is closer to the tension side, and therefore the tensile stress is less than the compressive stress. This is commonly the case with cast-iron sections formed into T shapes that are bent so that the tensile fibers are those closest to the neutral axis. Another solution is to form a composite section consisting of two materials so that the material which is stronger in tension helps the weaker one. Common examples of such composites are reinforced concrete beams, or wood beams reinforced with steel plates.

The analysis of nonhomogeneous beams, such as of composites mentioned above, may be made similar to that of homogeneous sections by transforming the composite section into an equivalent section composed of only one material. The nonhomogeneity, evident by the different moduli of elasticity of the various materials in the composites, is accounted for in the transformation approach. Thus, all the assumptions made in Section 3–4 that were used in analyzing homogeneous beams are assumed valid for the analysis of transformed sections.

In order to illustrate the procedure undertaken in the analysis of composite sections, let us consider the beam in Figure 3–15, subjected to pure bending. We shall assume that there is no slip between the steel plate and the wood. Hence, from the assumption that plane sections before bending remain plane after bending, and that the strain varies directly with the distance from the neutral axis, we conclude that at the point of contact of the steel and wood, the strain of the two materials has to be the same. The strain distribution over the section is shown in Figure 3–15c. From here we can develop a relationship between the strain in the wood and that in the steel: Let y_1 and y_2 represent the distance from the neutral axis to any fiber in the wood and steel, respectively. Then, from similar triangles, in Figure 3–5c, we get

$$\epsilon_1 = \epsilon_{St} = \left(\frac{y_2}{y_1}\right)\epsilon_{Wd} = \left(\frac{y_2}{y_1}\right)\epsilon_1 \qquad (a)$$

From $\sigma = \epsilon E$, Equation (a) becomes

$$\sigma_{St} = \left(\frac{y_2}{y_1}\right)\left(\frac{E_{St}}{E_{Wd}}\right)\cdot\sigma_{Wd} \qquad (b)$$

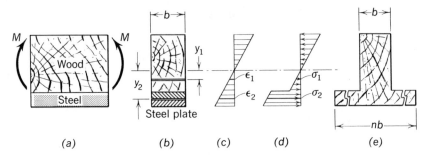

Figure 3–15 (b) Cross section. (c) Strain distribution. (d) Stress distribution. (e) Transformed section.

where E_{Wd} and E_{St} represent the moduli of elasticity of wood and steel, respectively. When y_1 equals y_2, at the junction point of the wood and plate, for example, the stress in the steel is greater than that in the wood by the factor $E_{St}/E_{Wd} = n$. The stress distribution over the section is shown in Figure 3–15d. Thus, at a common level y, the forces developed by the wood and steel over an area $dA = bdy$ are:

For wood,

$$dF_{Wd} = \sigma_{Wd} \cdot dA = \sigma_{Wd} \cdot bdy \qquad (c)$$

For steel,

$$dF_{St} = \sigma_{St} \cdot dA = (E_{St}/E_{Wd})b \cdot dy = \sigma_{Wd}(nbdy) \qquad (d)$$

Now let us examine expressions (c) and (d) above. Equation (d) shows that the force developed by the steel over an increment of the area y distance from the neutral axis is n times as large as the force developed by the wood over an increment of area ($b \cdot dy$) *equal* to that of the steel and located *also* at y distance from the neutral axis. Hence, we may transform the steel section into an equivalent wood section if we (*a*) multiply dA_{St} by n, and (*b*) keep the centroid of the transformed steel section at the same y distance from the neutral axis as for the steel section. These conditions fall within the basic requirements for static equilibrium applicable to our case; that is, (*a*) $\Sigma F_x = 0$ (keeping in mind that there are no externally applied *axial* forces, this step determines the location of the neutral axis), and (*b*) $\Sigma M_{n.a.} = 0$. (From this step we obtain the resisting moment.) In our case, the transformation is accomplished by merely obtaining a new width b for the steel section such that

$$b_1 = nb \qquad (e)$$

The transformed section is shown in Figure 3–15e. The transformation of wood to steel could have been made with equal facility. Also, if the composite beam consists of more than two materials, or two or more materials of various shapes, the transformation may be obtained in terms of one of these materials, although the procedure may be more cumbersome.

Once the transformed section is obtained, the beam of one material (but transformed shape) is treated on the basis of the elastic theory discussed in Section 3–4. The *strains and stresses* vary linearly from the neutral axis of the transformed section. Thus,

$$\sigma = \frac{My}{I}$$

applies. Here I represents the moment of inertia of the transformed section, and y represents the distance from the neutral axis of the transformed section to a particular fiber of the beam. The stress σ for the material of which the transformed section is made (in our case wood) is correct; for the other material (in our case steel), we multiply the stress computed by n to obtain the actual stress in that material. This concept is illustrated in the following example.

Example 3–14:

Given: A wood beam having a 8-in. × 10-in. rectangular cross section is reinforced with a steel plate 8 in. × ½ in. as shown in Figure *a*. The beam is subjected to a bending moment of 200,000 in.-lb. The moduli of elasticity of the steel and wood are $E_{St} = 30 \times 10^6$ psi and $E_{Wd} = 1.5 \times 10^6$ psi, respectively.
Find: The maximum bending stresses in the steel and wood.

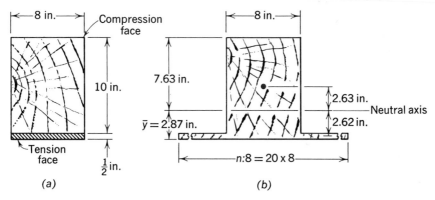

(*a*) Actual section. (*b*) Transformed section.

Procedure: We may transform either the wood into steel or steel into wood with equal ease. Assuming the latter, the transformed section is shown in Figure *b*, where

$$n = E_{St}/E_{Wd} = \frac{30 \times 10^6}{1.5 \times 10^6} = 20$$

For the transformed section, the neutral axis is located at y distance from the bottom where

$$\overline{y} = \frac{8 \times 10 \times 5.5 + 20 \times 8 \times \frac{1}{2} \times \frac{1}{4}}{8 \times 10 + 20 \times 8 \times \frac{1}{2}} = 2.87 \text{ in.}$$

The moment of inertia of the transformed section is

$$I = \frac{8 \times 10^3}{12} + 8 \times 10(2.63)^2 + \frac{20 \times 8 \times (\frac{1}{2})^3}{12} + 20 \times \frac{1}{2}(2.62)^2 \times 8$$

$$I = 1773 \text{ in.}^4$$

Wood, compression: $\sigma_{\max} = \dfrac{M \cdot c}{I} = \dfrac{(200,000)(7.63)}{1773}$

$$= \textbf{860 psi} \text{ compression}$$

Wood, tension: $\sigma_{\max} = \dfrac{(200,000)(2.37)}{1773} = \textbf{267 psi}$ tension

Steel, tension: $\sigma_{\max} = n(Mc/I) = 20\left(\dfrac{200,000 \times 2.87}{1773}\right)$

$$= \textbf{6450 psi} \text{ tension}$$

Example 3–15:

Given: The beam cross section shown in Figure *a*. The working stresses in bending for steel and wood are 18,000 psi and 1000 psi, respectively. $E_{\text{St}} = 30 \times 10^6$ psi and $E_{\text{Wd}} = 1.5 \times 10^6$ psi.
Find: The bending moment the section can safely carry.

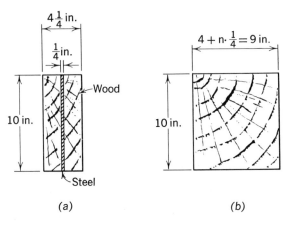

(a)   (b)

(*a*) Actual section. (*b*) Transformed section.

Procedure: The steel is assumed to be transformed into wood. The new width of the transformed steel section is

$$b_1 = nb = \left(\frac{30 \times 10^6}{1.5 \times 10^6}\right) \tfrac{1}{4} = 5 \text{ in.}$$

The transformed section is shown in Figure *b*. We must now decide which allowable working stress will govern. The allowable stress in the transformed steel section is now $18{,}000/n = 900$ psi < 1000 psi for wood, thus making the transformed steel stress as the governing criteria. The moment of inertia of the transformed section is

$$I = \frac{9 \times 10^3}{12} = 750 \text{ in.}^4$$

Then

$$M = \frac{\sigma \cdot I}{c} = \frac{900 \times 750}{5} = \textbf{135,000 in.-lb}$$

PROBLEMS

3–81 Determine the bending moments that the sections in Figures P 3–81*a* and P 3–81*b* safely carry if the working stresses in bending for the steel and wood are 18,000 psi and 1200 psi, respectively. Assume $E_{St} = 30 \times 10^6$ psi, and $E_{Wd} = 1.5 \times 10^6$ psi.

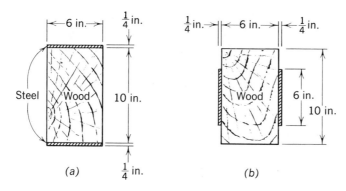

(a) (b)

Figure P 3–81

3–82 Determine the bending moments that the sections in Figures P 3–82*a* and P 3–82*b* can safely carry if the working stresses in bending for the steel and aluminum are 20,000 psi and 15,000 psi, respectively. $E_{St} = 30 \times 10^6$ psi, $E_{Al} = 10 \times 10^6$ psi.

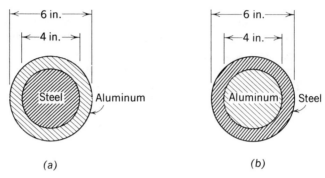

(a)   (b)

Figure P 3–82

3–83 Determine the bending moments that the sections in Figure P 3–83 can safely withstand if the working stresses for steel, wood, and aluminum are $\sigma_{St} = 20,000$ psi, $\sigma_{Wd} = 1200$ psi, and $\sigma_{Al} = 15,000$ psi. Assume $E_{St} = 30 \times 10^6$ psi, $E_{Wd} = 1.5 \times 10^6$ psi, and $E_{Al} = 10 \times 10^6$ psi.

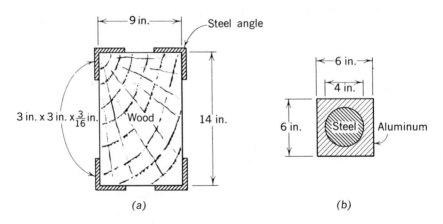

(a)   (b)

Figure P 3–83

3–84 Determine the bending moments that the sections in Figures P 3–84*a* and P 3–84*b* can safely withstand if the working stresses for oak and pine are 1200 psi and 1000 psi, respectively. Assume $E_O = 1.7 \times 10^6$ psi and $E_{Pi} = 1.0 \times 10^6$.

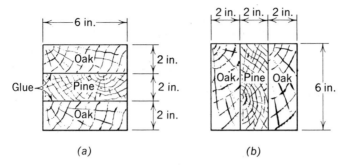

Figure P 3–84

3–85 A cantilever beam 12 ft long has the cross section as shown in Figure P 3–81*a*. For a uniform load of 300 lb/ft (including the weight of the beam), determine the maximum bending stresses in the wood and steel plates. Assume $n = 20$.

3–86 A simply supported beam 15 ft long has the cross section shown in Figure P 3–81*b*. For a uniform load of 275 lb/ft (including the weight of the beam), determine the maximum bending stresses in the wood and steel plates. Assume $n = 20$.

3–87 A cantilever beam 10 ft long has the cross section shown in Figure P 3–83*a*. For a uniform load of 250 lb/ft (including the weight of the beam), and a concentrated load 800 lb at the free end, determine the maximum bending stresses in the wood and steel angles. Assume $n = 20$.

3–88 Compute the uniform load, w lb/ft, that two simply supported beams of length *l*, having cross sections as shown in Figures P 3–84*a* and P 3–84*b*, safely carry if the working load for oak and pine are 1000 psi and 800 psi, respectively. Assume $n = 1.6$.

3—9 Reinforced Concrete Beams

Perhaps the most common example of a composite fabricated from two materials of widely different physical properties is a reinforced concrete member. Compressive strength, fire safety, economy, weather resistance, durability, and ease of fabrication are some of the factors which make concrete a very desirable building material. However, as mentioned in the preceding section, concrete has relatively little, if not actually negligible, tensile strength; hence, it can not be relied upon to take any tensile load. To overcome this deficiency, steel bars are embedded in the concrete to form a composite such that the concrete withstands compression, and the steel

withstands tension. Here, we shall limit our discussion to reinforced concrete beams; we shall cover even beams only briefly.[7]

The transformed-section approach presented in the previous section
may be used here to analyze reinforced concrete beams, with one notable
exception: *the concrete is assumed to take no tension.* In other words, the
concrete below the neutral axis (on the convex side) will crack well before
the steel (also on the convex side) approaches the yield stress, and because no
tensile force can be transmitted across a crack, all the tensile forces will
necessarily be transmitted to the steel. Just how far along the depth the
crack would propagate is debatable. It is, however, reasonable to assume
that the crack will reach the neutral axis; hence, our reasoning for the above
assumption.

It is customary to transform the reinforcing steel to concrete, assuming
the transformed area of the steel is to be concentrated at the centroid of the
steel bars. Hence, the resultant tensile force is assumed to act at this level.
Also, because strain varies linearly with the distance from the neutral axis,
it is convenient to assume that the stress also varies linearly. This is a reasonable assumption, although not quite correct because the stress-strain relationship for concrete does not have the pronounced linearity displayed by
steel.

Now let us consider the reinforced concrete beam in Figure 3–16a.

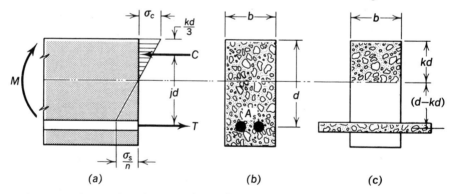

Figure 3–16 (*a*) Reinforced concrete beam. (*b*) Actual section. (*c*) Transformed section.

The pure moment M is resisted by the internal couple $C \cdot jd = T \cdot jd$. The
compressive force C is the product of the average compressive stress and the
area over which this stress acts, that is,

$$C = \left(\frac{\sigma_c}{2}\right) b \cdot kd \qquad (a)$$

---

[7] For a comprehensive coverage of reinforced concrete, see, for example, J. N. Cernica, *Fundamentals of Reinforced Concrete,* Addison-Wesley Publishing Company, Reading, Massachusetts, 1964.

The tensile force T is the product of the tensile stress in the bars and the areas of the bars; that is,

$$T = (\sigma_s)A_s \tag{b}$$

Based on maximum stresses, σ_c and σ_s, the corresponding resisting moments are

$$M_c = C \cdot jd = \tfrac{1}{2}\sigma_c kj \cdot bd^2 \tag{3-15}$$

and

$$M_s = T \cdot jd = \sigma_s A_s \cdot jd \tag{3-16}$$

The quantities k and j are interrelated. From Figure 3–16a, the resultant C is located at $kd/3$ from the extreme compression fiber. Thus, $jd = d - kd/3$, or

$$j = 1 - (k/3) \tag{3-17}$$

By locating the neutral axis we may easily determine k; that is,

$$\Sigma\, Ay = A\bar{y}: \qquad bkd(kd/2) = nA_s(d - kd) \tag{c}$$

(transformed section Figure 3-16c)

Dividing both sides by bd^2, we have

$$(\tfrac{1}{2})k^2 = n\left(\frac{A_s}{bd}\right)(1 - k) \tag{d}$$

Letting $p = A_s/bd$, and simplifying, we have

$$k^2 + (2np)k - 2np = 0 \tag{e}$$

Solving the quadratic for k, Equation (e) yields

$$k = \sqrt{2np + (pn)^2} - np \tag{3-18}$$

With the neutral axis located, the values of j and k can be substituted into Equations (3–15) and (3–16) where for a given moment the maximum stress in the concrete and steel may be determined; or for given maximum concrete and steel stresses, the maximum allowable moment may be determined.

The equation $\sigma = Mc/I$ may be employed for concrete beams as well. Here I represents the transformed moment of inertia[8] of the section, that is,

$$I = \frac{b(kd)^3}{3} + nA_s(d - kd)^2 \tag{3-19}$$

so that the maximum stresses are

$$\sigma_c = \frac{Mkd}{I} \tag{3-20}$$

and

$$\sigma_s = \frac{nM_s(d - kd)}{I} \tag{3-21}$$

---

[8] The moment of inertia of the steel rods with respect to their own centroidal axis is negligible.

We seldom confront a design situation where the allowable stresses in the concrete and steel are both reached. It is therefore prudent to check both stresses (concrete and steel) when designing.

Example 3–16:

Given: The cross section in the figure below is subjected to a bending moment of 500,000 in.-lb. Assume $(E_s/E_c) = n = 12$.
Find: The maximum bending stresses in the concrete and steel.

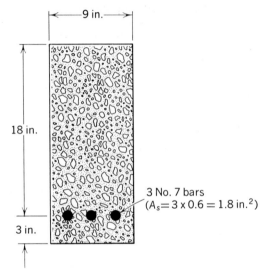

3 No. 7 bars
$(A_s = 3 \times 0.6 = 1.8 \text{ in.}^2)$

Procedure: We may use either Equations (3–15) and (3–16) or Equations (3–20) and (3–21) to determine the stresses. For comparison, we shall use both.

$$p = \frac{A_s}{bd} = \frac{1.8}{9 \times 18} = 0.0111$$

From Equation (3–18), we have

$$k = \sqrt{2np + (pn)^2} - np$$
$$= \sqrt{2(0.0111 \times 12) + (0.0111 \times 12)^2} - 0.0111 \times 12$$

$$k = 0.403$$

$$j = 1 - (k/3) = 1 - (0.403/3) = 0.866$$

From Equations (3–15) and (3–16) we have

$$\sigma_c = \frac{2M_c}{kjbd^2} = \frac{2 \times 500,000}{(0.403)(0.866)(9)(18)^2} = \textbf{990 psi}$$

$$\sigma_s = \frac{M_s}{A_s jd} = \frac{500,000}{(1.8)(0.866)(18)} = \textbf{17,700 psi}$$

Now, using Equations (3–20) and (3–21) to compare the results, we shall first determine I using Equation (3–19); that is,

$$I = \frac{b(kd)^3}{3} + nA_s(d - kd)^2$$

$$I = \frac{9(0.403 \times 18)^3}{3} + 12 \times 1.8(18 - 0.403 \times 18)^2 = 3650 \text{ in.}^4$$

Thus,

$$\sigma_c = \frac{M_c(kd)}{I} = \frac{500,000(0.403 \times 18)}{3650} = \mathbf{990 \text{ psi}}$$

and

$$\sigma_s = \frac{nM_s(d - kd)}{I} = \frac{12 \times 500,000(18 - 0.403 \times 18)}{3650} = \mathbf{17,700 \text{ psi}}$$

which yields the respective stresses that are identical from two somewhat different approaches.

Example 3–17:

Given: A bending moment of 250,000 in.-lb. Allowable stress are $\sigma_s = 20,000$ psi, $\sigma_c = 1000$ psi. Assume $n = 12$.

Find: A suitable cross section to carry the bending moment.

Procedure: Note that in Example 3–16 we were given a section and then investigated the stresses for a given moment. In other words, we *analyzed* the section for stress. In this example, we are to select an economical cross section for given allowable working stresses to carry the given bending moment. Thus, our problem here is one of *design*.

In order to ensure the maximum economy, it is desirable to reach both working stresses, concrete and steel, concurrently under the given moment. Ideally, the stress distribution would be as shown in Figure *a*.

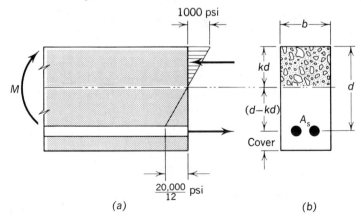

(a)   (b)

From calculations, $d = 13.75$ in., $b = 8$ in., $A_s = \frac{1}{2}$ in.^{2} (2 No. 7 bars), and cover = say, 3 in.

Using similar triangles, we may find the value of k (then j) for this "*balanced design*" situation; that is,

$$\frac{kd}{1000} = \frac{d - kd}{\dfrac{20,000}{12}}$$

from which

$$k = \frac{12(1 - k)}{20}$$

Solving,

$$k = 0.375, \quad j = 1 - (k/3) = 0.875$$

Using Equation (3–15), and rearranging terms

$$bd^2 = \frac{2M_c}{\sigma_c kj} = \frac{2 \times 250,000}{(1000)(0.375)(0.875)} = 1525 \text{ in.}^3$$

Now, because both b and d are unknowns, we must assume one and solve for the other. Lateral stability and efficiency usually calls for a ratio of $d/b = 1.75$ to 2. Assuming $d/b = 1.75$,

$$\frac{d^3}{1.75} = 1525 \qquad \text{or} \qquad d = (1525 \times 1.75)^{1/3} = 13.875 \text{ in., say } 13.9 \text{ in.}$$

and

$$b = \frac{13.9}{1.75} = 7.95 \text{ in., say } 8 \text{ in.}$$

From Equation (3–16) we have

$$A_s = \frac{M_s}{\sigma_s jd} = \frac{250,000}{20,000 \times 0.875 \times 13.9} = 1.03 \text{ in.}^2$$

We find that 2 No. 7 bars are satisfactory, having an area of $2(0.6) = 1.2$ in.$^2 > 1.03$ in.^{2} needed. The design (dimensions) is shown in Figure b. The 3-in. cover for the steel was selected arbitrarily to protect the steel from corrosion, fire, and so forth.

PROBLEMS

3–89 For the beam shown in Figure P 3–89 determine the compressive and tensile stresses σ_c and σ_s in the concrete and steel. Assume $n = 10$. The unit weight for reinforced concrete is assumed to be 150 lb/cu ft. The uniform load denoted by w_{LL} does not include the weight of the beam.

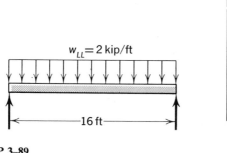

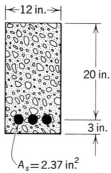

Figure P 3–89

$A_s = 2.37$ in.^{2}

3–90 For the beam in Problem P 3–89, determine the maximum uniform live load w_{LL} that the beam may safely carry if the stresses in the concrete and steel are not to exceed 1200 psi and 22,000 psi, respectively. Assume $n = 10$.

3–91 The beam in Figure P 3–91 is to carry a uniform load of 1.5 kips /ft (excluding the weight of the beam of 150 lb/cu ft) and a concentrated load of 5 kips at the center of the span. If the working stresses are $\sigma_c = 1000$ psi and $\sigma_s = 20,000$ psi, determine the maximum span l that the beam may have. Assume $n = 12$.

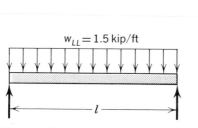

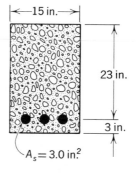

Figure 3–91

$A_s = 3.0$ in.^{2}

3–92 Compute the total load that may be concentrated at the third points of the beam shown in Figure P 3–92, if $n = 10$ and the allowable stresses are $\sigma_c = 1200$ psi and $\sigma_s = 20,000$ psi.

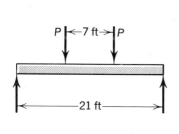

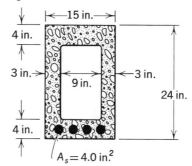

Figure P 3–92

$A_s = 4.0$ in.^{2}

3–93 Calculate the maximum uniform live load (the weight of the slab is considered dead load) that the slab in Figure P 3–93 can carry for a 12-ft span, given that $\sigma_c = 900$ psi, $\sigma_s = 18,000$ psi, $n = 12$. [Hint: Assume a strip to be 1-ft wide and treat this as a simply supported beam.]

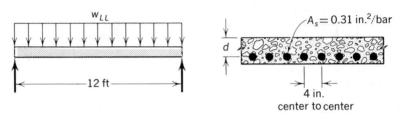

Figure P 3–93

3–94 The cross section shown in Figure P 3–94 is to carry a bending moment of 450,000 in.-lb. If the working stress are $\sigma_s = 20,000$ psi and $\sigma_c = 1000$ psi, determine the minimum value of b and A_s for $n = 12$.

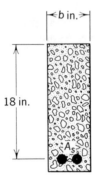

Figure P 3–94

3–95 Design a simply supported slab (that is, find d and A_s) to carry a uniform load of 150 lb/sq ft, besides its own weight, over a 14-ft span if $n = 12$, $\sigma_c = 1000$ psi, and $\sigma_s = 18,000$ psi. Assume $d/b = 1.75$.

3–96 Design a cantilever beam, 10 ft long, to carry a concentrated load of 8 kips at the free end, if $n = 12$, $\sigma_s = 20,000$ psi, and $\sigma_c = 1200$ psi. Assume a ratio of $d/b = 1.75$. [Note: The weight of the beam is considered uniform dead load.]

3–97 Design a simply supported beam, 18 ft long, that has a uniformly decreasing live load from W_0 at the left support to zero at the right support, if $n = 12$, $\sigma_c = 1000$ psi, and $\sigma_s = 20,000$ psi. Assume the depth of the beam to be constant over the span, and consider the weight of the beam as a dead load.

3—10 Bending Stress in Curved Members

The flexure formula derived in Section 3–4 for straight members may be seriously in error when applied to hooks, rings, U frames, chain links, or other components which have an initial curvature in the plane of loading. The amount of error depends upon the ratio $R/(c_1 + c_2)$, Figure 3–17. The stress distribution for curved beams is not linear as it is for a straight beam. Furthermore, the neutral axis of the curved member does not coincide with the centroid of the cross section as it does for the straight members. Here we are concerned with the development of expressions for the bending stresses of curved members.

Let us consider a segment of a curved beam (Figure 3–17a) subjected to a pure bending moment that is assumed to lie in the plane of symmetry of the cross section. As in the case of straight members, we will assume that transverse planes (normal to the center line of the curved member) before bending remain transverse planes after bending, and that the material is homogeneous, isotropic, and obeys Hooke's law. Also, we shall assume that there is no lateral pressure between longitudinal fibers.[9] Now, let $d\theta$ denote the angle between the two transverse planes at the ends of the curved segment, and $\Delta d\theta$ denotes the change in angle of these planes after the bending

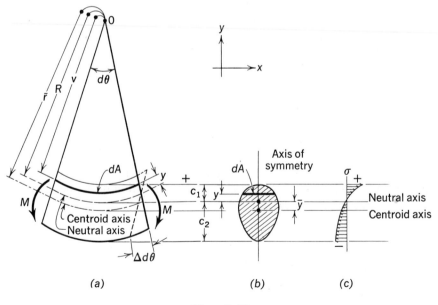

(a) (b) (c)

Figure 3–17

---

[9] For other than thin members, such stresses, although existant, are usually negligible.

is applied, as shown in Figure 3–17a. At a distance y from the neutral axis the strain is

$$\epsilon_y = \frac{y(\Delta d\theta)}{(R - y)d\theta} \qquad (a)$$

Assuming *elastic* deformation, that is, $\sigma/\epsilon = E$, we get

$$\sigma_y = E \cdot \epsilon_y = \frac{E \cdot y(\Delta d\theta)}{(R - y)d\theta} \qquad (b)$$

An inspection of Equation (b) shows that the stress is zero at the neutral axis (where $y = 0$) and varies hyperbolically as illustrated in Figure 3–17c.

In order to determine the location, and subsequently the shift of the neutral axis with respect to the centroidal axis, we refer to the stress distribution (Figure 3–17c) and the equilibrium requirements for the beam. (a) The sum of the normal forces over the cross section is zero, and (b) the moment of these normal forces equals the bending moment M. From requirement (a) and Equation (b) above, we have

$$\int_A \sigma_y \cdot dA = \int_A \frac{E \cdot y(\Delta d\theta)dA}{(R - y)d\theta} = \frac{E(\Delta d\theta)}{d\theta} \int_A \frac{ydA}{(R - y)} = 0 \qquad (c)$$

Because $E\Delta d\theta/d\theta \neq 0$, we conclude that

$$\int_A \frac{ydA}{(R - y)} = 0 \qquad (d)$$

In Figure 3–17a we may note again what the quantities $\bar{r}$, R, v, and y represent. From this we note that $y = R - v$. Substituting for y in Equation (d) we have

$$\int_A \frac{R - v}{v} dA = R \int_A \frac{dA}{v} - \int_A dA = 0$$

from which

$$R = \frac{A}{\displaystyle\int_A \frac{dA}{v}} \qquad (3\text{–}22)$$

which determines the location of the neutral axis with respect to the center of curvature. From this we can easily relate the location of the neutral axis to that of the centroidal axis, that is,

$$\bar{y} = \bar{r} - \frac{A}{\displaystyle\int_A \frac{dA}{v}} \qquad (3\text{–}23)$$

Where $\bar{r}$ is given in Figure 3–17a.

Now, from condition (b), that is, internal moment $= M$, we have

$$\int_A \sigma_y \cdot y \cdot dA = M \qquad (e)$$

When substituting for σ, Equation (e) becomes

$$\frac{E(\Delta d\theta)}{d\theta} \int_A \frac{y^2 dA}{(R-y)} = M \qquad (f)$$

The integral in Equation (f) may be simplified by adding and subtracting an R to one of the y's in the numerator. That is, let $y = R - (R-y)$ so that

$$\int_A \frac{y^2 dA}{(R-y)} = \int_A \frac{y[R-(R-y)]dA}{R-y} = R\int_A \frac{y\,dA}{R-y} - \int_A y\,dA \qquad (g)$$

The first integral on the right side of Equation (g) is equal to zero [see Equation(d)]. The second integral is the first moment of the cross-sectional area with respect to the neutral axis. Thus, if y is positive when measured from the neutral axis towards the center of curvature, $-\int_A y\,dA = -[A(-\bar{y})] = A\bar{y}$, yielding, from Equation (g),

$$\int_A \frac{y^2 dA}{R-y} = A\bar{y} \qquad (h)$$

Substituting in Equation (f)

$$\frac{E(\Delta d\theta)}{d\theta} \cdot A\bar{y} = M$$

which gives, when $[E(\Delta d\theta/d\theta)]$ is eliminated from Equation (b)

$$\sigma_y = \left(\frac{M}{A\bar{y}}\right)\left(\frac{y}{R-y}\right) \qquad (3\text{--}24)$$

Referring to Figure 3–17b, the maximum tensile (assumed positive) and compressive (negative) stresses occur at c_1 and c_2, respectively, from the neutral axis:

$$\sigma_{concave} = \left(\frac{M}{A\bar{y}}\right)\left(\frac{c_1}{R-c_1}\right) \qquad (3\text{--}25)$$

and

$$\sigma_{convex} = \left(\frac{M}{A\bar{y}}\right)\left(\frac{c_2}{R+c_2}\right) \qquad (3\text{--}26)$$

The discussion up to this point assumed that the member was subjected to pure bending only. More generally, for a curved member subjected to both bending and transverse forces, it is merely necessary to resolve the transverse force into a force acting through the *centroid* of the cross section, and into a couple acting in the plane of symmetry of the cross section. The moments may be then treated as before. The force, if not perpendicular to the cross section, is then resolved into components perpendicular and parallel to the cross section. This results in normal and shear stresses over the section, calculated by (P/A) and (VQ/Ib), respectively, as discussed previously. The normal and bending stress may then be added algebraically to obtain the net fiber stress acting over the section.

Example 3-18:

Given: A circular ring loaded and of the dimensions as shown in the figure.

Find: The maximum stresses at section *A–A*.

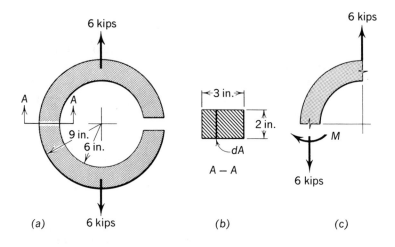

(a)　6 kips　　　　(b)　　　　(c)

Procedure: We note that we have here a direct force and a bending moment acting on section *A–A* as shown in Figure *c*. Thus, the maximum stresses will be

$$\sigma = \frac{P}{A} \pm \left(\frac{M}{A\bar{y}}\right)\left(\frac{y}{R - y}\right)$$

From Equation (3–22),

$$R = \frac{A}{\int_A \frac{dA}{v}} = \frac{(3 \times 2)}{\int_6^9 \frac{2dr}{r}} = \frac{6}{2(\ln 9 - \ln 6)}$$

$$R = \frac{3}{\ln (9/6)} = \frac{3}{0.4055} = 7.398 \text{ in.}$$

from which [Equation (3–23)]

$$\bar{y} = 7.5 - 7.398 = 0.102 \text{ in.}$$

Thus,

$$A = 6 \text{ in.}^2$$
$$R = 7.398 \text{ in.}$$
$$\bar{y} = 0.102 \text{ in.}$$
$$P = 6000 \text{ lb}$$

$$M = (6000)(7.5) = 45{,}000 \text{ in.-lb (positive)}$$

$$c_1 = 1.5 - 0.102 = 1.398 \text{ in. (Figure 3–17}b)$$

$$c_2 = 1.5 + 0.102 = 1.602 \text{ in. (Figure 3–17}b)$$

from which

Maximum tensile: $\quad \sigma_t = \dfrac{6000}{6} + \dfrac{45{,}000}{6 \times 0.102} \times \dfrac{1.398}{6} = \mathbf{18{,}132 \text{ psi}}$

Maximum compressive: $\sigma_c = \dfrac{6000}{6} - \dfrac{45{,}000}{6 \times 0.102} \times \dfrac{1.602}{9} = \mathbf{-12{,}100 \text{ psi}}$

For comparison, let us determine these bending stresses using the straight-beam formula, that is,

$$\sigma_{t,c} = \frac{P}{A} \pm \frac{Mc}{I}$$

$$\sigma_{t,c} = \frac{6000}{6} \pm \frac{45{,}000 \times 1.5}{\dfrac{2 \times (3)^3}{12}} = \begin{array}{l} 16{,}000 \text{ psi, tension} \\ -14{,}000 \text{ psi, compression} \end{array}$$

PROBLEMS

3–98 Determine the maximum stresses at section A–A, Figure P 3–98, for $P = 1$ kip.

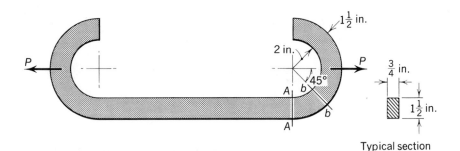

Typical section

Figure P 3–98

3–99 For the problem of Figure P 3–98, determine the maximum value that P may have as shown if the stress at section b–b is not to exceed 15,000 psi.

3–100 Determine the maximum stresses at section A–A of the crane hook shown in Figure P 3–100 if the value of $P = 100$ kips.

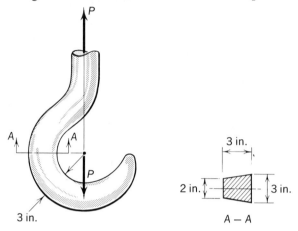

Figure P 3–100

3–101 Determine the maximum P that the crane hook in Figure P 3–101 may safely withstand if the stress at section A–A is not to exceed 16,000 psi.

3–102 Determine the maximum stresses at section A–A in the U frame of Figure P 3–102 if the load $P = 4$ kips.

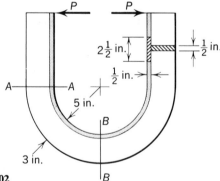

Figure P 3–102

3–103 Determine the maximum stresses at a section B–B of the U frame of Figure P 3–102 if the load $P = 4$ kips.

3–104 Determine the maximum P that the U frame in Figure P 3–102 can safely withstand at section B–B if the stress at that section is not to exceed 15,000 psi.

3—11 Principal Stresses in Beams

In Section 3–4, we established a procedure for determining the bending (normal) stresses for beams. In Section 3–5, we developed a method for finding shearing stresses in beams. Using these stresses, we will proceed here to determine the principal stresses and maximum shearing stresses in beams.

Let us consider the beam in Figure 3–18a. The bending and shear-stress distributions over an arbitrary section a–a are shown in Figures 3–18b and 3–18c, respectively. As previously noted, the bending stress is a maximum

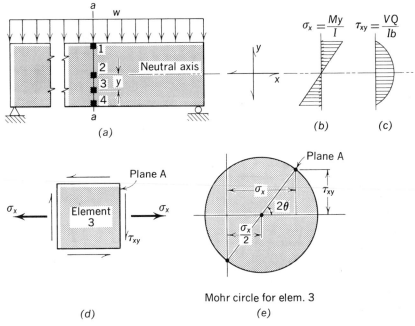

$$\sigma_x = \frac{My}{I} \qquad \tau_{xy} = \frac{VQ}{Ib}$$

(a)

(b)

(c)

(d)

Mohr circle for elem. 3

(e)

Figure 3–18

at the extreme fibers and zero at the neutral axis; however, the shear stress is zero at the extreme fibers and a maximum at the neutral axis. Thus, the small elements 1 and 4 in Figure 3–18a experience only bending stresses oriented in the x direction, whereas element 2 is subjected to only pure shear. More generally, however, element 3 at y distance from the neutral axis experiences both normal (in the x direction only) and shear stresses, as shown in Figure 3–18d. From the Mohr circle for this general stress condition, shown in Figure 3–18e, we can easily obtain the maximum (or minimum) normal or shear stress; that is,

$$\sigma_{\max,\min} = \frac{\sigma_x}{2} \pm \sqrt{\frac{(\sigma_x)^2}{4} + (\tau_{xy})^2} \qquad (3\text{--}27)$$

and

$$\tau_{\max} = \sqrt{\frac{(\sigma_x)^2}{4} + (\tau_{xy})^2} \qquad (3\text{--}28)$$

The direction of the principal stresses (maximum and minimum normal, and zero shear) may be spotted easily from the Mohr circle. Keeping in mind that counterclockwise rotation is positive on the Mohr circle, relative to an original reference plane, say plane A,

$$\tan 2\theta = -\frac{\tau_{xy}}{(\sigma_x/2)} = -\frac{2\tau_{xy}}{\sigma_x} \qquad (3\text{--}29)$$

The direction (as well as the magnitude) of the principal stresses may prove helpful to a designer. As an example, because of its inability to take tensile load, concrete has to be properly reinforced with steel bars or inclined bars (called *stirrups*) to take care of "diagonal" tension. For their most effective use, the bars or stirrups are inclined perpendicular to a "potential" crack. Hence, if we knew the direction of the principal stress (maximum tension for our example) we could design appropriately. Because σ and τ vary with both distances x along the beam, and y over the depth (M and V vary), the direction of the principal stress, given Equation (3–29), varies. At the extreme fibers where $\tau = 0$ the principal stresses ($\sigma_{\max}$ and $\sigma_{\min}$) are oriented at zero slope all along the beam, which is the same slope as the bending stresses. In other words, the boundary lines give the direction of the principal stresses for the fibers at those levels. We may, in this manner, determine θ for various points defined by coordinates x and y, and form a family of curves, called *stress trajectories*, whose slopes indicate the direction of the principal stresses. Such trajectories are illustrated in Figure 3–19 for the beam in Figure 3–18*a*. The directions of the tensile stress are given by the solid lines, those for the compressive stress are given by the broken lines. The two types intersect at 90-degree angles, as evident by inspecting the Mohr circle (the two principal stresses occur on planes 90 degrees with each other).

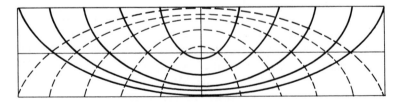

Figure 3–19

The maximum stresses (normal or shear) are not readily obtainable for a loaded beam. The maximum flexure stresses are readily obtained at the most extreme fiber where the moment is a maximum; the maximum shear stress is, in most instances, at the neutral axis (rectangular, I shapes, T shapes, and so forth). But the use of Equations (3–27) and (3–28) may result in stresses which are in excess of those computed on the basis of $M \cdot c/I$ and VQ/Ib. A close examination of various points along the beam and at various depths is advisable for cases where the maximum is not readily evident.

Example 3–19:

Given: A cantilever beam loaded as shown in Figure a.
Find: The maximum normal stresses and shearing stress in the beam at the junction of the flange and the web.

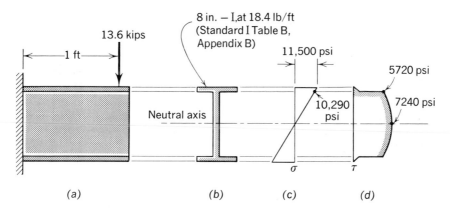

(a)　　　　　(b)　　　　　(c)　　　　　(d)

Procedure: In this case, the maximum bending moment is at the fixed end: $M_{max} = 1 \times 13{,}600 \times 12 = 163{,}500$ in.-lb. The shear V is constant $V = 13{,}600$ lb over the entire beam (neglecting the weight of the beam). The maximum bending stress is, therefore,

$$\sigma_{max} = \frac{Mc}{I} = \frac{163{,}500 \times 4}{56.9} = 11{,}500 \text{ psi}$$

at the extreme fiber, as shown in Figure c. At the junction of the web and flange, the stress is, by proportion,

$$\sigma_w = \left(\frac{4 - 0.425}{4}\right) \cdot 11{,}500 = 10{,}290 \text{ psi}$$

Based on $\tau = VQ/Ib$, the shear-stress distribution over the section is as shown in Figure d. At the neutral axis, $Q = 8.16$ in.^{3}, $b = 0.270$ in., $\tau = 7240$ psi. At the junction of flange-web, $Q = 6.44$, $b = 0.270$

in., $\tau = 5720$ psi. From Equation (3–27), the normal stress at the flange-web junction is

$$\sigma_w = \frac{10{,}290}{2} + \sqrt{\frac{(10{,}290)^2}{4} + (5720)^2} = \mathbf{17{,}980\ psi}$$

and from Equation (3–28),

$$\tau_{\max} = \mathbf{7690\ psi}$$

from which we notice that both the shear and normal stresses based on combined stresses are larger than those based on only shear and only moment, respectively.

PROBLEMS

[*Note:* For properties of I sections, see Table B-2, Appendix B.]

3–105 Determine the principal stresses and maximum shear stress at point *A* in the beam shown in Figure P 3–105.

10-in. I, at 25.4 lb/ft (Standard I)

Figure P 3–105

3–106 Determine the principal stresses and maximum shear stress at point *A* in the beam shown in Figure P 3–106.

10-in. I, at 25.4 lb/ft (Standard I)

Figure P 3–106

3–107 Determine the principal stresses and maximum shear stress at point *A* in the box-beam section shown in Figure P 3–107.

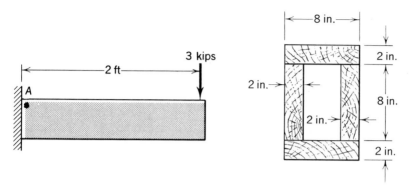

Figure P 3–107

3–108 Determine the principal stresses and maximum shear stress at point *A* in the box-beam section shown in Figure P 3–108.

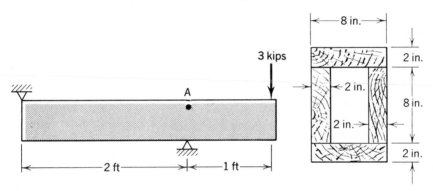

Figure P 3–108

3–109 Determine the maximum concentrated load that the beam in Figure P 3–105 may safely carry if the maximum normal and shear stresses at point *A* are not to exceed 20,000 psi and 15,000 psi, respectively.

3–110 Determine the maximum concentrated load that the beam in Figure P 3–106 may safely carry if the maximum normal and shear stresses at point *A* are not to exceed 24,000 psi and 16,000 psi, respectively.

3–111 Determine the maximum concentrated load that the beam in Figure P 3–107 may safely carry if the maximum normal and shear stresses at point *A* are not to exceed 1400 psi and 200 psi, respectively.

3—12 Stresses due to Bending and Torsion

The discussion of Section 3–11 where we considered principal stresses resulting from shear and bending may be extended one step further to also include torsion. However, because noncircular members are rarely designed to carry bending *and* torsion (possibly because of the more complicated analysis described in Section 2–5), we shall consider here only circular shafts.

The circular shaft in Figure 3–20a is subjected to a torque T and a concentrated load P. An arbitrary section a–a experiences a shear force V, a bending moment M, and a torque T. The corresponding stresses may be

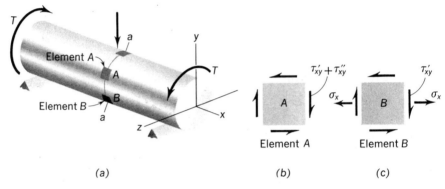

(a)   (b)   (c)

Figure 3–20

determined using the conventional stress formulas derived in Chapters 2 and 3, that is,

due to V: $\tau''_{xy} = \dfrac{VQ}{Ib}$

due to M: $\sigma_x = \dfrac{My}{I}$

and due to T: $\tau'_{xy} = \dfrac{Tr}{J}$

Element A, assumed to be at the level of the neutral axis of the shaft (Figure 3–20a), has zero bending stress (for our loading) but a maximum shear stress τ''_{xy}, caused by V, and a torsional stress τ'_{xy}, caused by T (Figure 3–20b). Their effect on a typical element is that of the algebraic sum of the two shears, τ''_{xy} and τ'_{xy}. The element may then be treated in pure shear, as discussed in Sections 1–11 and 1–12, to find the magnitudes and orientation of the principal stresses. However, usually (but not always) the stress caused by V is small compared to those caused by T or M; hence, such analysis is less important. The more critical case usually includes bending and torsion. Element B, located on the extreme tensile surface of the shaft, experiences a maximum tensile stress σ_x and a shear stress τ'_{xy}. The shear stress τ''_{xy} caused by V is zero. The stresses are shown on the enlarged element B, Figure 3–20c. From a plot of the Mohr circle (see Section 1–12) of the stresses for this element we can then determine a general expression for principal stresses and maximum shearing stresses for shafts under bending and torsion (neglecting the effect of V):

$$\sigma_{max,min} = \frac{\sigma_x}{2} \pm \sqrt{\frac{(\sigma_x)^2}{4} + (\tau_{xy})^2} \qquad (3\text{--}30)$$

and

$$\tau_{max} = \pm \sqrt{\frac{(\sigma_x)^2}{4} + (\tau_{xy})^2} \qquad (3\text{--}31)$$

The above equations may be written in terms of shaft diameter d, bearing in mind that $I = \pi d^4/64$ and $J = 2I$. From

$$\sigma_x = \frac{My}{I} = \frac{M \cdot d/2}{\pi d^4/64} = \frac{32M}{\pi d^3}$$

and

$$\tau_{xy} = \frac{Tr}{J} = \frac{T \cdot d/2}{\pi d^4/32} = \frac{16T}{\pi d^3}$$

Equations (3–30) and (3–31) become [10]

$$\sigma_{max,min} = \frac{16}{\pi d^3}(M \pm \sqrt{M^2 + T^2}) \qquad (3\text{--}30')$$

and

$$\tau_{max} = \frac{16}{\pi d^3}\sqrt{M^2 + T^2} \qquad (3\text{--}31')$$

Example 3–20:

Given: The shaft loaded as shown in the figure.

Find: The maximum normal and shearing stresses in the shaft.

Procedure: We note that P may be resolved into a horizontal component of $\frac{1}{2}(400) = 200$ lb, and a vertical component of $0.866(400) = 346.5$ lb. The corresponding moments at the fixed end (point of maximum moments) are

$$M_y = \quad 200 \times 8 \text{ ft} = 1600 \text{ ft-lb} = 19,200 \text{ in.-lb}$$

$$M_z = 346.5 \times 8 \text{ ft} = 2771 \text{ ft-lb} = 33,250 \text{ in.-lb}$$

The maximum moment M_{max} on the shaft is the vector sum of M_y and M_z.

---

[10] Note that these equations only apply to solid shafts. Equations (3–30) and (3–31) may be used for hollow shafts, or use Mohr's circle.

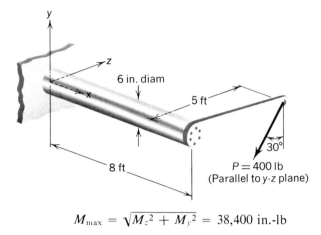

$$M_{max} = \sqrt{M_z^2 + M_y^2} = 38{,}400 \text{ in.-lb}$$

The maximum torque is

$$T_{max} = 346.5 \times 5 \text{ ft} = 1732 \text{ ft-lb} = 20{,}800 \text{ in.-lb}$$

Using Equations (3–30′) and (3–31′), we get

$$\sigma_{max,min} = \frac{16}{\pi \times 216} [38{,}400 \pm \sqrt{(38{,}400)^2 + (20{,}800)^2}] = \begin{matrix} +1930 \\ -1225 \end{matrix} \text{ psi}$$

and

$$\tau_{max} = \frac{16}{\pi \times 216} \sqrt{(38{,}400)^2 + (20{,}800)^2} = 1030 \text{ psi}$$

An investigation will show that the effect of V is not a governing factor here.

PROBLEMS

[*Note:* In the following problems neglect the direct shear stress caused by the transversed shear force V.]

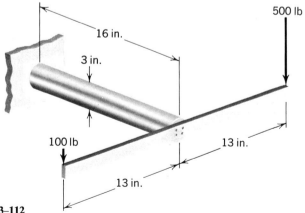

Figure P 3–112

3–112 Determine the principal stress and maximum shear stress for the shaft in Figure P 3–112.

3–113 Determine the principal stress and maximum shear stress for the shaft in Figure P 3–113.

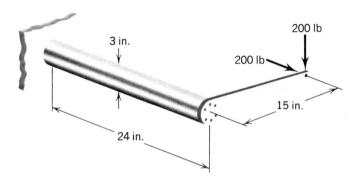

200 lb

3 in.

200 lb

15 in.

24 in.

Figure P 3–113

3–114 Determine the principal stress and maximum shear stress for the shaft in Figure P 3–114. The bearings on the shaft supports are assumed to be frictionless. The shaft is maintained in rotational equilibrium by the torque R.

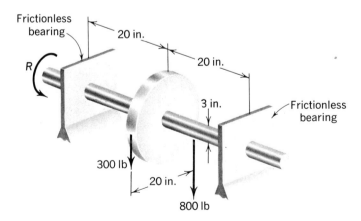

Frictionless bearing

20 in.

20 in.

R

3 in.

Frictionless bearing

300 lb

20 in.

800 lb

Figure P 3–114

3–115 Determine the principal stress and maximum shear stress for the shaft in Figure P 3–115. The bearings are assumed to be frictionless.

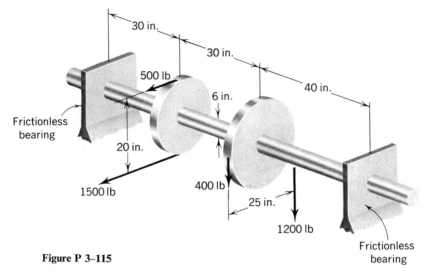

Figure P 3–115

3–116 Determine the principal stress and maximum shear stress for the shaft in Figure P 3–116.

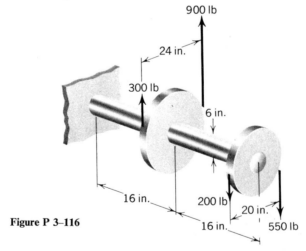

Figure P 3–116

3–117 Determine the minimum diameter that the shaft in Figure P 3–112 may have if the normal and shear stresses are not to exceed 12,000 psi and 9000 psi, respectively.

3–118 Determine the minimum diameter that the shaft in Figure P 3–113 may have if the normal and shear stresses are not to exceed 10,000 psi and 8000 psi, respectively.

3–119 Determine the minimum diameter that the shaft in Figure P 3–114 may have if the normal and shear stresses are not to exceed 9000 psi and 6000 psi, respectively.

3–120 Determine the minimum diameter that the shaft in Figure P 3–115 may have if the normal and shear stresses are not to exceed 11,000 psi and 8000 psi, respectively.

3–121 Determine the minimum diameter the shaft in Figure P 3–116 may have if the normal and shear stresses are not to exceed 12,000 psi and 8000 psi, respectively.

3–122 The element shown in Figure P 3–122c is taken from the shaft near the support as shown in Figure P 3–122a. Determine (a) the length L which satisfies the stress condition given for the element, and (b) the principal stresses.

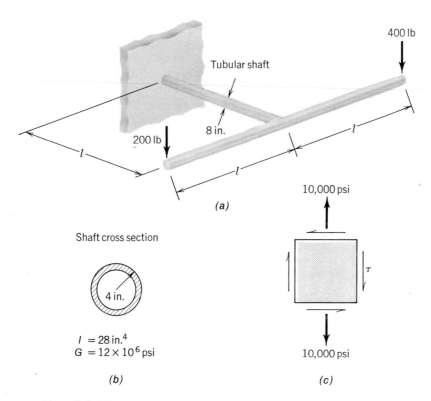

400 lb

Tubular shaft

8 in.

200 lb

10,000 psi

(a)

Shaft cross section

4 in.

$I = 28\,\text{in.}^4$
$G = 12 \times 10^6\,\text{psi}$

(b)

10,000 psi

τ

10,000 psi

(c)

Figure P 3–122

3–123 The sign post in Figure P 3–123 is subjected to a twisting and bending moment stress caused by wind blowing perpendicularly to the plane of the sign. The post has an outside diameter of 6 in. and an inside diameter of 5½ in. For a wind pressure of 30 lb/sq ft, determine (*a*) the maximum twisting stress in the post, and (*b*) the maximum normal stress on an element near the support (use Mohr's circle).

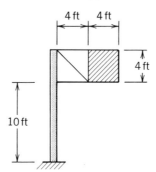

Figure P 3-123

3—13   Elastic Bending of Nonsymmetrical Sections

One of the assumptions made in the development of the bending stress formula (Section 3–4) was that the loads were applied in a plane of symmetry of the beam; hence, the beam would be subject to no torque. We will now proceed to develop a more general approach encompassing a general case of loading and a general cross section.

Figure 3–21 shows an arbitrary section of a beam subjected to a pure bending moment M. In our case, we shall assume the principal plane, in which M acts, to be x–y and the principal axes of the cross section to be y and z. We also assume the beam to be homogeneous and isotropic, and that we stay within the elastic limit of the material. Thus, plane sections will remain plane, and the bending strains and stresses will vary linearly from the neutral axis. Now let us resolve M into rectangular components M_y and M_z acting in the xz and xy planes, respectively (y and z are assumed to be principal axes).

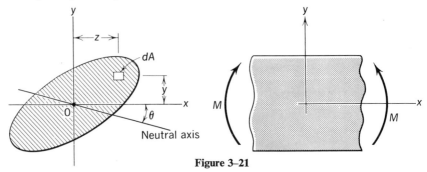

Figure 3–21

Observing that the deformation is elastic, we write

$$\sigma_y = ky$$

from which $k = M_z/I_z$. The stress then becomes

$$\sigma_y = \frac{M_z y}{I_z} \qquad (3\text{-}32)$$

By a similar analysis, the stress σ_z is found to be

$$\sigma_z = \frac{M_y z}{I_y} \qquad (3\text{-}33)$$

Thus, when the bending moments do not act in the principal planes, we may resolve the moments into components which would coincide with the principal planes. Then the total stress is the algebraic sum (or difference) of the two stresses produced by the individual components of M. Example 3–21 illustrates the procedure.

Example 3–21:

Given: The section in Figure *a* subjected to a bending moment of $M = 50{,}000$ in.-lb acting in plane xy.
Find: Stress at point 1.

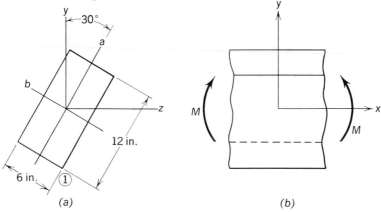

(a)   (b)

Procedure: $M_b = (50{,}000) \cos 30° = 43{,}300$ in.-lb

$M_a = (50{,}000) \sin 30° = 25{,}000$ in.-lb

$$I_b = \frac{6(12)^3}{12} = 864 \text{ in.}^4$$

$$I_a = \frac{12(6)^3}{12} = 216 \text{ in.}^4$$

Thus,

$$\sigma_1 = \frac{(43{,}300)(6)}{864} + \frac{(25{,}000)(3)}{216} = 300 + 347 = \mathbf{647\ psi}$$

Chapter 4 ◆ DEFLECTION OF BEAMS

4—1 Introduction

In Chapter 3 we developed relationships between transverse loads and internal deformations in beams. The resultant effect of the accumulations of these internal deformations, assumed elastic within the range of loading, is an external deformation. Although a beam deforms laterally (according to the Poisson-ratio concept), as well as in the plane of loading, for our purpose here only the latter will be considered. The distance a point on the neutral surface of an initially straight beam travels in the plane of loading during bending is called the *deflection* of the point. A line in the plane of loading coinciding with the deformed neutral axis is called the *elastic curve* or *deflection curve* of the beam. Thus, the deflection of the beam at any point along its length is merely the distance from the original location of the neutral surface to the elastic curve. Also, the slopes of the tangents to the elastic curves at various points are considered to be the slopes of the beam at the respective points along the length of the beam.

Both the deflections and the slopes of bent beams are important criteria, and must be considered in engineering design. Excessive deflections may cause cracks, and perhaps spalling, of plastered ceilings for example; also, the esthetic devaluation of the structure is possible. In the field of machine design or dynamic analysis, excessive deflection may obstruct proper functioning of a machine part, for example. For this reason, building codes specify maximum deflections of anywhere from 1/360 to 1/240 of the span, depending on the type and use of a structure; the deflection of a machine part must frequently be limited if the machine is to function properly. As a tool in the analysis of statically indeterminate structures, the slope and deflection relationships provide the means by which we may set up additional equations, compatible with the geometry of the structure, and independent of those based on statics, to lead to a solution.

250

Several methods are available for determining beam deflection, each with merit and special usefulness for particular cases. Related to the principles on which they are derived, these methods may be separated into two general categories, *energy methods* and *geometry of deformation methods*. The energy methods discussed in this text are found in Chapter 6 and, therefore, will not be mentioned at this time. The procedures based on geometry of deformation include (*a*) the *integration* method, (*b*) the *moment-area* methods and its rather closely related *conjugate-beam* and *elastic weights* methods, and (*c*) the *superposition* methods.

4—2   Differential Equation of the Elastic Curve

In Section 3–4 we observed that the curvature of a beam, for the assumed conditions stated in that section, was directly related to the quantities M, I, and E as expressed by Equation (3–5). Rewritten to express curvature $1/\rho$, Equation (3–5) takes the form

$$\frac{1}{\rho} = \frac{M}{EI} \qquad (a)$$

Because M generally varies along the span of the beam, the curvature will, obviously, also vary. Hence, it would be quite difficult, and cumbersome to determine the complete shape of the elastic curve under such circumstances. It is therefore almost necessary to express the shape of the elastic

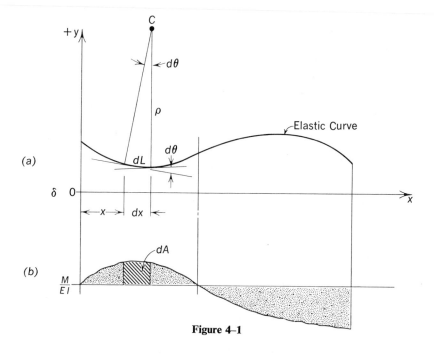

Figure 4–1

curve in terms of its rectangular coordinates x and y, if we are to put the expressions of slope and deflections to good use.

Let us consider the curve of Figure 4–1a, assumed to represent a segment of the elastic line of a beam. At a distance x from a reference point, say point O, the support, an increment of length dL, will have a change of slope from one end to the other by an amount $d\theta$. Thus, $dL = \rho \, d\theta$ from which

$$\frac{d\theta}{dL} = \frac{1}{\rho} \tag{b}$$

For small angles (that is, small deflections), $dy/dx = \tan \theta = \theta$, and $dL \approx dx$. Thus, $d\theta/dL \approx d\theta/dx = d^2y/dx^2$. Hence, Equation ($b$) becomes

$$\frac{d^2y}{dx^2} = \frac{1}{\rho} \tag{c}$$

Combining Equations (a) and (c), we have

$$EI \frac{d^2y}{dx^2} = M \tag{4–1}$$

which is the differential equation of the elastic curve of the beam.

This equation might have also been obtained by making use of the radius of curvature as expressed in any calculus textbook, that is,

$$\frac{1}{\rho} = \frac{d^2y/dx^2}{[1 + (dy/dx)^2]^{3/2}} \tag{d}$$

As mentioned previously, dy/dx is quite small, making $(dy/dx)^2$ in Equation (d) insignificant. Thus, Equation (d) yields expression (c) above.

Careful consideration should be given to sign conventions. In Chapter 3 the moment was assumed positive when the beam (elastic curve for our purpose) was concave upward (for example, the left segment of the elastic curve in Figure 4–1). Thus, corresponding to the *positive moment* we have the *positive curvature* $1/\rho$, which, in turn, also makes d^2y/dx^2 positive. Also, for the portion of positive moment, there is a positive *change* in the *slope;* that is $d\theta/dx$ is positive when M is positive. A similar analysis for a portion that is concave downward (associated with a negative bending moment) will indicate that d^2y/dx^2 and $d\theta/dx$ will be negative because M is negative. Hence, we conclude that d^2y/dx^2 and $d\theta/dx$ assume the same sign as the quantity M.[1]

Now we shall proceed to interrelate more meaningfully the quantities y, θ, M, V, and L representing deflection, slope, moment, shear, and load, respectively. From Section 3–2, $dM/dx = V$ and $dV/dx = w$. In the present

---

[1] For other choices of coordinate systems, this is not the case.

case, we found that $d\theta/dx = M/EI$ and $dy/dx = \theta$. Hence, we may relate all these quantities in terms of x and y as follows:

Deflection:
$$\delta = y \qquad (a)$$

Slope:
$$\theta = \frac{dy}{dx} \qquad (b)$$

Moment:
$$M = \frac{d\theta}{dx} EI = \frac{d^2y}{dx^2} EI \quad (c) \qquad\qquad (4\text{-}2)$$

Shear:
$$V = \frac{dM}{dx} = \frac{d^3y}{dx^3} EI \quad (d)$$

Load:
$$w = \frac{dV}{dx} = \frac{d^4y}{dx^4} EI \qquad (e)$$

We have already shown in Chapter 3 (Sections 3–2 and 3–3) that the relationships between L, V, and M were indeed helpful to draw the shear and moment diagrams. The same relative use of the expressions in Equations (4–2) may be made in obtaining two new diagrams, the slope diagram and deflection diagram, the importance of which has already been mentioned, but which shall be evident later as additional material is covered. From Equations (4–2c) and (b) we have

$$\theta = \int \frac{M}{EI} \cdot dx + C_1 \qquad (4\text{-}3)$$

and

$$\delta = y = \int \theta dx + C_2 \qquad (4\text{-}4)$$

In a specific interval between two points, $a < x < b$, Equations (4–3) and (4–4) take the general form

$$\theta_b - \theta_a = \int_a^b \frac{M}{EI} \cdot dx \qquad (4\text{-}5)$$

and

$$\delta_b - \delta_a = \int_a^b \theta dx \qquad (4\text{-}6)$$

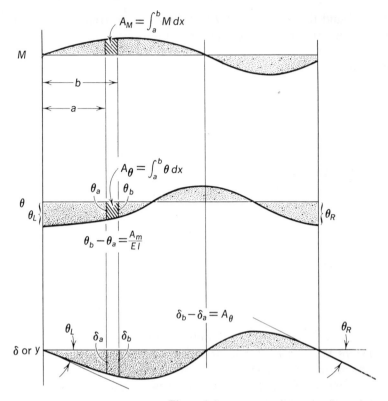

$$A_M = \int_a^b M\,dx$$

$$A_\theta = \int_a^b \theta\,dx$$

$$\theta_b - \theta_a = \frac{A_m}{EI}$$

$$\delta_b - \delta_a = A_\theta$$

Figure 4–2

Now to observe the usefulness of these equations in drawing the θ and δ diagrams, we have only to study the operations involved. We note that the difference in slopes $(\theta_b - \theta_a)$ at points $x = a$ and $x = b$ is equal to the area under the moment diagram, divided by EI, between these two points. Similarly, from Equation (4–6), we note that the difference in deflections $(\delta_b - \delta_a)$ at points $x = a$ and $x = b$ is equal to the area under the θ diagram between these points. Also, from the expression in Equation (4–2), we note that the value of moment M at a point is the *slope* of the curve at that point; the *value* of the slope θ at a specific location is the slope of the δ curve at that point. This is illustrated using the series of curves in Figure 4–2. From Equation (4–2) it is obvious that if the equation of the elastic curve δ is known, then θ, M, V, and w can be determined through a series of successive differentiations. Conversely, one may surmise that from the load function we may obtain the values of V, M, θ, and δ through a series of successive integrations. Symbolically, these are

$$V = \int w\,dx + C_1 \quad (a)$$

$$M = \int V dx + C_2 \quad (b)$$

$$\theta = \int \frac{M}{EI} dx + C_3 \quad (c) \qquad\qquad (4\text{--}7)$$

$$\delta = \int \theta dx + C_4 \quad (d)$$

The constants of integration (C_1, C_2 C_3, C_4) are evaluated by considering the boundary conditions with which the structure must comply. These operations are perhaps best illustrated in the examples of the next section. Thus, we may derive a series of curves (curves for V, M, θ, and δ), interrelated to one another, such that the shape of one is useful in obtaining another adjacent (consecutive) to it, as is explained in the examples of the next section.

The product EI, commonly referred to as the *flexural rigidity*, is frequently a constant throughout the length of beams that are usually confronted in engineering problems. When EI is a variable, it should also be expressed as a function of x.

We must not lose sight of the implied assumptions on which the above expressions were derived: The elastic curve was assumed to be a continuous curve with no abrupt changes in either ordinates or slopes. This is a reasonable and widely accepted supposition. Also, it is assumed that the deformation due to shear is negligible. This, also, is a reasonable assumption, especially for slender members, where bending stresses usually govern. The deflection due to shear is, however, discussed in Section 4–6.

4—3 The Integration Method

This method consists of performing two successive integrations of the moment equation in order to get the elastic curve. The bending moment equations may be derived directly from the load and shear functions, as described in Chapter 3. From this a complete relationship is developed between w, V, M, θ, and δ as expressed by Equation (4–7). Or, the bending moment equation may be given, and the two integrations are performed to obtain the δ curve. Whatever the situation, it should be noted that integration constants always occur in the process, and therefore must be accounted for as dictated by the boundary conditions.

Although we assume the elastic curve to be a "smooth" curve, the mathematical expression for this curve is not always a continuous function; in fact, in most instances it is not. Abrupt changes in loading will cause abrupt changes in the V and M curves (either changes in ordinates or changes in slopes, or both), requiring separate equations for these curves between the points where such changes occur. Consequently, slope and deflection equations must be written to cover these particular intervals. Note that because the elastic curve is assumed smooth, neither the deflection nor the slope curves can have abrupt changes anywhere over the length of a beam which is continuous over its length (this condition does not hold for hinged or semihinged beams). Hence, both deflection and slope curves have the *same* respective values at a small increment of distance on either side of the point at which the abrupt change in loading takes place. This is an important aid in evaluating the constants of integration.

Drawing V, M, θ, and δ diagrams, in that order, and placing them directly below the beam should prove indeed helpful in constructing these curves and when writing the corresponding equations (especially for evaluating the constants of integration). Similarly, writing the equation for any one curve, or portion of a curve, "near" the curve, rather than, say, on a separate sheet, makes for a clear and effective presentation. It minimizes the need for elaborate explanations of the procedure and, quite likely, minimizes the possibility of errors. The procedure is somewhat self-explanatory in the following examples. Briefly, it consists of plotting the values of V, M, θ, and δ on the vertical scale, positive values plotted above, negative below a reference line, for each value of x along the length of the beam. This is merely an expansion of the steps already described in Chapter 3 for the construction of the V and M diagrams.

Example 4–1:

Given: A simply supported beam of length L and constant EI, supports a uniform load w lb/ft, as shown in Figure a.

Find: (a) The V, M, θ, and δ diagrams and equations, (b) the maximum deflection (at the middle of the span), and (c) the slope of the elastic curve at the left support.

Procedure: Because w is assumed to act down, it shall be assumed *negative*. Thus, a load diagram would represent the w load below and the reactions above a reference line, as shown in Figure b. The constants of integration may now be obtained from the boundary conditions as follows (see Figures c, d, e, and f):

At $x = 0$,

$$V = +\frac{wl}{2}; \text{ hence, } C_1 = +\frac{wl}{2}$$
$$M = 0; \text{ hence, } C_2 = 0$$
$$\theta_A = C_3/EI$$
$$\delta = 0; \text{ hence, } C_4 = 0$$

At $x = l$,

$$\delta = 0; \text{ hence, } C_3 = -\frac{wl^3}{24}$$

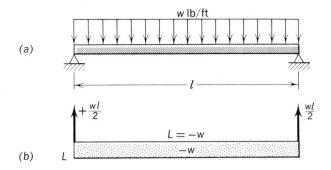

(a)

(b) L

$V = -wx + C_1$

(c) V

$M = \dfrac{-wx^2}{2} + C_1 x + C_2$

(d) M

$\theta = \dfrac{1}{EI}\left(\dfrac{-wx^3}{6} + \dfrac{C_1 x^2}{2} + C_2 x + C_3\right)$

(e) θ θ_A

$\delta = \dfrac{1}{EI}\left(\dfrac{-wx^4}{24} + \dfrac{C_1 x^3}{6} + \dfrac{C_2 x^2}{2} + C_3 x + C_4\right)$

(f) δ θ_A

Equations: Constants evaluated

$$V = w(-x + l/2)$$

$$M = w\left(\dfrac{-x^2}{2} + \dfrac{l}{2}x\right)$$

$$\theta = \dfrac{w}{EI}\left(\dfrac{-x^3}{6} + \dfrac{l}{4}x^2 - \dfrac{l^3}{24}\right)$$

$$\delta = \dfrac{w}{EI}\left(\dfrac{-x^4}{24} + \dfrac{l}{12}x^3 - \dfrac{l^3}{24}x\right)$$

Substituting the values for these constants in the respective equations directly above the diagrams which they represent, we obtain the series of equations underneath the diagrams.

To determine the maximum deflection (in our case obviously at midspan), we merely substitute $x = l/2$ in the δ equation and get

$$\delta_{max} = \delta_{x=l/2} = -\frac{5wl^4}{384EI} \qquad \text{(negative indicates the deflection to be below the reference line)}$$

Similarly, the value of θ at the left support is $\theta_A = C_3/EI$; that is,

$$\theta_A = -\frac{wl^3}{24EI}$$

Example 4—2:

Given: A cantilever beam, l ft long, loaded as shown in Figure a. EI is constant.

Find: (a) The V, M, θ, and δ diagram and corresponding equations, (b) the maximum deflection, and (c) the maximum slope of the elastic curve.

Procedure: In order to satisfy equilibrium, the reaction must provide a vertical force equal and opposite to the load, $w_0l/2$, and a counterclockwise moment to counteract the rotational tendency. These reactions and loads, plotted according to our sign convention (positive above and negative below the reference axis) are shown in Figure b. Subsequently, we proceed to construct the V, M, θ, and δ diagrams.

The constants of integration are determined from the boundary conditions. (See Figures b, c, d, e, and f.)

At $x = 0$,

$$V = + w_0l/2; \text{ hence, } C_1 = + w_0l/2$$

$$M = - w_0l^2/3; \text{ hence, } C_2 = - w_0l^2/3$$

$$\theta = 0; \text{ hence, } C_3 = 0$$

$$\delta = 0; \text{ hence, } C_4 = 0$$

The maximum deflection is obviously at the free end, where $x = l$. Substituting $x = l$ into the δ equation we get

$$\delta_{max} = \frac{1}{EI}\left(-\frac{w_0l^4}{120} + \frac{w_0l^4}{12} - \frac{w_0l^4}{6}\right) = -\frac{11\,w_0l^4}{120EI}$$

Similarly, the maximum slope of the elastic line (θ_{max}) also occurs at the free end. When $x = l$ in the θ equation, we have

$$\theta_{max} = \frac{1}{EI}\left(-\frac{w_0l^3}{24} + \frac{w_0l^3}{4} - \frac{w_0l^3}{3}\right) = -\frac{w_0l^3}{8EI}$$

(a)

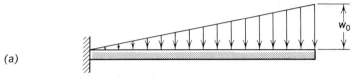

(b) L $\dfrac{w_0 l^2}{3}$ $\dfrac{w_0 l}{2}$ $L = -\dfrac{w_0}{l} x$

(c) V $\dfrac{w_0 l}{2}$ $V = \dfrac{-w_0}{2l} x^2 + C_1$

(d) M $\dfrac{w_0 l^2}{3}$ $M = \dfrac{-w_0}{6l} x^3 + C_1 x + C_2$

(e) θ $\theta = \dfrac{1}{EI}\left(\dfrac{-w_0}{24l} x^4 + \dfrac{C_1}{2} x^2 + C_2 x + C_3\right)$

(f) δ $\delta = \dfrac{1}{EI}\left(\dfrac{-w_0}{120l} x^5 + \dfrac{C_1}{6} x^3 + \dfrac{C_2}{2} x^2 + C_3 x + C_4\right)$

Equations: Constants evaluated

$$L = -\frac{w_0}{l}\, x$$

$$V = -\frac{w_0}{l}\frac{x^2}{2} + \frac{w_0 l}{2}$$

$$M = -\frac{w_0}{l}\frac{x^3}{6} + \frac{w_0 l}{2} x - \frac{w_0 l^2}{3}$$

$$\theta = \frac{w_0}{EI}\left(\frac{-x^4}{24l} + \frac{l x^2}{4} - \frac{l^2}{3} x\right)$$

$$\delta = \frac{w_0}{EI}\left(\frac{-x^5}{120l} + \frac{l x^3}{12} - \frac{l^2 x^2}{6}\right)$$

Example 4–3:

Given: The cantilever beam shown in Figure a, loaded with a con-centrated load P at the free end. The left half of the beam has an I twice that on the right half.

Find: (a) The V, M, θ, and δ diagrams and equations, (b) the slope of the elastic line at $l/2$ and at the free end (in terms of EI), and (c) the deflection of the elastic line at $l/2$ and at the free end (in terms of EI).

Procedure: Note that although the M curve is continuous, M/EI is not. Hence, a separate θ and δ equation is required for each half-section. θ_1 and δ_1 represent the equations of the respective curves for the left half and θ_2 and δ_2 for the right half. Because the elastic curve is a smooth continuous curve, $\theta_1 = \theta_2$ and $\delta_1 = \delta_2$ at $x = l/2$, a con-dition which provides the means for evaluating the constants C_3 and C_5 (see Figures d, e, f, and g):

At $x = 0$,

$$M = -Pl; \text{ hence, } C_1 = -Pl$$
$$\theta_1 = 0; \text{ hence, } C_2 = 0$$
$$\delta_1 = 0; \text{ hence, } C_4 = 0$$

At $x = l/2$,

$$\theta_1 = \theta_2; \text{ hence, } C_3 = -(3/16)Pl^2$$
$$\delta_1 = \delta_2; \text{ hence, } C_5 = -(5/96)Pl^3$$

The equations with the constants evaluated are shown following the diagrams on the next page. Now, to determine θ and δ at $l/2$ and at l we shall use the equations derived above, substituting for x an $l/2$,

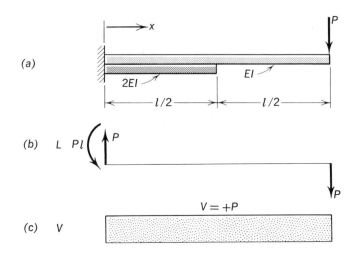

(a)

2EI   EI

l/2   l/2

(b)   L   Pl   P

V = +P

(c)   V

(d) M

$$M = Px + C_1$$

(e) $\frac{M}{EI}$

$$M_1 = \frac{P}{2}x + \frac{C_1}{2}$$

$$M_2 = P\left(x - \frac{l}{2}\right) + \frac{C_1}{2}$$

(f) θ

$$\theta_1 = \frac{1}{EI}\left(\frac{Px^2}{4} + \frac{C_1}{2}x + C_2\right)$$

$$\theta_2 = \frac{1}{EI}\left[\frac{P}{2}\left(x - \frac{l}{2}\right)^2 + \frac{C_1}{2}\left(x - \frac{l}{2}\right) + C_3\right]$$

(g) δ

$$\delta_1 = \frac{1}{EI}\left(\frac{Px^3}{12} + \frac{C_1}{4}x^2 + C_2 x + C_4\right)$$

$$\delta_2 = \frac{1}{EI}\left[\frac{P}{6}\left(x - \frac{l}{2}\right)^3 + \frac{C_1}{4}\left(x - \frac{l}{2}\right)^2 + C_3\left(x - \frac{l}{2}\right) + C_5\right]$$

Equations: Constants evaluated

$$V = +P$$

$$M = Px - Pl$$

$$M_1 = \frac{P}{2}x - \frac{Pl}{2}$$

$$M_2 = P\left(x - \frac{l}{2}\right) - \frac{Pl}{2}$$

$$\theta_1 = \frac{P}{EI}\left(\frac{x^2}{4} - \frac{l}{2}x\right)$$

$$\theta_2 = \frac{P}{EI}\left[\frac{(x - l/2)^2}{2} - \frac{l}{2}\left(x - \frac{l}{2}\right) - \frac{3}{16}l^2\right]$$

$$\delta_1 = \frac{P}{EI}\left(\frac{x^3}{12} - \frac{l}{4}x^2\right)$$

$$\delta_2 = \frac{P}{EI}\left[\frac{1}{6}\left(x - \frac{l}{2}\right)^3 - \frac{l}{4}\left(x - \frac{l}{2}\right)^2 - \frac{3}{16}l^2\left(x - \frac{l}{2}\right) - \frac{5Pl^3}{96}\right]$$

and *l*, in the appropriate equations:

$$\theta_{x=l/2} = \frac{1}{EI}\left(\frac{Pl^2}{16} - \frac{Pl^2}{4}\right) = -\frac{3}{16}\frac{Pl^2}{EI} = C_3$$

As a check, note that the area under the M/EI diagram between $x = 0$ and $x = l/2$ gives $\theta_{x=l/2} - \theta_{x=0}$, and because $\theta = 0$ at $x = 0$

$$\theta_{x=l/2} = \left(\Sigma\frac{M}{EI}\right) = -\frac{1}{EI}\left(\frac{Pl}{2} + \frac{Pl}{4}\right)\frac{l}{4} = \left(-\frac{3}{16}\right)\left(\frac{Pl^2}{EI}\right)$$

Similarly, the total area under the M/EI diagram gives θ at free end; that is,

$$\theta_{x=l} = -\left(\Sigma\frac{M}{EI}\right) = -\frac{1}{EI}\left[\left(\frac{3}{16}Pl^2 + \frac{Pl}{2}\times\frac{l}{2}\times\frac{1}{2}\right)\right] = -\frac{5}{16}Pl^2/EI$$

or evaluating θ_2 at $x = l$, from the θ_2 equation we get

$$\theta_{x=l} = \frac{1}{EI}\left[\frac{P}{2}\left(l - \frac{l}{2}\right)^2 - \frac{Pl}{2}\left(l - \frac{l}{2}\right) - \frac{3Pl^2}{16}\right] = -\frac{5}{16}Pl^2/EI.$$

From the δ equations, using δ_1 evaluated at $x = l/2$

$$\delta_{x=l/2} = \frac{1}{EI}\left(\frac{Pl^3}{96} - \frac{Pl^3}{16}\right) = -\frac{5Pl^3}{96EI}$$

and from the δ_2 equation, evaluated at $x = l$, we obtain the maximum δ_l at the free end.

$$\delta_{\text{max}} = \frac{1}{EI}\left[\frac{P}{6}\left(\frac{l}{2}\right)^3 - \frac{Pl}{4}\left(\frac{l}{2}\right)^2 - \frac{3}{16}Pl^2\left(\frac{l}{2}\right) - \frac{5Pl^3}{96}\right] = -\frac{3Pl^3}{16EI}$$

Example 4–4:

Given: A simply supported steel beam 16 ft long, loaded as shown in Figure *a*. Assume that EI is uniform over the length of the beam.

Find: An economical wide-flange section such that the maximum deflection of the beam is about equal to 1/240 of the span.

Procedure: The sketching of the V, M, θ, and δ may facilitate the writing of equations, especially for evaluating the constants of integration. We shall work the problem in terms of general quantities w, l, E, and I, then evaluate these quantities for the final answer.

The maximum deflection occurs where the slope of the elastic curve is zero. Setting $\theta = 0$ and solving for x, we get

$$\theta = 0 = \frac{w_0}{EI}\left(\frac{lx^2}{12} - \frac{x^4}{24l} - \frac{7}{360}l^3\right) = \frac{w_0}{12EI}\left(lx^2 - \frac{x^4}{2l} - \frac{7}{30}l^3\right)$$

Simplifying and rearranging, we have

$$x^4 - 2l^2x^2 + \frac{14}{30}l^4 = 0$$

$$x^2 = l^2 \pm \frac{\sqrt{4l^4 - (28/15)l^4}}{2} = (l^2 \pm l^2\sqrt{8/15})$$

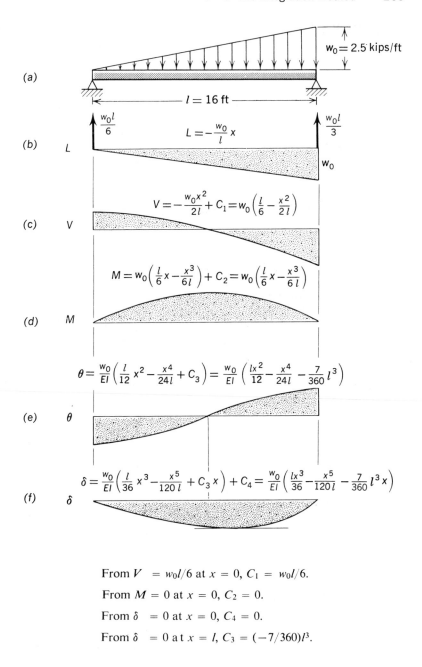

(a)

(b) L

(c) V

(d) M

(e) θ

(f) δ

$w_0 = 2.5$ kips/ft

$l = 16$ ft

$\dfrac{w_0 l}{6}$ $L = -\dfrac{w_0}{l}\, x$ $\dfrac{w_0 l}{3}$

w_0

$V = -\dfrac{w_0 x^2}{2l} + C_1 = w_0\left(\dfrac{l}{6} - \dfrac{x^2}{2l}\right)$

$M = w_0\left(\dfrac{l}{6}x - \dfrac{x^3}{6l}\right) + C_2 = w_0\left(\dfrac{l}{6}x - \dfrac{x^3}{6l}\right)$

$\theta = \dfrac{w_0}{EI}\left(\dfrac{l}{12}x^2 - \dfrac{x^4}{24l} + C_3\right) = \dfrac{w_0}{EI}\left(\dfrac{lx^2}{12} - \dfrac{x^4}{24l} - \dfrac{7}{360}l^3\right)$

$\delta = \dfrac{w_0}{EI}\left(\dfrac{l}{36}x^3 - \dfrac{x^5}{120\,l} + C_3 x\right) + C_4 = \dfrac{w_0}{EI}\left(\dfrac{lx^3}{36} - \dfrac{x^5}{120\,l} - \dfrac{7}{360}l^3 x\right)$

From $V = w_0 l/6$ at $x = 0$, $C_1 = w_0 l/6$.

From $M = 0$ at $x = 0$, $C_2 = 0$.

From $\delta = 0$ at $x = 0$, $C_4 = 0$.

From $\delta = 0$ at $x = l$, $C_3 = (-7/360)l^3$.

For $\delta_{\max}$: $x = 0.519l$. Substituting this in the δ equation gives

$$\delta_{\max} = \dfrac{0.0065 w_0 l^4}{EI}$$

Correcting for units gives

$$\delta_{max} = \frac{0.0065w_0l^4}{EI} \times 1728 \text{ in.}^3/\text{ft}^3$$

Now we evaluate

$$\delta_{max} = \frac{1}{240} \times l = \frac{0.0065 \times 1728 \times 2500 \times l^3 \cdot l}{30 \times 10^6 I}$$

Solving for I, we get

$$I = \frac{0.0065 \times 1728 \times 2500 \times (16)^3 \times 240}{30 \times 10^6} = 76.6 \text{ in.}^4$$

From Appendix B, we note that an 8 in. WF 24 lb/ft section will do; $I = 82.5 > 76.6$; therefore, our solution is satisfactory.

PROBLEMS

For the following problems assume EI = constant along the length of the member, unless specified otherwise. Assume the origin for x to be at the extreme left of the beam when writing the equations for the deflection curve. V, M, θ, and δ diagrams will prove helpful in solving the problems.

4-1 Derive the deflection equation and determine the maximum deflection for the member in Figure P 4-1 in terms of EI, P, and l.

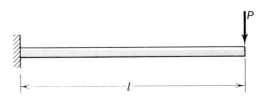

Figure P 4-1

4-2 Derive the deflection equation and determine the maximum deflection for the member in Figure P 4-2 in terms of EI, P, a, and l.

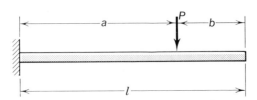

Figure P 4-2

4–3 Derive the deflection equation and determine the maximum deflection for the member in Figure P 4–3 in terms of EI, w, and l.

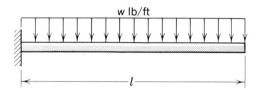

w lb/ft

l

Figure P 4–3

4–4 Derive the deflection equation and determine the maximum deflection for the member in Figure P 4–4 in terms of EI, w_0, and l.

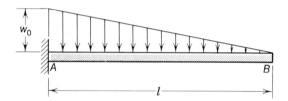

w_0

A

B

l

Figure P 4–4

4–5 Derive the deflection equation and determine the maximum deflection for the beam in Figure P 4–5 in terms of W, EI, and l.

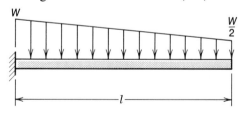

W

$\dfrac{W}{2}$

l

Figure P 4–5

4–6 Derive the deflection equation and determine the maximum deflection for the beam in Figure P 4–6 in terms of W, EI, and l.

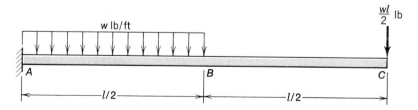

$\dfrac{wl}{2}$ lb

w lb/ft

A

B

C

$l/2$

$l/2$

Figure P 4–6

4–7 Derive the deflection equation and determine the maximum deflection for the beam in Figure P 4–7 in terms of EI, M, and l.

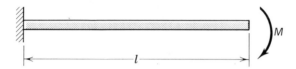

Figure P 4–7

4–8 Derive the deflection equation and determine the deflection at points B, C, and D. Assume $EI = 130 \times 10^7$ lb-in.^{2}

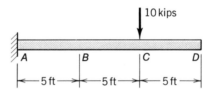

Figure P 4–8

4–9 Derive the deflection equation and determine the maximum deflection for the beam in Figure P 4–9 in terms of EI, P, and l.

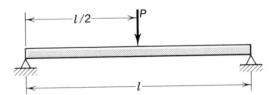

Figure P 4–9

4–10 Derive the deflection equation and determine the maximum deflection for the beam in Figure P 4–10 in terms of EI, M, and l.

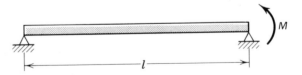

Figure P 4–10

4–11 Derive the deflection equation and determine the maximum deflection for the beam in Figure P 4–11 in terms of EI, w_B, and l.

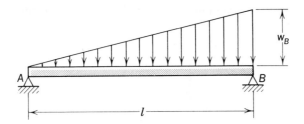

Figure P 4–11

4–12 Determine the maximum deflection, in terms of *EI*, for the beam in Figure P 4–12.

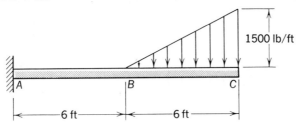

1500 lb/ft

A   B   C

6 ft   6 ft

Figure P 4–12

4–13 Derive the deflection equations for the beam in Figure P 4–13 and determine the deflection at point *D*, in terms of *P*, *EI*, and *l*.

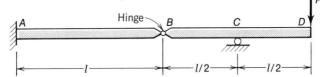

Hinge   B   C   D   P

A

l   *l*/2   *l*/2

Figure P 4–13

4–14 Determine the maximum deflection, in terms of *EI*, for the beam in Figure P 4–14.

5 kips

1 kip/ft

0.5 kip/ft

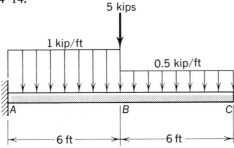

A   B   C

6 ft   6 ft

Figure P 4–14

4–15 Determine the maximum deflection, in terms of EI, for the beam in Figure P 4–15.

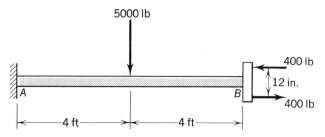

Figure P 4–15

4–16 Determine the maximum deflection, in terms of EI, for the beam in Figure P 4–16.

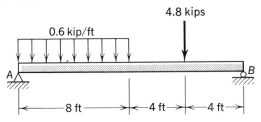

Figure P 4–16

4–17 Determine the maximum deflection, in terms of EI, for the beam in Figure P 4–17.

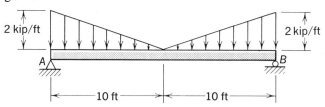

Figure P 4–17

4–18 Determine the maximum deflection, in terms of EI, for the beam in Figure P 4–18.

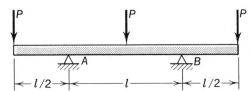

Figure P 4–18

4–19 Determine the maximum deflection, in terms of *EI*, for the beam in Figure P 4–19.

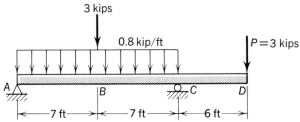

Figure P 4–19

4–20 Determine the maximum deflection, in terms of *EI*, for the beam in Figure P 4–20.

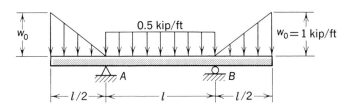

Figure P 4–20

4–21 Determine the maximum deflection, in terms of *EI*, for the beam in Figure P 4–21.

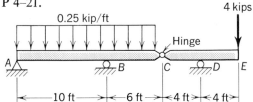

Figure P 4–21

4–22 Determine the minimum moment of inertia that the beam in Figure P 4–14 must have if the maximum deflection is not to exceed 0.4 in. Assume $E = 30 \times 10^6$ psi.

4–23 Determine the minimum moment of inertia that the beam in Figure P 4–15 must have if the maximum deflection is not to exceed 0.5 in. Assume $E = 30 \times 10^6$ psi.

4–24 Determine the minimum moment of inertia that the beam in Figure P 4–16 must have if the maximum deflection is not to exceed 0.8 in. Assume $E = 30 \times 10^6$ psi.

4–25 Determine the minimum moment of inertia that the beam in Figure P 4–17 must have if the maximum deflection is not to exceed 0.6 in. Assume $E = 30 \times 10^6$ psi.

4–26 Determine the minimum moment of inertia that the beam in Figure P 4–19 must have if the maximum deflection is not to exceed 0.75 in. Assume $E = 30 \times 10^6$ psi.

4–27 Determine the minimum moment of inertia that the beam in Figure P 4–21 must have if the maximum deflection is not to exceed 1 in. Assume $E = 30 \times 10^6$ psi.

4–28 Determine the maximum deflection in terms of W, EI, and l for the beam in Figure P 4–28.

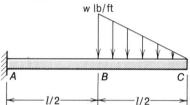

Figure P 4–28

4–29 Determine the slope and deflection in terms of EI, w, and l, at the overhanging end of the beam in Figure P 4–29.

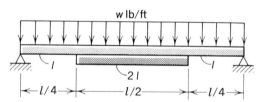

Figure P 4–29

4–30 Determine the maximum deflection in terms of EI, w, and l, for the beam in Figure P 4–30.

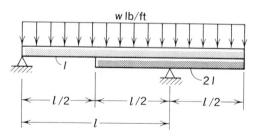

Figure P 4–30

4-31 Determine the slope and deflection in terms of EI, P, and l, at the overhanging end of the beam in Figure P 4-31.

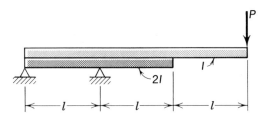

Figure P 4-31

4—4 The Moment-Area Method

The integration method becomes rather cumbersome and laborious when the deflection is caused by concentrated rather than by distributed loads. This results in discontinuities in the V, M, θ, and δ diagrams. Similar discontinuities, and consequently similar inconveniences, are encountered when the moment of inertia of the beam varies along the length of the beam. The *moment-area method* is a semigraphical, much faster method for determining deflections in such cases. It was developed by Professor Charles E. Green of the University of Michigan in 1873. This method relates the M/EI diagram to the deflection and to the slopes of the elastic curve, giving a physical depiction of the approach and establishing a strong basis for developing a tool to solve many structural problems, including rigid frames and indeterminate beams.

Let us re-examine the meaning of Equation (4–5), slightly modified here as Equation (4–8), and relate its physical significance to Figure 4–3. Integrating the moment diagram between points A and B, we get the angle between the tangents to the elastic curve at A and B. Thus, we have

$$\theta_{AB} = \theta_B - \theta_A = \int_A^B \frac{M}{EI} dx \qquad (4\text{–}8)$$

The physical interpretation of Equation (4–8) may be given in Figure 4–3. Line AB represents a segment of the elastic curve, and correspondingly we have the M/EI diagram for the section of a straight beam between points A_0 and B_0. It is important to realize that the deflection has been greatly exaggerated (for convenience). Actually, the angles are quite *small*, such that the angle is approximately equal to its tangent and sine. Hence, just as the change in slope $d\theta$ between the tangents at points C and D on the elastic line is equal to the area under the M/EI diagram between points C and D, we note that the change in slope between the tangents at points A and B, much farther apart on the elastic curve than C and D, is equal to

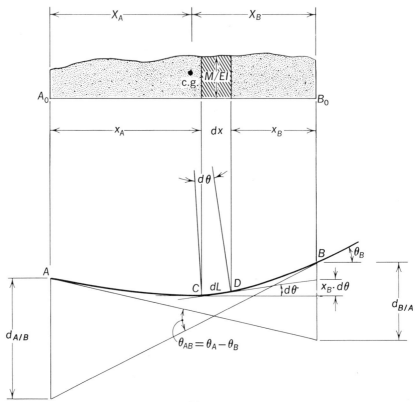

Figure 4–3

the area under the M/EI curve bounded by A and B. This leads us to the *first moment-area theorem:*

> THEOREM 1. The angle between tangents to the elastic curve at points A and B, on the elastic curve, is equal to the area under the M/EI diagram between points A and B.

Let us now consider the relationship between deflection and tangential deviation — defined as the vertical distance (perpendicular to the beam) between a certain point on the deflection curve and the tangent line to the deflection curve at some other point (distances $d_{A/B}$ and $d_{B/A}$ in Figure 4–3 are tangential deviations). In Figure 4–3 tangent lines are drawn to points C and D on the deflection curve. Relative to point B, the curvature of length dL (distance CD) contributes an increment of deviation $x_B d\theta$. Thus, for small angles, and assuming a positive direction of x_B, the distance $d_{B/A}$ may be approximated by

$$d_{B/A} = \int_B^A x_B d\theta = \int_B^A (M/EI)\, x_B dx_B \qquad (4\text{–}9)$$

Note that in the notation $d_{B/A}$, point B represents the point on curve where the deviation is sought, and A represents the point of tangency; the reverse is true for $d_{A/B}$.

In a similar manner the deviation $d_{A/B}$ may be obtained:

$$d_{A/B} = \int_A^B x_A d\theta = \int_A^B (M/EI) x_A dx_A \qquad (4\text{--}10)$$

In Equations (4–9) and (4–10) the term $(M/EI \cdot dx)$ = increment of area under the M/EI diagram. Thus, $(M/EI \cdot dx) \cdot x$ represents the *first moment* (statical moment) with respect to B, in case of Equation (4–9), and with respect to A in the case of Equation (4–10), of the total area under the M/EI curve between points A and B. Hence, we state the *second moment-area theorem* as follows:

> THEOREM 2. The tangential deviation of a point B located on the elastic curve to the tangent at another point A is equal to the first moment, with respect to B, of the area under the M/EI diagram between points A and B.

Caution must be exercised on the sign convention related to negative and positive moments, such as shown in Figures 4–4a and 4–4b. For a positive moment the tangential deviation is considered positive, and point B is above the tangent drawn at A (Figure 4–4a). For a negative moment, the tangential deviation is negative, and point B is *below* the tangent drawn at A (Figure 4–4b). Hence, if the moment changes from positive to negative between points A and B, the deviation with respect to B is merely the difference between the first moments with respect to B of the positive and negative segments of the M/EI diagram. Also, for such a case, the angle between the tangents at A and B is the *difference* (*net*) of positive and negative areas under the M/EI diagram.

By now we have perhaps concluded that neither of the above theorems gives us, directly, the slope or deflection of a point on the beam. This conclusion is generally correct except for special cases, such as cantilever beams or special symmetrically loaded simple beams. However, by suitable

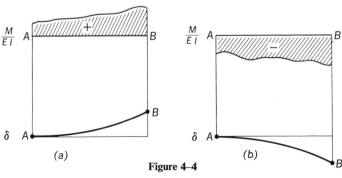

(a)   (b)

Figure 4–4

manipulations and geometrical considerations, the theorems provide a powerful analysis tool. This may become evident after one studies the examples following this discussion.

Lastly, one must realize that not all problems are most easily solved by this method. For some problems this method becomes somewhat tedious, especially when the area under the M/EI diagram and its centroid are not easily determined. Fortunately most problems we deal with either have M/EI diagrams which are simple rectangles, triangles or parabolas, or geometric shapes which may be resolved into rectangles, triangles or parabolas.[2]

Example 4–5:

Given: The cantilever beam shown in Figure *a*. (It is the same beam that was analyzed in Example 4–3 by the integration method.)

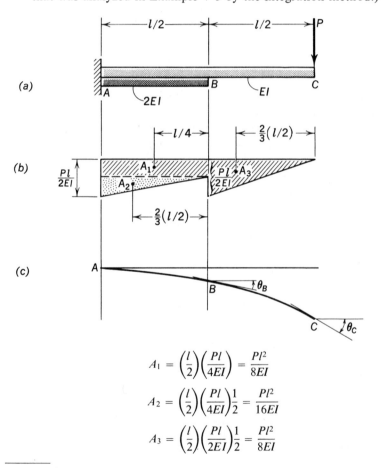

$$A_1 = \left(\frac{l}{2}\right)\left(\frac{Pl}{4EI}\right) = \frac{Pl^2}{8EI}$$

$$A_2 = \left(\frac{l}{2}\right)\left(\frac{Pl}{4EI}\right)\frac{1}{2} = \frac{Pl^2}{16EI}$$

$$A_3 = \left(\frac{l}{2}\right)\left(\frac{Pl}{2EI}\right)\frac{1}{2} = \frac{Pl^2}{8EI}$$

---

[2] For the properties of various geometrical shapes see Table B–1 in Appendix B.

Find: (*a*) The slope of the elastic line at *B* and at *C* in terms of *E*, *I*, *P*, and *l*; (*b*) the deflection of the beam at *B* and at *C* in terms of *E*, *I*, *P*, and *l*.

Procedure: The area under the *M/EI* diagram is conveniently divided into composite areas (rectangle and triangles) for which the areas and centroids may be readily determined, as shown in Figure *b*. Because the beam has zero slope at the support, the tangent to the elastic line at *A* is horizontal. Thus, the slope of the elastic line at a point is merely the area under the *M/EI* diagram between that point and point *A* (Theorem 1), and the deflection at a point is equal to the static moment, about that point, of the area under the *M/EI* diagram between point *A* and that specific point. (The negative slope and deflection are readily evident from Figure *c*.) Thus,

At *B*,

$$\theta_B = -(A_1 + A_2) = -\left(\frac{Pl^2}{8EI} + \frac{Pl^2}{16EI}\right) = -\frac{3Pl^2}{16EI}$$

$$\delta_B = -\left[\left(\frac{Pl^2}{8EI} \times \frac{l}{4}\right) + \left(\frac{Pl^2}{16EI} \times \frac{l}{3}\right)\right] = -\frac{5Pl^3}{96EI}$$

At *C*,

$$\theta_C = -(A_1 + A_2 + A_3) = -\left(\frac{Pl^2}{8EI} + \frac{Pl^2}{16EI} + \frac{Pl^2}{8EI}\right) = -\frac{5Pl^2}{16EI}$$

$$\delta_C = -\left[\frac{Pl^2}{8EI}\left(\frac{l}{2} + \frac{l}{4}\right) + \frac{Pl^2}{16EI}\left(\frac{l}{2} + \frac{l}{3}\right) + \frac{Pl^2}{8EI}\left(\frac{l}{3}\right)\right] = -\frac{3Pl^3}{16EI}$$

Example 4—6:

Given: The cantilever beam in Figure *a*. Assume *EI* is constant for the length of the beam.

Find: The deflection and slopes of the beam at points *B* and *C*, in terms of *w*, *a*, *b*, and *EI*.

Procedure: The moment diagram is parabolic in shape. The area under the curve and the centroid of the area may be obtained from Table B–1, Appendix B. Because the tangent at *A* to the elastic curve is horizontal, the slope of θ_B is the area under the *M* diagram between *A* and *B*, divided by *EI*. Thus,

$$\theta_B = -wa^3/6EI = \theta_C$$

because portion *B–C* does not bend and, from Theorem 2,

$$\delta_B = -\frac{wa^3}{6EI}\left(\frac{3}{4}a\right) = -\frac{wa^4}{8EI}$$

and

$$\delta_C = -\frac{wa^3}{6EI}\left(\frac{3}{4}a + b\right)$$

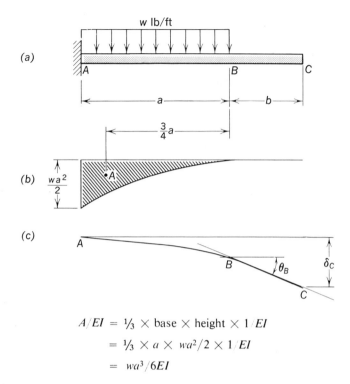

$$A/EI = \tfrac{1}{3} \times \text{base} \times \text{height} \times 1/EI$$
$$= \tfrac{1}{3} \times a \times wa^2/2 \times 1/EI$$
$$= wa^3/6EI$$

Example 4–7:

Given: The beam shown in Figure *a*. *EI* is constant for the whole length of the beam.

Find: The deflections of the beam at *B* and *D* in terms of *w*, *l*, and *EI*.

Procedure: First let us draw the tangent line to point *A*. Then, the tangential deviation at *C*, is, by virtue of Theorem 2,

$$d_C = -\frac{A_1}{EI} \times \frac{l}{3} = -\frac{wl^3}{16EI} \times \frac{l}{3} = -\frac{wl^4}{48EI}$$

Point *C* is below the tangent drawn at *A* (Figure 4–7); hence, the negative sign. Similarly,

$$d_D = -\frac{1}{EI}\left[A_1\left(\frac{l}{2}+\frac{l}{3}\right) + A_2 \times \frac{3}{8}l\right]$$

$$d_D = -\frac{1}{EI}\left(\frac{wl^3}{16} \times \frac{5}{6}l + \frac{wl^3}{48} \times \frac{3}{8}l\right) = -\frac{23wl^4}{384EI}$$

Therefore,

$$\delta_D = d_D - \frac{3}{2}d_C = \frac{11wl^4}{384EI}$$

(below the reference axis, see Figure *c*.)

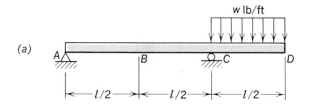

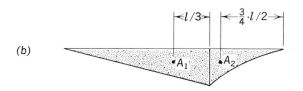

$$A_1 = \frac{1}{2}\left(\frac{wl^2}{8}\right)l = \frac{wl^3}{16}$$

$$A_2 = \frac{1}{3} \times \frac{l}{2} \times \frac{wl^2}{8} = \frac{wl^3}{48}$$

At B:

$$d_B = -\text{(area between } A \text{ and } B)\cdot(1/3)(l/2)$$

$$d_B = -\frac{1}{EI}\left(\frac{wl^2}{16} \times \frac{l}{2} \times \frac{1}{2}\right)\frac{l}{6} = -\frac{wl^4}{384EI}$$

$$\delta_B = \frac{1}{2}d_C - d_B = \frac{wl^4}{96EI} - \frac{wl^4}{384EI} = \frac{wl^4}{128EI}$$

Example 4—8:

Given: The beam in Figure *a* of the previous example.
Find: (a) The angle θ_A between the tangent at A and the horizontal line, and (b) the maximum upward deflection of the beam between supports.
Procedure: The moment and deflection diagrams for this beam are shown in Figures *a* and *b*. From Figure *b*, we note that

$$\theta_A = \frac{d_C}{l} = \frac{wl^4/48EI}{l} = \frac{wl^3}{48EI}$$

At the point of maximum deflection, the tangent line to the elastic

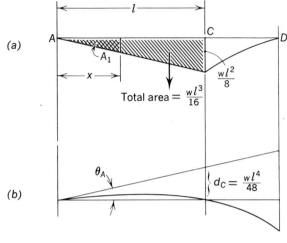

(a) M diagram.   (b) δ diagram.

curve is horizontal. From Theorem 1, θ_A, from above, equals the area under the M/EI diagram, which in our case may be expressed in terms of x. From the above observation,

$$\theta_A = \frac{wl^3}{48EI}$$

The area A, under the M/EI diagram between point A and that at maximum deflection is

$$A_1 = \frac{1}{EI}\left(\frac{wl}{8}\cdot x\right)\cdot x\cdot\frac{1}{2} = \frac{wlx^2}{16EI}$$

Equating θ_A to A_1

$$\frac{wl^3}{48EI} = \frac{wlx^2}{16EI}$$

$$x = \sqrt{\frac{l^2}{3}} = \frac{l}{\sqrt{3}}$$

Thus, the area $A_1 = wl^3/48EI$, and it has its centroid at $(\frac{2}{3})x$ from point A. If we determined the static moment of this area, about A, the result (by virtue of Theorem 2) is

$$\delta_{\max} = \frac{wl^3}{48EI}\frac{2}{3}\cdot x = \frac{wl^4}{72\sqrt{3EI}}$$

PROBLEMS

Use the moment-area method to solve the following problems. The slopes of the elastic line and deflections of the beams at a point will be represented by θ and δ, respectively, with the appropriate subscript representing the point in question.

4–32 Determine θ_B and δ_B of the beam in Figure P 4–32 in terms of P, l, and EI.

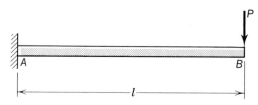

Figure P 4–32

4–33 Determine θ_B and θ_C, and δ_B and δ_C of the beam in Figure P 4–33 in terms of P, a, b, and EI.

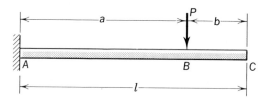

Figure P 4–33

4–34 Determine θ_B and δ_B of the beam in Figure P 4–34 in terms of w, l, and EI.

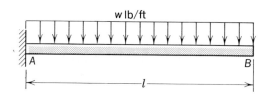

Figure P 4–34

4–35 Determine θ_B and θ_C, δ_B, and δ_C of the beam in Figure P 4–35 in terms of w_0, l, and EI. (Point C is located at $l/2$ from the support.)

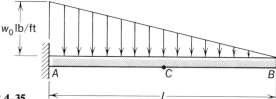

Figure P 4–35

4-36 Determine θ_B and θ_C, and δ_B and δ_C of the beam in Figure P 4-36 in terms of M_B, l, and EI. (Point C is located at $l/2$ from the support.)

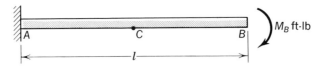

Figure P 4-36

4-37 Determine θ and δ under the load P of the beam in Figure P 4-37 in terms of P, l, and EI.

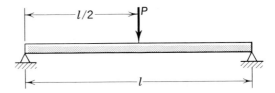

Figure P 4-37

4-38 Determine the maximum deflection of the beam in Figure P 4-38 in terms of M_B, l, and EI. (See Example 4-8.)

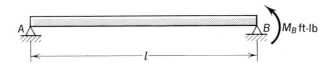

Figure P 4-38

4-39 Determine the maximum deflection of the beam in Figure P 4-39 in terms of w_B, l, and EI. (See Example 4-8.)

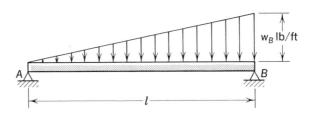

Figure P 4-39

4-40 Determine θ_B and θ_C, and δ_B and δ_C of the beam in Figure P 4-40 in terms of EI.

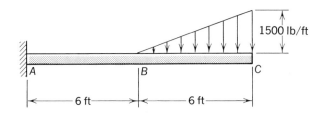

Figure P 4–40

4–41 Determine θ_B and θ_C and δ_B and δ_C of the beam in Figure P 4–41 in terms of EI.

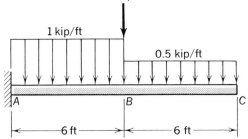

Figure P 4–41

4–42 Determine θ_B and δ_B of the beam in Figure P 4–42 in terms of EI.

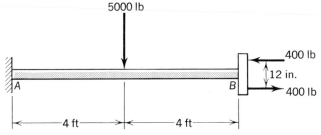

Figure P 4–42

4–43 Determine θ_C and δ_C of the beam in Figure P 4–43 in terms of EI.

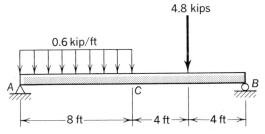

Figure P 4–43

4–44 Determine the maximum deflection of the beam in Figure P 4–44 in terms of *EI*.

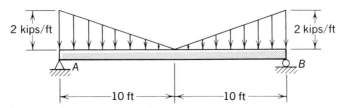

Figure P 4–44

4–45 Determine the deflection of one of the cantilever ends of the beam in Figure P 4–45 in terms of *P*, *l*, and *EI*.

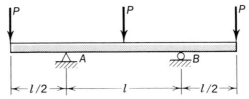

Figure P 4–45

4–46 Determine θ_D and δ_D of the beam in Figure P 4–46 in terms of *EI*.

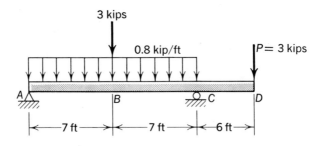

Figure P 4–46

4–47 Determine θ_D and δ_D of the beam in Figure P 4–47 in terms of *w*, *l*, and *EI*.

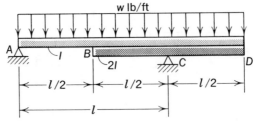

Figure P 4–47

4–48 Determine the deflection of the center of the span of the beam in Figure P 4–48 in terms of w, l, and EI.

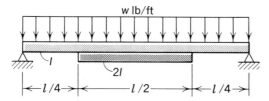

w lb/ft

2l

←l/4→|←l/2→|←l/4→|

Figure P 4–48

4–49 Determine θ_C and θ_D, and δ_C and δ_D of the beam in Figure P 4–49 in terms of P, l, and EI.

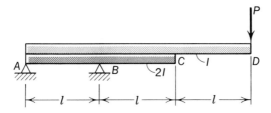

P

A   B 2l C l D

←l→|←l→|←l→|

Figure P 4–49

4–50 Determine θ_A and θ_B for the beam in Figure P 4–49 in terms of P, l, and EI.

4—5 Deflection by Superposition

The principle of superposition, first mentioned in Section 1–10, will be restated here: *If several causes act simultaneously on a system and each effect is directly proportional to its cause, the total effect is the sum of the individual effects when considered separately, provided that Hooke's law is valid within the range of these effects either individually or combined.* As related here, we might think of the causes as the loads applied to the beam (P, w, or couples) and of the effects as the deformations of the beam (slope and deflection). Thus, the total slope and deflection at a point in a beam is the algebraic sum of the slopes and deflections produced by each load, provided we do not exceed the proportional limit of the material. Hence, we have the basic idea for deflection by superposition. The superposition method uses results for slopes and deflections for certain simple loading conditions (some cases are given in Table 4–1) determined by integration, moment-area or other methods. It combines these results to get the total effect.

TABLE 4-1 Beam Deflection Formulas

Type of Load	Slope at Free End	Maximum Deflection	Deflection Equations [3]
1. Cantilever beam	Concentrated load at the free end		
	$\theta = \dfrac{Pl^2}{2\,EI}$	$\delta = \dfrac{Pl^3}{3\,EI}$	$\delta = \dfrac{Px^2}{6\,EI}(3l - x)$
2. Cantilever beam	Concentrated load at any point of the beam		
	$\theta = \dfrac{Pa^2}{2\,EI}$	$\delta = \dfrac{Pa^2}{6\,EI}(3l - a)$	$\delta = \dfrac{Px^2}{6\,EI}(3a - x) \quad \text{for } 0 < x < a$ $\delta = \dfrac{Pa^2}{6\,EI}(3x - a) \quad \text{for } a < x < l$
3. Cantilever beam	Bending moment applied at the free end		
	$\theta = \dfrac{Ml}{EI}$	$\delta = \dfrac{Ml^2}{2\,EI}$	$\delta = \dfrac{Mx^2}{2\,EI}$
4. Cantilever beam	Uniformly distributed load		
	$\theta = \dfrac{wl^3}{6\,EI}$	$\delta = \dfrac{wl^4}{8\,EI}$	$\delta = \dfrac{wx^2}{24\,EI}(x^2 + 6l^2 - 4lx)$
5. Cantilever beam	Uniformly distributed load on only part of beam		
	$\theta = \dfrac{w}{6\,EI}(l^3 - a^3)$	$\delta = \dfrac{w}{24\,EI}$ $(3l^4 - 4a^3 l + a^4)$	$\delta_1 = \dfrac{w\,(l-a)x^2}{12\,EI}\left[3(l+a) - 2x\right]$ $\text{for } 0 < x < a$ $\delta_2 = \dfrac{w}{24\,EI}\left[6(l^2 - a^2)x^2\right.$ $\left. - 4(l-a)x^3 + (x-a)^4\right]$ $\text{for } a < x < l$
6. Cantilever beam	Uniformly decreasing load		
	$\theta = \dfrac{wl^3}{24\,EI}$	$\delta = \dfrac{wl^4}{30\,EI}$	$\delta = \dfrac{wx^2}{120\,l \cdot EI}$ $\times \left[10l^3 - 10l^2 x + 5lx^2 - x^3\right]$

[3] An upward deflection is considered positive while a downward one is negative. (For load an upward acting load is positive and a downward one is negative.)

Type of Load	Slope as Shown	Maximum Deflection	Deflection Equation
7. Cantilever beam		Uniformly increasing load	
w lb per unit length	$\theta = \dfrac{wl^3}{8EI}$	$\delta = \dfrac{11\,wl^4}{120\,EI}$	$\delta = \dfrac{wx^2}{120\,l\cdot EI}(x^3 - 10l^2x + 20l^3)$
8. Simply supported beam		Concentrated load at the center of the span	
	$\theta_L = \theta_R = \dfrac{Pl^2}{16EI}$	$\delta = \dfrac{Pl^3}{48\,EI}$	$\delta = \dfrac{Px}{12\,EI}\left(\dfrac{3}{4}l^2 - x^2\right)$ for $0 < x < \dfrac{l}{2}$
9. Simply supported beam		Concentrated load at any point	
	$\theta_L = \dfrac{Pb\,(l^2 - b^2)}{6\,l\cdot EI}$	$\delta = \dfrac{Pb\,(l^2 - b^2)^{3/2}}{9\sqrt{3}\,l\cdot EI}$	$\delta_1 = \dfrac{Pbx}{6\,l\,EI}(l^2 - x^2 - b^2)$ for $0 < x < a$
	$\theta_R = \dfrac{Pab\,(2l - b)}{6\,l\,EI}$	at $x = \sqrt{\dfrac{l^2 - b^2}{3}}$	$\delta_2 = \dfrac{Pb}{6\,l\,EI}\left[\dfrac{l}{b}(x - a)^3 + (l^2 - b^2)x - x^3\right]$ for $a < x < l$
10. Simply supported beam		Uniformly distributed load	
w lb per unit length	$\theta_L = \theta_R = \dfrac{wl^3}{24\,EI}$	$\delta = \dfrac{5\,wl^4}{384\,EI}$	$\delta = \dfrac{wx}{24\,EI}(l^3 - 2lx^2 + x^3)$
11. Simply supported beam		Uniformly increasing load	
w lb per unit length	$\theta_L = \dfrac{7\,wl^3}{360\,EI}$	$\delta = \dfrac{2.5\,wl^4}{384\,EI}$	$\delta = \dfrac{wx}{360\,EI}(7l^4 - 10l^2x^2 + 3x^4)$
	$\theta_R = \dfrac{8\,wl^3}{360\,EI}$	at $x = 0.519\,l$	
12. Simply supported beam		Load increasing uniformly to center	
w lb per unit length	$\theta_L = \theta_R = \dfrac{5\,wl^3}{192\,EI}$	$\delta = \dfrac{wl^4}{120\,EI}$	$\delta = \dfrac{wx}{960\,l\cdot EI}(5l^2 - 4x^2)^2$ for $0 < x < l/2$

Type of Load	Slope as Shown	Maximum Deflection	Deflection Equation
13. Simply supported beam		Load decreasing uniformly to center	
w lb per unit length	$\theta_L=\theta_R=\dfrac{wl^3}{64\,EI}$	$\delta=\dfrac{3\,wl^4}{640\,EI}$	$\delta=\dfrac{wx}{960\,l\cdot EI}$ $\times(15l^4-40l^2x^2+40lx^3-16x^4)$ for $0<x<l/2$
14. Simply supported beam		Uniform load partially distributed at one end	
w lb per unit length	$\theta_L=\dfrac{w(l-a)^2}{24\,l\cdot EI}$ $(l^2+2al-a^2)$	(when $a>\frac{l}{2}$) $\delta_c=\dfrac{w(l-a)^2}{96\,EI}(l^2+4al-2a^2)$	
	$\theta_R=\dfrac{w}{24\,l\cdot EI}$ $\times(l^2-a^2)^2$	(when $a<\frac{l}{2}$) $\delta_c=\dfrac{w}{384\,EI}(5l^4-12a^2l^2+8a^4)$	
15. Simply supported beam		Bending moment applied at one end	
$\big)M$	$\theta_L=\dfrac{Ml}{6\,EI}$	$\delta=\dfrac{Ml^2}{9\sqrt{3}\,EI}$ at $x=l/\sqrt{3}$	$\delta=\dfrac{Mlx}{6\,EI}\left(1-\dfrac{x^2}{l^2}\right)$
	$\theta_R=\dfrac{Ml}{3\,EI}$		
16. Overhanging beam		Concentrated load at the end of overhang	
P	$\theta_L=\dfrac{Pal}{6\,EI}$	$\delta=\dfrac{Pal^2}{9\sqrt{3}\,EI}$ (between supports)	$\delta_1=\dfrac{Pax}{6\,l\,EI}(l^2-x^2)(0<x<l)$
	$\theta_R=\dfrac{Pal}{3\,EI}$	$\delta=\dfrac{Pa^2}{3\,EI}(a+l)$ (at free end)	$\delta_2=\dfrac{P(x-l)}{6\,EI}$ $\big[(x-l)^2-a(l-3x)\big]$ for $l<x<l+a$
17. Overhanging beam		Uniformly distributed load on overhang	
w lb per unit length	$\theta_L=\dfrac{wla^2}{12\,EI}$	$\delta=\dfrac{wl^2a^2}{32\,EI}$ (between supports)	$\delta_1=\dfrac{wa^2x}{12\,l\cdot EI}(l^2-x^2)\,0<x<l$
	$\theta_R=\dfrac{wla^2}{6\,EI}$	$\delta=\dfrac{wa^3}{24\,EI}(3a+4l)$	$\delta_2=\dfrac{w(x-l)}{24\,EI}\big[2a^2(3\bar{x}-l)$ $+(x-l)^2(x-l-4a)\big]$ for $l<x<l+a$

The following examples attempt to illustrate the concept and procedure of superposition.

Example 4—9:

Given: For the beam shown in Figure *a*, *EI* is constant along its full length. Joints *B* and *C* are assumed rigid; that is, the angles between adjacent members remain right angles.

Find: The vertical deflection of point *B*.

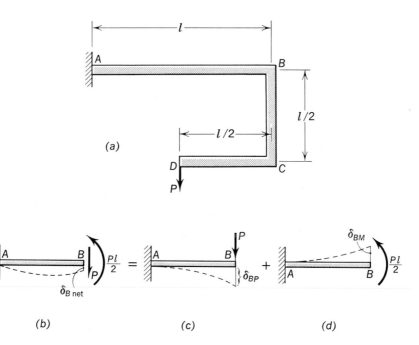

(a)

(b)   (c)   (d)

Procedure: From equilibrium (that is, $\Sigma F_y = 0$ and $\Sigma M_B = 0$), we find that at joint *B* there is a shear force *P* and a bending moment $Pl/2$ acting as shown in the figure. Hence, at point *B* the net deflection is the resultant deflection of the load *P* and the moment $Pl/2$. Assuming $\delta_{B \text{ net}}$ to be downward, as indicated by the dotted line, then we can say

$$\delta_{B \text{ net}} = \delta_{BP} - \delta_{BM}$$

This assumes that $\delta_{BP} > \delta_{BM}$. If our assumption is incorrect, the sign for $\delta_{B \text{ net}}$ will turn out negative. Visualizing the deflections by means of sketches is indeed advisable.

From case 1, Table 4–1, the δ_B caused by *P* is

$$\delta_{BP} = \frac{Pl^3}{3EI} \quad \text{(downward)}$$

From case 3, δ_B caused by M is

$$\delta_{BM} = \frac{Ml^2}{2EI} = \frac{\frac{1}{2}(Pl) \cdot l^2}{2EI} = \frac{Pl^3}{4EI} \quad \text{(upward)}$$

Because $\delta_{BP} > \delta_{BM}$, the sign will be positive, indicating a correct assumption. Thus, we have

$$\delta_{B \text{ net}} = \frac{Pl^3}{3EI} - \frac{Pl^3}{4EI} = \frac{Pl^3}{12 \cdot EI} \quad \text{(downward)}$$

Example 4–10:

Given: The overhanging beam in Figure *a*. Assume that E for the beam is the same as for the rod and that I is uniform along the entire length of the beam.
Find: The vertical deflection of the beam at point C.

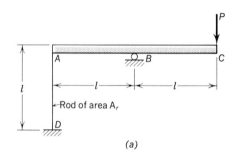

(a)

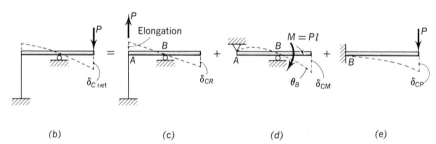

(b) (c) (d) (e)

Procedure: By summing moments about support B we find that the axial force in the rod AD is P. This results in an elongation, as shown in Figure *c*. Thus, we have

$$\delta_{AD} = \frac{Pl}{A_r E} = \delta_{CR} \quad \text{(by proportion)}$$

The overhang BC transmits a moment to segment AB which bends AB and results in the angle θ_B at support B as shown in Figure *d*. Using case 15, Table 4–1, we have

$$\delta_{CM} = \theta_B \times l = \frac{Ml}{3EI} \cdot l = \left(\frac{Pl \cdot l}{3EI}\right) \cdot l = \frac{Pl^3}{3EI}$$

Finally, because of P, the length BC bends resulting in δ_{CP}. Using case 1, Table 4–1, we get

$$\delta_{CP} = \frac{Pl^3}{3EI}$$

Hence,

$$\delta_{C\ net} = \delta_{CR} + \delta_{CM} + \delta_{CP}$$

$$\delta_{C\ net} = \frac{Pl}{E}\left(\frac{1}{A_r} + \frac{l^2}{3I} + \frac{l^2}{3I}\right) \text{(downward)}$$

Example 4—11:

Given: The U frame shown in Figure a. Assume that the EI for the vertical members is the same as for the horizontal beam, and that joints B and C are rigid.
Find: The horizontal travel of point A relative to D.

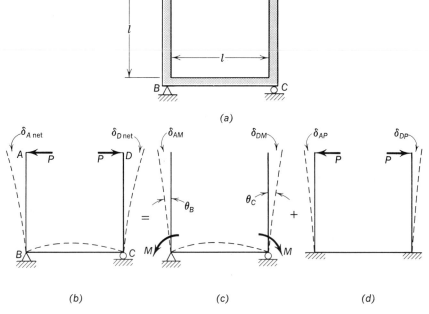

(a)

(b)   (c)   (d)

Procedure: The dashed lines in Figure b show (exaggerated) the deflected shape of the structure. At points B and C, the moment causes bending in beam BC which in turn causes δ_{DM}, as shown in Figure c.

Using case 15, Table 4–1, with a moment applied at each end, we get

$$\theta_C = \frac{Ml}{6EI} + \frac{Ml}{3EI} = \frac{Ml}{2EI} \quad \begin{array}{l}\text{(same moment at}\\ \text{end } B \text{ and } C)\end{array}$$

and

$$\delta_{DM} = \theta_C \cdot l = \frac{Ml^2}{2EI} = \frac{(Pl) \cdot l^2}{2EI} = \frac{Pl^3}{2EI}$$

The cantilever action, Figure d, gives δ_{DP}. Using case 1, Table 4–1, we have

$$\delta_{DP} = \frac{Pl^3}{3EI}$$

Hence,

$$\delta_{D \text{ net}} = \delta_{DP} + \delta_{DM} = \frac{5Pl^3}{6EI}$$

Relative to A, point D moves $2 \times \delta_{D \text{ net}} = \dfrac{5Pl^3}{3EI}$ because,

$$\delta_{D \text{ net}} = \delta_{A \text{ net}}.$$

PROBLEMS

Unless otherwise specified, the EI is to be assumed uniform along the length of the beams.

4–51 Determine the deflection at point C of the cantilever beam in Figure P 4–51 in terms of EI.

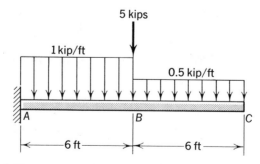

Figure P 4–51

4–52 Determine the deflection at point *B* of the cantilever beam in Figure P 4–52 in terms of *EI*.

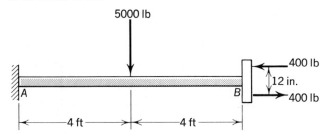

Figure P 4–52

4–53 Determine the deflection at the center of the span of the beam in Figure P 4–53 in terms of *EI*, *M*, *l*, and w_B.

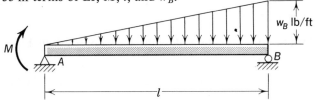

Figure P 4–53

4–54 Determine the deflection at the center of the span of the beam in Figure P 4–54 in terms of *EI*.

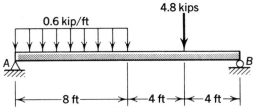

Figure P 4–54

4–55 Determine (*a*) the slope of the elastic curve at support *B*, and (*b*) the deflection at one of the overhanging ends of the beam in Figure P 4–55 in terms of *P*, *l*, and *EI*.

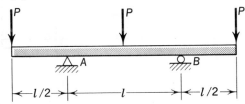

Figure P 4–55

4–56 Determine the deflection at point D of the overhanging beam in Figure P 4–56 in terms of EI.

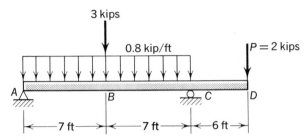

Figure P 4–56

4–57 For the beam in Figure P 4–56, determine the slope of the elastic curve at support C in terms of EI.

4–58 Determine the deflection at point B of the beam in Figure P 4–56 in terms of EI.

4–59 Determine the deflection at the center of the span of the beam in Figure P 4–59, in terms of l and EI.

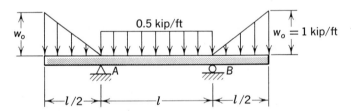

Figure P 4–59

4–60 Determine the deflection of one of the overhanging ends of the beam in Figure P 4–59 in terms of l and EI.

4–61 Determine the deflection at the hinge, point C, of the beam in Figure P 4–61 in terms of EI.

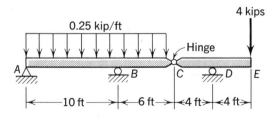

Figure P 4–61

4-62 For the beam in Figure P 4–61, determine the deflection of the over-hanging end, point E, in terms of EI.

4-63 For the beam in Figure P 4–61, determine the slope of the elastic curve at support B in terms of EI.

4-64 For the beam in Figure P 4–61, determine the slopes of beams ABC and CDE, at point C, in terms of EI. Before you calculate, would you say the slopes would be the same or different?

4-65 Determine the deflection at point C of the beam in Figure P 4–65.

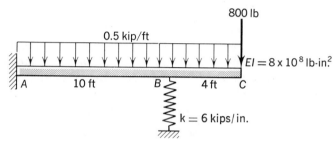

Figure P 4–65

4-66 Determine the compressive force in the spring of Figure P 4–65.

4-67 Determine the horizontal deflection of point C in Figure P 4–67 in terms of I and EI.

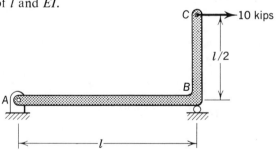

Figure P 4–67

4—6 Deflection due to Shearing Stress

So far, the discussion of deflection neglected the effects of shear forces; it assumed the deflection to be caused only by bending. For most cases the deflection due to shear is negligible, but for beams which are short and subjected to high shear forces, this may not be true. Here we shall consider the deflection of beams produced by shear stresses.

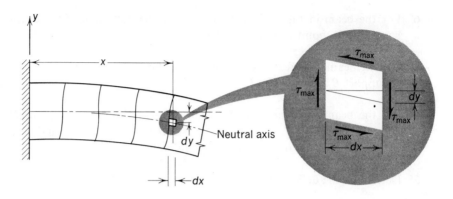

Figure 4–5

For an approximate evaluation of shearing-stress deflection, let us refer to Figure 4–5. An element on the neutral axis, originally rectangular, now changes into a rhombus. Thus, we have an estimate of the slope of the elastic line:

$$\frac{dy}{dx} = 2 = \frac{\tau_{\max}}{G} = \frac{1}{G}\left(\frac{V_x Q}{Ib}\right) = \frac{kV_x}{GA} \tag{4–11}$$

where V_x/A represents the average shear stress, and k is a numerical factor by which the average shear stress must be multiplied in order to give $\tau_{\max}$. For a rectangular cross-section, for example, $k = 3/2$ [see Equation (3–10)].

While this is a common and acceptable approach for most needs, it is not totally correct: (1) it neglects the rotation of the element, and (2) it ignores the varying distribution of shear stresses over the section. Hence, using $\tau_{\max}$ yields a deflection which is slightly larger than the actual. A more accurate $k = 6/5$ may be obtained via the energy approach.

$$\left[\text{Hint: From Example 6–7}, U = \frac{V^2 h^5 bl}{240 GI^2} = \frac{3}{5}\frac{V^2 l}{GA}. \text{ Also, from Equation (6–1)},\right.$$

$$U = \frac{Vy}{2}. \text{ Thus, } y = \frac{6}{5}\frac{V^2 l}{GA}. \text{ From Equation (4–11) } y = \int_0^l \frac{kV_x}{GA}\,dx = \frac{kV_x l}{GA}.$$

$$\left.\text{Hence, } k = \frac{6}{5}\right]$$

Example 4–12:

Given: The simply supported beam loaded as shown in Figure *a*. Assume the ratio $E/G = 2.5$ and neglect the weight of the beam. $k = \frac{3}{2}$. *Find:* (*a*) The maximum deflection of the beam due to shear, and (*b*) the deflection due to both shear and bending for a ratio of $d/l = 1/10$ and $1/20$.

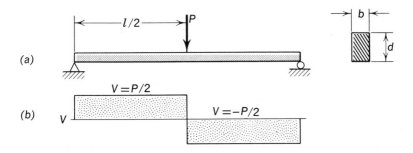

(a)

(b)

Procedure: From Figure b, $V_x = P/2$ for $0 < x < l/2$. Thus, from Equation (4–11)

$$\frac{dy}{dx} = -\frac{kV}{GA} = -\frac{3/2(P/2)}{GA}$$

where $k = \frac{3}{2}$ for rectangular beams. The integral of the above equation with respect to x gives us the expression for deflection, in terms of x,

$$\delta_v = y_v = -\left(\frac{3P}{4GA}x + C\right)$$

When $x = 0$ (support), $y = 0$, and therefore, $C = 0$. When $x = l/2$ the deflection is a maximum, so that

$$\delta_{\text{max } v} = -\frac{3P}{4GA}\left(\frac{l}{2}\right) = -\frac{3}{8}\frac{Pl}{GA}$$

From Table 4–1, case 8, $\delta_{\text{max } m}$ from bending alone is

$$\delta_{\text{max } m} = -\frac{Pl^3}{48EI}, \text{ also at } l/2$$

Thus, the total deflection is

$$\delta_{\text{tot}} = -\frac{Pl}{8}\left(\frac{3}{GA} + \frac{l^2}{6EI}\right)$$

But, $G = (E/2.5)$. Also, $A = I/r^2$, where r = radius of gyration = $d/\sqrt{12}$ for a rectangular section. Thus,

$$\delta_{\text{tot}} = -\frac{Pl^3}{48EI}\left[\frac{7 \cdot 5}{2}\left(\frac{d}{l}\right)^2 + 1\right]$$

For $d/l = 1/10$

$$\delta_{\text{tot}} = -\frac{Pl^3}{48EI} \times (0.0375 + 1)$$

For $d/l = 1/20$

$$\delta_{\text{tot}} = -\frac{Pl^3}{48EI} \times (0.00938 + 1)$$

Hence, we note that the deflection produced by the shear is 3.75 per-cent in the case of $(d/l) = 1/10$, and 0.94 percent for $(d/l) = 1/20$, of the deflection caused by bending alone. In both instances the shear deflection is rather negligible compared to the bending deflection. However, as (d/l) becomes larger, say $d/l = \frac{1}{2}$, the importance of the shear deflection increases.

Example 4–13:

Given: The simply supported beam loaded with a uniform load w, as shown in Figure a. Assume the ratio $E/G = 2.5$ and neglect the weight of the beam $k = \frac{3}{2}$.

Find: (a) The deflection of the beam due to shear. and (b) the deflec-tion due to both shear and bending.

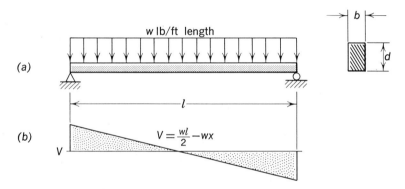

(a)

(b)

$$V = \frac{wl}{2} - wx$$

Procedure: From Figure b, $V_v = wl/2 - wx$. For a rectangular sec-tion, $k = \frac{3}{2}$. Thus, from Equation (4–11) we have

$$\frac{dy}{dx} = -\frac{3/2(wl/2 - wx)}{GA}$$

from which

$$\delta = y = -\int \frac{3/2(wl/2 - wx)}{GA} \cdot dx = -\frac{3/2}{GA}\left(\frac{wl}{2}x - \frac{wx^2}{2}\right) + C$$

When $x = 0$, $\delta = 0$, from which $C = 0$ and the equation for δ due to shear is

$$\delta_v = -\frac{3wx}{4GA}(l - x)$$

The maximum δ_v occurs when $x = l/2$:

$$\delta_{\max v} = -\frac{3wl^2}{16GA}$$

The maximum δ produced by bending alone is found in Table 4–1, case 10, also at the center of the span

$$\delta_{\max\ m} = -\frac{5wl^4}{384EI}$$

Combining the two deflections

$$\delta_{tot} = -\frac{wl^2}{16}\left(\frac{3}{GA} + \frac{5l^2}{24EI}\right)$$

As was assumed, $G = (E/2.5)$; $A = I/r^2 = 12I/d^2$. Thus,

$$\delta_{tot} = -\frac{wl^2}{384EI}(15d^2 + 5l^2)$$

or

$$\delta_{tot} = -\frac{5wl^4}{384EI}\left[3\left(\frac{d}{l}\right)^2 + 1\right]$$

For a ratio of $d/l = 1/10$, the total deflection is

$$\delta_{tot} = -\frac{5wl^4}{384EI} \times (0.03 + 1)$$

From which we note that the deflection produced by the shear is about 3 percent of that produced by bending alone. If $d/l = \frac{1}{3}$,

$$\delta_{tot} = -\frac{5wl^4}{384EI}(\frac{1}{3} + 1)$$

indicating that the deflection caused by shear is about 33 percent of that caused by bending alone.

PROBLEMS

For the following problems assume that $E/G = 2.5$.

4–68 Determine the deflection at point C of the beam in Figure P 4–68 in terms of w, EI, and l.

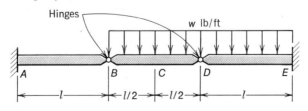

Figure P 4–68

4-69 Determine the vertical and horizontal deflection of point B, in terms of W, EI, and l. All members have the same EI, and joint C is rigid.

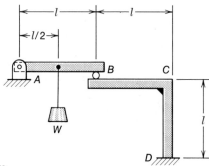

Figure P 4-69

4-70 In Figure P 4-70, find the vertical deflection at point C, in terms of l, EI, and w. AB and BC have the same EI, and joint B is to be assumed rigid.

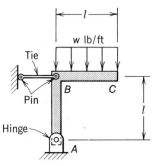

Figure P 4-70

4-71 Determine the deflection at midpoint of member BC in Figure P 4-71, in terms of W, EI, and l. Joints B and C are rigid, and EI is the same for all members.

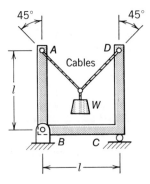

Figure P 4-71

4–72 Determine the horizontal travel of point E in Figure P 4–72, in terms of P, EI, and l. EI is the same for all members, and joints B, C, and D are rigid.

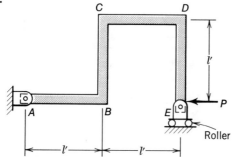

Figure P 4–72

4–73 Determine the value of EI for beam CDE in Figure P 4–73 if the deflection at point E is 3 times the elongation of the spring at C. Assume the beam is steel.

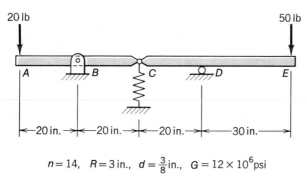

$n = 14$, $R = 3$ in., $d = \frac{3}{8}$ in., $G = 12 \times 10^6$ psi

Figure P 4–73

4–74 Determine the vertical deflection of the hinge in Figure P 4–74 if $EI = 60 \times 10^7$ lb-in.^{2}

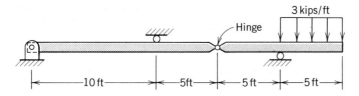

Figure P 4–74

4–75 Determine the vertical deflection of point E in Figure P 4–75 in terms of P, EI, and l. EI is constant throughout, and joints C and D are rigid.

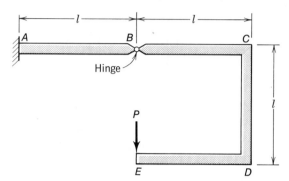

Figure P 4–75

4–76 Determine the deflection of point E in Figure P 4–76 in terms of w, EI, and l.

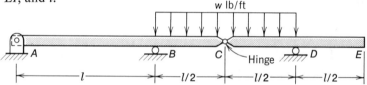

Figure P 4–76

4–77 Determine the deflection of point C in Figure P 4–77 in terms of w, EI, and l.

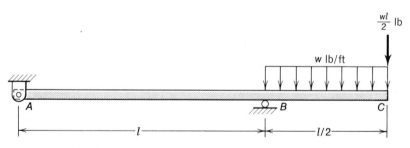

Figure P 4–77

4–78 Determine the deflection at point D in Figure P 4–78 in terms of w, EI, and l.

4–79 Determine the deflection and slope of point A in Figure P 4–79 in terms of w, EI, and l. EI is the same for AB and BC.

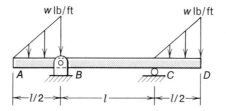

Figure P 4-78

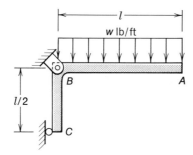

Figure P 4-79

4-80 Determine the deflection and slope at point *B* in Figure P 4-80 in terms of *P*, *EI*, and *l*. All members have the same *EI*.

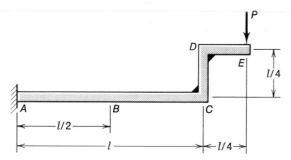

Figure P 4-80

4-81 Determine the deflection and slope at point *C* of beam *BC* in Figure P 4-81 in terms of *w*, *EI*, and *l*.

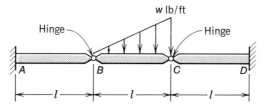

Figure P 4-81

4–82 Determine the deflection and slope at point D in Figure P 4–82 in terms of P, EI, and l.

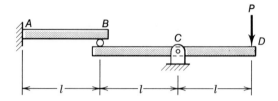

Figure P 4–82

4–83 For the steel beam in Figure P 4–83 determine (a) the deflection at the free end caused by bending only, (b) the deflection at the free end caused by shear, and (c) the percentage of the total deflection produced by shear.

Figure P 4–83

4–84 For the steel beam in Figure P 4–84 determine (a) the deflection at the free end caused by bending only, (b) the deflection at the free end caused by shear, and (c) the percentage of the total deflection produced by shear.

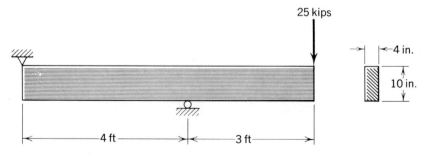

Figure P 4–84

4-85 For the beam in Figure P 4–85 determine (a) the deflection at the center of the span caused by bending only, (b) the deflection at the center of the span by shear, and (c) the percentage of the total deflection produced by shear, for a ratio of $d/l = 1/10$.

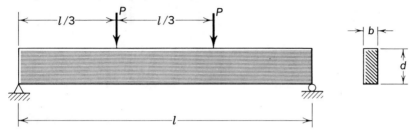

Figure P 4–85

4-86 For the steel cantilever shaft shown in Figure P 4–86, determine (a) the total deflection caused by shear and bending at the free end, and (b) the percentage of total deflection produced by shear.

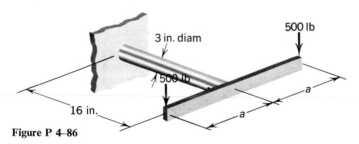

Figure P 4–86

4-87 For the steel cantilever shaft shown in Figure P 4–87, determine (a) the total deflection caused by shear and bending at the free end, and (b) the percentage of total deflection produced by shear.

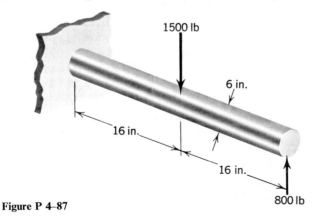

Figure P 4–87

Chapter 5 ◆ STATICALLY INDETERMINATE BEAMS

5—1 Introduction

The beams considered thus far were statically determinate; that is, the equations of static equilibrium were sufficient to solve for the external reactions and thus permit us to analyze the beam for stresses and deformation. Although statically determinate beams, within our assumptions, are the more common of the two classes, statically indeterminate beams frequently provide advantages in economy, function or esthetics. Their analysis is, therefore, a necessary part in the study of beams.

The concept of indeterminateness was introduced in Chapter 1 (Section 1–9) for axially loaded members, and in Chapter 2 (Section 2–3) for members in torsion. The basic concepts introduced there hold in this chapter for beams. When the equations of equilibrium are not sufficient to determine all the reactions, we are confronted with a statically indeterminate condition. We then supplement the equations of equilibrium with additional equations that are based on the deformation of the member. In the case of beams, the deformations we refer to are the slopes and deflections, compatible with the restraints of the supports. In other words, the supports provide certain restraints, either partial or complete, against translation or rotation (or both) of the beam. From this we derive other equations to supplement those of equilibrium.

With the boundary conditions established (either assumed or given), any of the methods presented in Chapter 4 may be used to determine the unknown reactions. Frequently one method of analysis has an advantage over another, depending upon the beams and loads. Therefore, several methods will be presented here (Sections 5–2 through 5–6). However, once the reactions are determined, the method for determining stresses and deflections is essentially the same as that presented in Chapters 3 and 4.

5—2 The Integration Method

The integration method introduced in Section 4–3 for statically determinate beams may also be used to solve statically indeterminate beams.
As such, it is a means by which we obtain additional equations to supplement those from statics. They are necessary to solve an indeterminate-beam
problem. Hence, by evaluating the slope and deflection equations, as dictated by the beam's boundary conditions (slopes and deflections at supports), the number of equations to meet the number of unknowns, moments
and/or forces, can be written. Once the unknown quantities are resolved,
the procedure usually reduces to that of determining either stresses or deflections, as presented in Chapters 3 and 4. Here we shall stress merely the
method for setting up the equations and solving for these unknowns.

The following two examples illustrate the method.

Example 5—1:

Given: The statically indeterminate beam shown in Figure *a*.

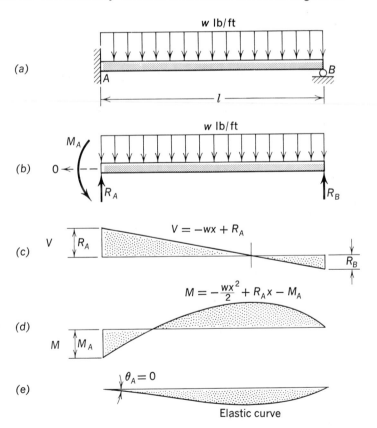

Find: (*a*) The vertical reactions and moment, and (*b*) the equation of the elastic line.

Procedure: We note that the horizontal reaction in the wall is zero, because there is no horizontal component of load. This leaves two independent equations of equilibrium applicable ($\Sigma F_y = 0$, and $\Sigma M_0 = 0$) and three unknowns, R_A, R_B, and M_A, which are assumed to act as shown in Figure *b*. Hence, one equation supplementing the two independent equations of equilibrium will be written based on deformation and boundary conditions, that is, at $x = 0$, $y_A = 0$, and $\theta_A = 0$. Also, at $x = l$, $y_B = 0$.

The moment equation, Figure *d*,

$$EI \frac{d^2 y}{dx^2} = -\frac{wx^2}{2} + R_A x - M_A$$

from which, by integration, we get the slope equation:

$$EI \frac{dy}{dx} = -\frac{wx^3}{6} + \frac{R_A x^2}{2} - M_A x + C_1$$

and because at $x = 0$,

$$EI \frac{dy}{dx} = 0$$

we obtain $C_1 = 0$.

Integrating again we get the deflection equation

$$EIy = -\frac{wx^4}{24} + \frac{R_A x^3}{6} - \frac{M_A x^2}{2} + C_2$$

$C_2 = 0$, from $EIy = 0$ at $x = 0$. Also, at $x = l$, $EIy = 0$, which yields

$$0 = -\frac{wl^4}{24} + \frac{R_A l^3}{6} - \frac{M_A l^2}{2} \qquad (a)$$

giving a relationship between R_A and M_A. A second relationship may be obtained by summing moments about point B. We then have

$$\Sigma M_B = 0: \qquad \frac{wl^2}{2} + M_A - R_A l = 0 \qquad (b)$$

Equations (*a*) and (*b*) are sufficient to yield the values of M_A and R_A in terms of w and l:

$$\frac{1}{2}\left(\frac{wl^2}{2} + M_A - R_A l \right) = 0$$

$$\frac{-\dfrac{wl^2}{24} - \dfrac{M_A}{2} + \dfrac{R_A l}{6} = 0}{+\dfrac{5wl^2}{24} - \dfrac{R_A l}{3} = 0}$$

from which $R_A = 5wl/8$ (upward as assumed).

Substituting into Equation (*b*) above, we get

$$\frac{wl^2}{2} + M_A - \frac{5wl^2}{8} = 0$$

$$M_A = \frac{wl^2}{8} \text{ (counterclockwise as assumed)}$$

From $\Sigma F_y = 0 = R_B + R_A - wl$,

$$R_B = \tfrac{3}{8}wl \text{ (upward as assumed)}$$

Now that the values for R_A and M_A are known, we may substitute into the y equation above:

$$EIy = -\frac{wx^4}{24} + \frac{5wlx^3}{48} - \frac{wl^2x^2}{16}$$

which is the deflection equation for our statically indeterminate member.

Example 5—2:

Given: The beam in Figure *a*.

Find: The reactions and moments at the two fixed supports.

Procedure: The common assumption when analyzing such beams is that one end is fixed against rotation and vertical translation but permits horizontal translation, as shown in Figure *b*. Otherwise, a consideration of axial deformation of the beam would be necessary for a solution of the horizontal reactions at A and B. For small deflections, however, because such reactions would not alter the vertical deflection to any measurable degree, the above assumption will be considered reasonable. Thus, we note that the beam is statically indeterminate to the second degree (two more unknowns than equations of static equilibrium). The boundary conditions dictate that, at $x = 0$,

$$\theta_A = 0 \qquad \text{and} \qquad y_A = 0$$

And at $x = l$,

$$\theta_B = 0 \qquad \text{and} \qquad y_B = 0$$

Thus, from Figure *d*,

$$EI\frac{d^2y}{dx^2} = -M_A + R_Ax - \frac{w_0x^2}{2} + \frac{w_0x^3}{6}$$

$$EI\frac{dy}{dx} = -M_Ax + R_A\frac{x^2}{2} - \frac{w_0x^3}{6} + \frac{w_0x^4}{24l} + C_1$$

$$EIy = -M_A\frac{x^2}{2} + R_A\frac{x^3}{6} - \frac{w_0x^4}{24} + \frac{w_0x^5}{120l} + C_1x + C_2$$

From $\theta_A = 0$, $y_A = 0$ at $x = 0$, $C_1 = 0$ and $C_2 = 0$. Thus,

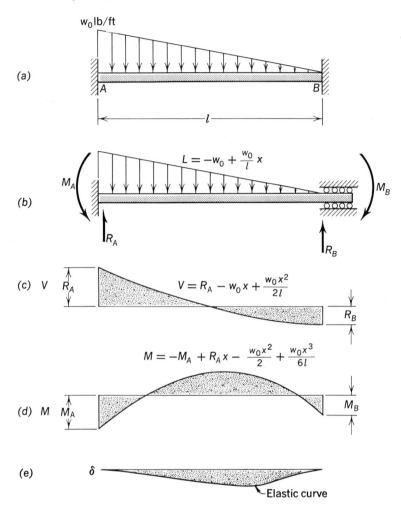

$$Ely = -M_A\frac{x^2}{2} + R_A\frac{x^3}{6} - \frac{w_0 x^4}{24} + \frac{w_0 x^5}{120l}$$

When $x = l$, $EI \cdot y = 0$ and $EI(dy/dx) = 0$, yielding a relationship between R_A and M_A:

$$0 = -\frac{M_A l^2}{2} + R_A\frac{l^3}{6} - \frac{w_0 l^4}{24} + \frac{w_0 l^4}{120} \qquad (a)$$

$$0 = -M_A l + R_A\frac{l^2}{2} - \frac{w_0 l^3}{6} + \frac{w_0 l^3}{24} \qquad (b)$$

Solving Equations (*a*) and (*b*) simultaneously, we get

$$-\frac{M_A}{2} + \frac{R_A l}{6} - \frac{w_0 l^2}{24} + \frac{w_0 l^2}{120} = 0$$

$$+\frac{M_A}{2} - R_A \frac{l}{4} + \frac{w_0 l^2}{12} - \frac{w_0 l^2}{48} = 0$$

$$-\frac{1}{12} R_A + \frac{7 w_0 l}{240} = 0$$

$$R_A = \frac{7}{20}(w_0 l) \quad \text{(upward as shown)}$$

Substituting into (*a*) gives

$$-\frac{M_A}{2} + \frac{7}{120} w_0 l^2 - \frac{w_0 l^2}{24} + \frac{w_0 l^2}{120} = 0$$

$$M_A = \frac{w_0 l^2}{20} \quad \text{(counterclockwise as shown)}$$

From $\Sigma\, M_B$, we get M_B:

$$M_B + R_A l - M_A - \left(\frac{w_0 l}{2}\right)\left(\frac{2}{3} l\right) = 0$$

$$M_B = \frac{w_0 l^2}{3} + \frac{w_0 l^2}{20} - \left(\frac{7}{20}\right) w_0 l^2 = \frac{8}{30} w_0 l^2 \quad \text{(clockwise as shown)}$$

From $\Sigma\, M_A$:

$$M_A - M_B - \frac{w_0 l^2}{6} + R_B l = 0$$

$$\frac{w_0 l^2}{20} - \frac{8}{30} \cdot w_0 l^2 - w_0 l^2 + R_B l = 0$$

hence,

$$R_B = \frac{3}{20} \times w_0 l \quad \text{(upward as shown)}$$

As a check, $\Sigma F y = 0$,

$$R_A + R_B = \frac{w_0 l}{2}$$

$$\frac{7}{20} w_0 l + \frac{3}{20} w_0 l = \frac{w_0 l}{2}$$

$$\frac{w_0 l}{2} = \frac{w_0 l}{2}$$

PROBLEMS

Unless specified otherwise, assume that EI is constant along the length of the beam.

5-1   Determine the vertical reactions and the moment at the fixed support for the beam in Figure P 5–1, in terms of P, b, a, and EI.

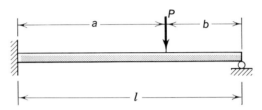

Figure P 5–1

5-2 Determine the vertical reactions and the moment at the fixed support for the beam in Figure P 5–2, in terms of P, l, and EI.

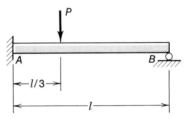

Figure P 5–2

5-3 Determine the vertical reactions and the moment at the fixed support for the beam in Figure P 5–3, in terms of P, l, and EI.

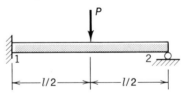

Figure P 5–3

5-4 Determine the vertical reactions and the moment at the fixed support for the beam in Figure P 5–4, in terms of P, l, a, and EI.

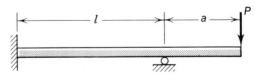

Figure P 5–4

5–5 For the problem in Figure P 5–4, determine the deflection at the free end in terms of P, a, l, and EI.

5–6 Determine the vertical reactions and the moment at the fixed support for the beam in Figure P 5–6 in terms of P, l, and EI.

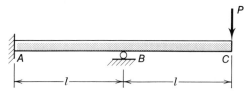

Figure P 5–6

5–7 Determine the vertical reactions and the moment at the fixed support for the beam in Figure P 5–7, in terms of w_0, l, and EI.

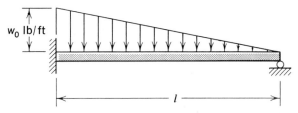

Figure P 5–7

5–8 Determine the vertical reactions and the moment at the wall support for the beam in Figure P 5–8, in terms of w_0, l, and EI.

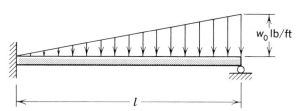

Figure P 5–8

5–9 Determine the vertical reactions and the moment at the wall support for the beam in Figure P 5–9, in terms of M, l, and EI.

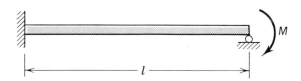

Figure P 5–9

5–10 Determine the vertical reactions and the moment at the wall support for the beam in Figure P 5–10, in terms of EI.

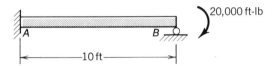

Figure P 5–10

5–11 Determine the maximum deflection of the beam in Figure P 5–9 in terms of M, l, and EI.

5–12 Determine the vertical reactions and the moment at the wall support for the beam in Figure P 5–12, in terms of w, a, l, and EI.

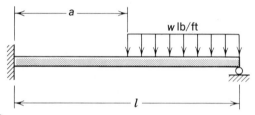

Figure P 5–12

5–13 Draw the shear and moment diagram for the beam in Figure P 5–12.

5–14 Determine the vertical reactions and the moment at the fixed support for the beam in Figure P 5–14, in terms of w_0, l, and EI.

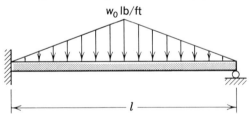

Figure P 5–14

5–15 Determine the vertical reactions and the moment at the fixed support for the beam in Figure P 5–15, in terms of w_0, l, and EI.

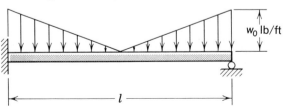

Figure P 5–15

5–16 Determine the vertical reactions and the moment at the fixed support for the beam in Figure P 5–16, in terms of w, l, a, and EI.

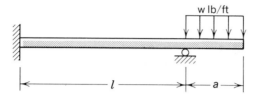

w lb/ft

l a

Figure P 5–16

5–17 Determine the vertical reactions and the moment at the fixed support for the beam in Figure P 5–17 in terms of EI.

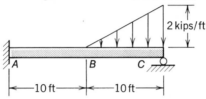

2 kips/ft

A   B   C

10 ft   10 ft

Figure P 5–17

5–18 Determine the deflection of the free end of the beam in Figure P 5–16, in terms of w, a, l, and EI.

5–19 Determine the deflection of the free end of the beam in Figure P 5–19, in terms of P, w, a, l, and EI.

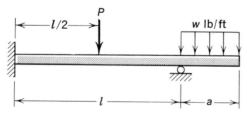

P

$l/2$ w lb/ft

l a

Figure P 5–19

5–20 Determine the vertical reactions and moments at the fixed supports for the beam in Figure P 5–20, in terms of P, b, a, and EI.

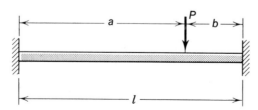

a   P   b

l

Figure P 5–20

5–21 Determine the vertical reactions and moments at the fixed supports for the beam in Figure P 5–21, in terms of w, l, and EI.

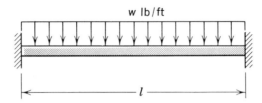

w lb/ft

l

Figure P 5–21

5–22 Determine the vertical reactions and moments at the fixed supports for the beam in Figure P 5–22, in terms of w_0, l, and EI.

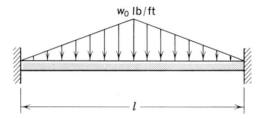

w_0 lb/ft

l

Figure P 5–22

5–23 Determine the vertical reactions and moments at the fixed supports for the beam in Figure P 5–23, in terms of w_0, l, and EI.

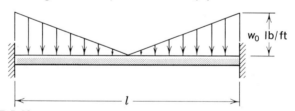

w_0 lb/ft

l

Figure P 5–23

5–24 Determine the vertical reactions and moments at the wall supports for the beam in Figure P 5–24, in terms of w, a, l, and EI.

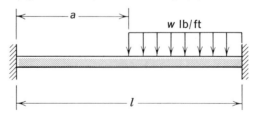

a

w lb/ft

l

Figure P 5–24

5–25 Determine the vertical reaction and moment at the wall support for the beam in Figure P 5–25, in terms of w, l, and EI.

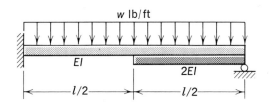

Figure P 5–25

5–26 Determine the vertical reactions and moments at the wall supports for the beam in Figure P 5–26, in terms of w, l, and EI.

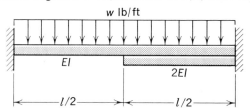

Figure P 5–26

5–27 Write the equation of the elastic line and determine the maximum deflection for the problem in Figure P 5–21 in terms of w, l, and EI.

5–28 Write the equation of the elastic line and determine the maximum deflection for the problem in Figure P 5–22 in terms of w_0, l, and EI.

5—3 The Superposition Method

The method of superposition as applied to statically determinate beams was introduced in Section 4–5. In essence, the principle permits the effects of several rather "simple" loads to be added algebraically in order to get the net effect of the more "complicated" loading. By adhering to this basic approach, and complying with the conditions consistent with the deformation of the beam, we may proceed to write additional equations to supplement those from equilibrium and ultimately solve a statically indeterminate problem. More specifically, for indeterminate beams, we may remove as many redundant restraints as necessary to make the beam statically determinate. Hence, the statically determinate beam will be deflected by the applied loads. We evaluate the deformation (deflection or slope) at points where the redundant restraints are removed. Then we apply the restraint back with such magnitude and orientation so as to make the combined deformation from loads and restraints consistent with the boundary conditions of the original beam.

Let us consider Figure 5–1a. The beam is statically indeterminant to the first degree. It has three forces and one moment resisting the effects of the load P as shown. Because statics provides us with only three independent equations of equilibrium, we must provide another equation based on deformation. We start by noting that the beam has zero vertical deflection at supports A and B. Furthermore, at support A the beam is restrained from rotation, in which case the slope of the elastic curve at this point is zero. Symbolically, we can write, at $x = 0$,

$$\theta_A = 0 \qquad \text{and} \qquad \delta_A = 0$$

and at $x = l$, $\delta_B = 0$

The "estimated" deflection, consistent with the above condition, is shown in Figure 5–1d (dashed line). Now, to make the beam statically de-

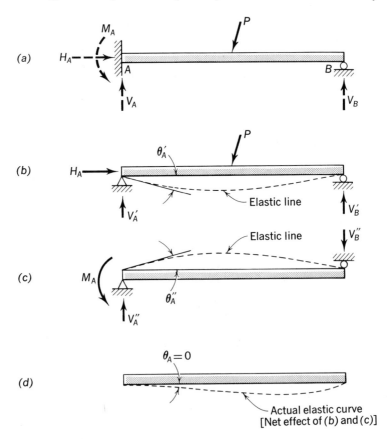

Figure 5–1

terminate, we remove the redundant M_A; hence, we are left with a simply supported beam loaded and deflecting as shown in Figure 5–1b. Because of this load, we recognize a slope at support A equal to θ'_A, expressed in terms of P, l, and EI (see Table 4–1, case 9). This is not consistent with our "original" boundary conditions, and therefore, the moment M_A must nullify this rotation Applying the moment as shown in Figure 5–1c, we get the slope θ''_A, expressed in terms of M, l, and EI (Table 4–1, case 15), that is,

$$\theta'_A = f_1(P, l, E, I)$$
$$\theta''_A = f_2(M_A, l, E, I)$$

and from $\theta'_A = \theta''_A$, we get an added equation, based on deformation where M_A may be expressed as a function of P and l, to supplement the equilibrium equations and subsequently determine the unknowns.

We could have also made the beam statically determinate by removing the support V_B. We would then have been left with a cantilever beam that would have had a vertical deflection at B, caused by the applied load P. To be consistent with the original structure, this deflection would have had to be nullified by a vertical reaction V_B. Thus, equating the two deflections at point B, a new equation, expressing V_B as a function of P and l, could have been obtained. Examples 5–3 and 5–4 will illustrate the basic approach.

Example 5—3:

Given: The beam shown in the Figure a.
Find: The reactions V_A, V_B, and M_A, in terms of w_0 and l.
Procedure: The beam in Figure a is statically indeterminate to the first degree. By removing the redundant M_A, we get a statically determinate "simple" beam. By removing the reaction V_B, we get a statically determinate cantilever beam. Choosing the latter, the beam is deflected by the load an amount δ'_B, as shown in Figure b. To satisfy the boundary conditions, we must nullify this deflection with the vertical support at B, V_B, as shown in Figure c. Thus, $\delta'_B = \delta''_B$.

$$\delta'_B = \frac{w_0 l^4}{30EI} \quad \text{(Table 4–1, case 6)}$$

$$\delta''_B = \frac{V_B l^3}{3EI} \quad \text{(Table 4–1, case 1)}$$

Equating the two deflections and solving for V_B gives

$$\frac{V_B l^3}{3EI} = \frac{w_0 l^4}{30EI}$$

$$V_B = \frac{w_0 l}{10} \quad \text{(upward as shown)}$$

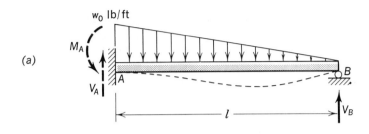

(a)

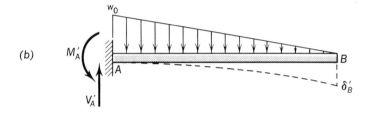

(b)

(c)

Summing moments about A, we get

$$M_A + \left(\frac{w_0 l}{10}\right) \cdot l = \left(\frac{w_0 l}{2}\right) \cdot \frac{l}{3}$$

$$M_A = \frac{w_0 l^2}{15} \quad \text{(counterclockwise as shown)}$$

Summing forces in the vertical direction, we get

$$V_A + V_B = \frac{w_0 l}{2}$$

$$V_A = \frac{w_0 l}{2} - V_B = \frac{w_0 l}{2} - \frac{w_0 l}{10} = \frac{2}{5} w_0 l \quad \text{(upward)}$$

Example 5—4:

Given: The load and beam arrangement shown in Figure *a*. Assume that the two beams have the same *EI* over their lengths.
Find: The deflection of the beams at point *C* in terms of *w*, *l*, and *EI*. Assume unyielding supports at points *A*, *B*, and *D*.
Procedure: The structure is statically indeterminate to the first degree. The deflected shape of the two beams is shown by dashed lines in

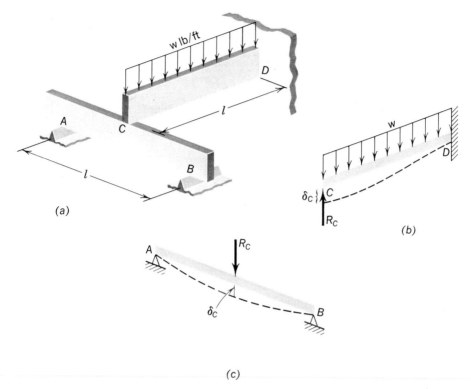

(a)

(b)

(c)

Figures b and c. From here we note that at point C the deflections of the two beams are the same. Thus, we can make the system statically determinate by first removing the reaction at C, then applying it back to each beam with a magnitude so that they both deflect the same amount. Referring to Figure b, beam $C–D$, we have by superposition, for the assumed orientation of R_C,

$$\delta_C = \delta_w - \delta_R = \frac{wl^4}{8EI} - \frac{R_C l^3}{3EI} \quad \text{(cases 4 and 1, Table 4–1)}$$

and for beam $A–B$, we have

$$\delta_C = \frac{R_C l^3}{48EI} \quad \text{(case 8, Table 4–1)}$$

Equating the two expressions for δ_C and solving, we find R_C:

$$\frac{R_C l^3}{48EI} = \frac{wl^4}{8EI} - \frac{R_C l^3}{3EI}$$

from which

$$R_C = \frac{6}{17} wl \quad \text{(as shown)}$$

Thus,

$$\delta_C = \frac{R_C l^3}{48EI} = \frac{w l^4}{136EI}$$

Example 5–5:

Given: The beam in Figure *a*.
Find: The shear and moment diagram for the beam. Assume unyielding supports.

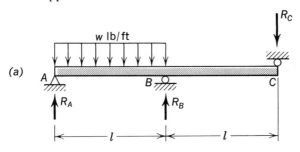

(a)

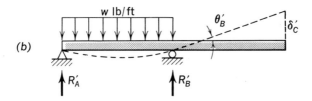

(b)

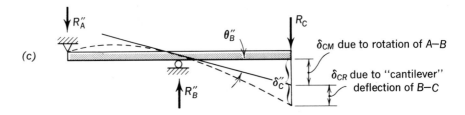

(c)

δ_{CM} due to rotation of A–B

δ_{CR} due to "cantilever" deflection of B–C

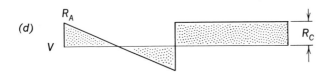

(d)

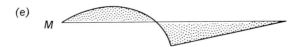

(e)

Procedure: One approach in making the structure in Figure *a* statically determinate is to remove one of the reactions. We note that the redundant reactions must then be placed back so as to make the deflection zero at that point. Removing R_C, the beam will deflect at point C by an amount δ'_C as shown in Figure *b*; that is,

$$\delta'_C = \theta'_B l = \frac{wl^3}{24EI} \times l = \frac{wl^4}{24EI} \quad \text{(see case 10, Table 4–1)}$$

When applying R_C as shown in Figure *c*, such that the net deflection at C is zero, we have

$$\delta'' = \delta_{CM} + \delta_{CR} = \theta''_B l + \frac{R_C l^3}{3EI} \quad \text{(see cases 15 and 1, Table 4–1)}$$

$$\delta''_C = \frac{Ml}{3EI} \times l + \frac{R_C l^3}{3EI} = \frac{R_C l^2}{3EI} \cdot l + \frac{R_C l^3}{3EI} = \frac{2R_C l^3}{3EI}$$

Letting $\delta'_C = \delta''_C$, we solve for R_C

$$\frac{2R_C l^3}{3EI} = \frac{wl^4}{24EI}$$

$$R_C = \frac{wl}{16} \quad \text{(downward as assumed)}$$

From $\Sigma M_A = 0$, we have

$$R_B = \frac{(wl/16) \times 2l + (wl)\frac{1}{2}l}{l} = \frac{wl}{8} + \frac{wl}{2} = \frac{5\,wl}{8} \quad \text{(upward)}$$

and from $\Sigma F_y = 0$, we get

$$R_A = wl - \frac{5wl}{8} + \frac{wl}{16} = \frac{7wl}{16} \quad \text{(upward)}$$

The shape of V and M diagrams are shown in Figures *d* and *e*, respectively.

PROBLEMS

Unless specified otherwise, assume that EI is constant along the length of the beam.

5–29 Determine the vertical reactions and the moment at the fixed support for the beam in Figure P 5–29, in terms of P, b, a, and EI.

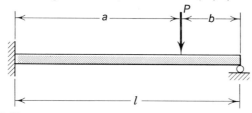

Figure P 5–29

5–30 Determine the vertical reactions and the moment at the fixed support for the beam in Figure P 5–30, in terms of P, l, a, and EI.

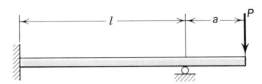

Figure P 5–30

5–31 Determine the vertical reactions and the moment at the fixed support for the beam in Figure P 5–31, in terms of w_0, l, and EI.

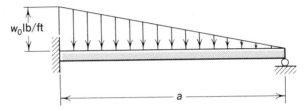

Figure P 5–31

5–32 Determine the vertical reactions and the moment at the fixed support for the beam in Figure P 5–32, in terms of M, l, and EI.

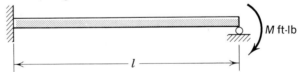

Figure P 5–32

5–33 Determine the maximum deflection of the beam in Figure P 5–32, in terms of M, l, and EI.

5–34 Determine the vertical reactions and the horizontal deflection of point C for the frame in Figure P 5–34 in terms of P, l, and EI.

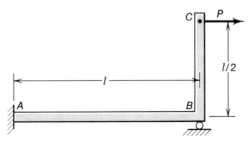

Figure P 5–34

5–35 Determine the vertical reactions and the moment at the fixed support for the beam in Figure P 5–35, in terms of w, a, l, and EI.

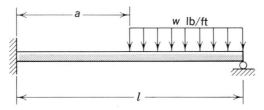

Figure P 5–35

5–36 Determine the vertical reactions and the moment at the fixed support for the beam in Figure P 5–36, in terms of w, l, and EI.

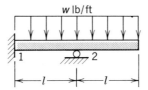

Figure P 5–36

5–37 Determine the vertical reactions and the moment at the fixed support for the beam in Figure P 5–37, in terms of w_0, l, and EI.

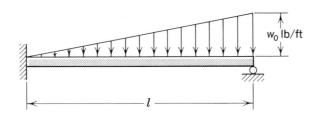

Figure P 5–37

5–38 Determine the vertical reactions for the beam in Figure P 5–38, in terms of P, l, and EI.

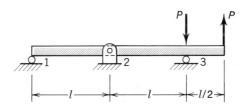

Figure P 5–38

5–39 Determine the vertical reactions and the moment at the fixed support for the beam in Figure P 5–39, in terms of w_0, l, and EI.

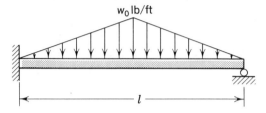

wₒ lb/ft

Figure P 5–39

5–40 Determine the vertical reactions and the moment at the fixed support in terms of w, l, and EI.

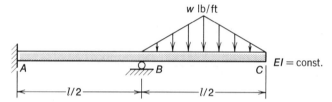

w lb/ft

EI = const.

Figure P 5–40

5–41 Determine the vertical reactions and the moment at the fixed support for the beam in Figure P 5–41, in terms of w, l, a, and EI.

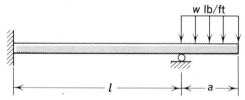

w lb/ft

Figure P 5–41

5–42 Determine the vertical reaction at a and the moment and horizontal reaction at c for the frame in Figure P 5–42, in terms of w, l, and EI.

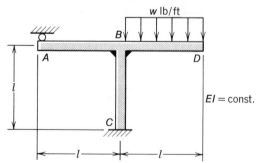

w lb/ft

EI = const.

Figure P 5–42

5-43 Determine the vertical reactions and moments at the fixed supports for the beam in Figure P 5–43, in terms of P, b, a, and EI.

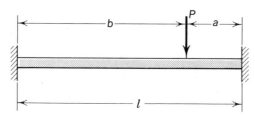

Figure P 5–43

5-44 Determine the vertical reactions and the moment at the fixed support for the beam in Figure P 5–44, in terms of P, l, and EI. ·

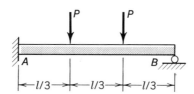

Figure P 5–44

5-45 Determine the vertical reactions and moments at the fixed supports for the beam in Figure P 5–45, in terms of w, l, and EI.

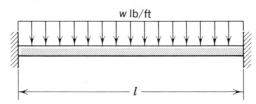

Figure P 5–45

· 5-46 Determine the vertical reactions and moments at the fixed supports for the beam in Figure P 5–46, in terms of w_0, l, and EI.

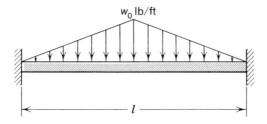

Figure P 5–46

5–47 Determine the vertical reactions and moments at the fixed supports for the beam in Figure P 5–47, in terms of w_0, l, and EI.

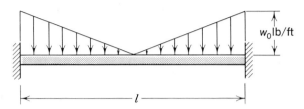

Figure P 5–47

5–48 Determine the vertical reactions and moments at the fixed supports for the beam in Figure P 5–48, in terms of w, a, l, and EI.

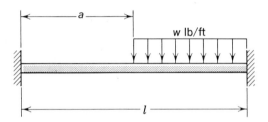

Figure P 5–48

5–49 Determine the reactions and moments for the beam in Figure P 5–49 in terms of w, l, and EI.

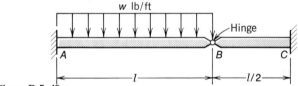

Figure P 5–49

5–50 Determine the vertical reactions and moment at the fixed support for the beam in Figure P 5–50, in terms of w, l, and EI.

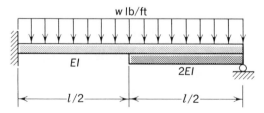

Figure P 5–50

5–51 Determine the vertical reactions and the moments at the fixed supports for the frame in Figure P 5–51 in terms of *EI*. Assume *EI* is constant for all beam sections, neglect any elongation in the tie.

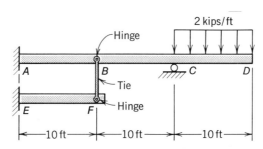

Figure P 5–51

5–52 Determine the vertical reactions and moments at the fixed supports for the beam in Figure P 5–52, in terms of *w*, *l*, and *EI*.

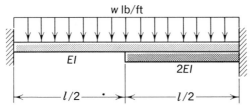

Figure P 5–52

5–53 Determine the reactions (including moments) for the beam in Figure P 5–53, in terms of *w*, *l*, and *EI*.

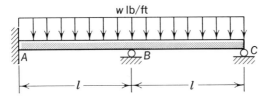

Figure P 5–53

5–54 Determine the vertical reactions for the beam in Figure P 5–54 in terms of *EI*.

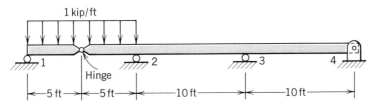

Figure P 5–54

5-55 Determine the vertical reactions for the beam in Figure P 5-55 in terms of *EI*.

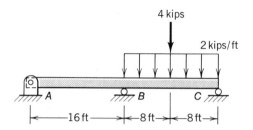

Figure P 5-55

5-56 Determine the reactions for the beam in Figure P 5-56. Also determine the deflection at point *C*. Neglect the weight of the beam.

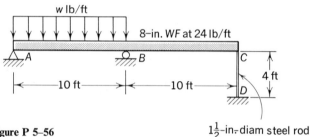

Figure P 5-56

5-57 Determine the reactions (including moments) for the beams in Figure P 5-57 in terms of *w*, *l*, and *EI*. *EI* is the same for both beams.

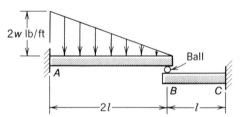

Figure P 5-57

5-58 Determine the reactions (including moments) for the beams in Figure P 5-58, in terms of *P*, *l*, *a*, and *EI*. Assume that *EI* is the same for both beams.

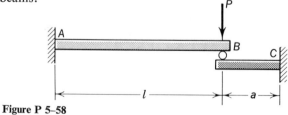

Figure P 5-58

5–59 Determine the reactions (including moments) for the beams in Figure P 5–59, in terms of P, l, and EI.

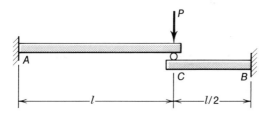

Figure P 5–59

5–60 Determine the reactions for the beam in Figure P 5–60, in terms of P, l, and EI. Assume that EI is the same for both beams.

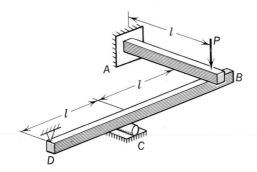

Figure P 5–60

5–61 Determine the reactions for beam AC in Figure P 5–61, and draw shear and moment diagrams for the beam.

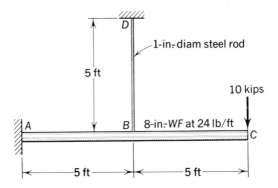

Figure P 5–61

5–62 Determine the reactions for the beam in Figure P 5–62, and draw shear and moment diagrams for the beam.

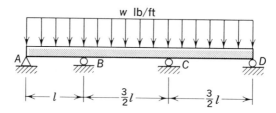

Figure P 5–62

5—4   The Moment-Area Method

The moment-area method presented in Chapter 4 for statically determinate beams may also be used for statically indeterminate beams. It serves as a tool for writing as many additional equations (to supplement those of equilibrium) as necessary in order to solve a statically indeterminate problem. The boundary conditions (slopes and deflections) of the indeterminate beam provide the conditions which subsequently permit the formulation of the supplementary equations. For example, the beam in Figure 5–2, statically indeterminate to the first degree, requires one equation in addition to those of equilibrium for the solution. For convenience, the moment diagram shown in Figure 5–2b is broken into the two moment diagrams, the negative moment caused by M_A and the positive moment caused by the load w. We can now use Theorem 2 to set up a relationship between M_A, w, and l; that is, the moment of the area under the moment diagram about point B must be zero because the slope at A is zero and the deflection at B is zero (see Section 4-4). Thus, we have

$$-\frac{M_A l}{2} \times \frac{2}{3}l + \frac{wl^3}{12} \times \frac{l}{2} = 0$$

from which

$$M_A = \frac{wl^2}{8}$$

(The positive sign here merely indicates that our assumption of the orientation of M_A was correct.)

By summing moments about A we get

$$-\frac{wl^2}{8} + \frac{wl^2}{2} = R_B l$$

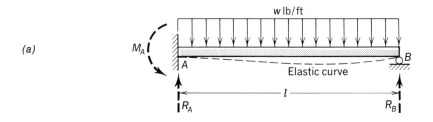

(a)

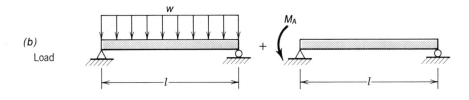

(b) Load

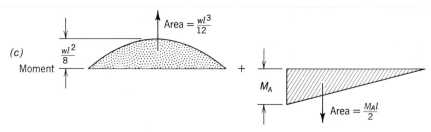

(c) Moment

Figure 5–2

which gives

$$R_B = \tfrac{3}{8}wl$$

and

$$R_A = \tfrac{5}{8}wl \quad (\text{from } \Sigma F_y = 0)$$

If additional restraints were present, additional equations may be written by observing the deflection and slope conditions at such restraints.

Example 5—6:

Given: The beam in Figure *a*. *EI* is constant along the length of the beam.

Find: The reactions and moments at the supports (compare answers with Example 5–2).

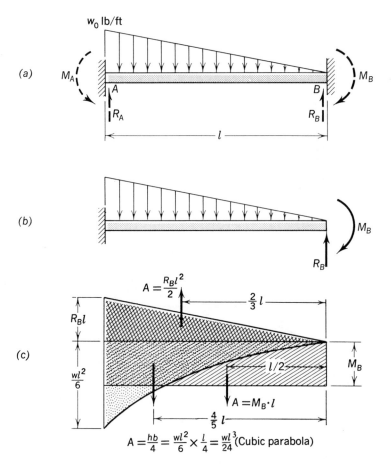

(a)

(b)

(c)

$$A = \frac{hb}{4} = \frac{wl^2}{6} \times \frac{l}{4} = \frac{wl^3}{24} \text{(Cubic parabola)}$$

Procedure: The beam is statically indeterminate to the second degree, thus requiring two "supplementary" equations.

Note that the deflections and slopes of the elastic curve at both supports *A* and *B* are zero. Thus, arbitrarily, we select M_B and R_B as the redundants (Figure *b*) and proceed by using the two moment-area theorems. The separate moment diagrams for the load and the two redundants are shown in Figure *c*. The areas under the various *M* diagrams are as shown in Figure *c*.

Using Theorem 2 and $\Sigma\, M_A = 0$, we get

$$\left(\frac{w_0 l^3}{24}\right)\frac{l}{5} + (M_B l)\cdot\frac{l}{2} - \left(R_B \frac{l^2}{2}\right)\frac{l}{3} = 0 \qquad (a)$$

Using Theorem 1, we get

$$-\frac{w_0 l^3}{24} - M_B l + \frac{R_B l^2}{2} = 0 \qquad (b)$$

Rewriting and solving Equations (a) and (b) simultaneously, we have

$$\frac{w_0 l^2}{24 \times 5} + \frac{M_B}{2} - \frac{R_B}{6} l = 0 \qquad (a')$$

$$-\frac{w_0 l^2}{24 \times 2} - \frac{M_B}{2} + \frac{R_B}{4} l = 0 \qquad (b')$$

$$\frac{w_0 l^2}{24}\left(\frac{1}{5} - \frac{1}{2}\right) + R_B l\left(\frac{6 - 4}{24}\right) = 0$$

from which $R_B = 3/20\, w_0 l$ (oriented upward as assumed).
Substituting in Equation (a') we get the value for M_B:

$$\frac{w_0 l^2}{120} + \frac{M_B}{2} - \frac{3 w_0 l^2}{120} = 0$$

from which

$$M_B = \frac{1}{30}\cdot w_0 l^2 \quad \text{(oriented as assumed)}$$

From $\Sigma\, M_A = 0$, we get:

$$M_A + R_B l - \frac{w_0 l}{2} \times \frac{l}{3} - M_B = 0$$

$$M_A = \frac{1}{30}\, w_0 l^2 + \frac{w_0 l^2}{6} - \frac{3}{20}\cdot w_0 l^2 = \frac{1}{20}\, w_0 l^2 \quad \text{(oriented as assumed)}$$

From $\Sigma\, F_y = 0$, we have:

$$R_A = \frac{w_0 l}{2} - R_B = \frac{7}{20}\cdot w_0 l \quad \text{(oriented as assumed)}$$

PROBLEMS

Unless specified otherwise, assume that *EI* is constant along the length of the beam. Use the moment-area method.

5–63 Determine the vertical reactions and the moment at the fixed support for the beam in Figure P 5–63, in terms of *P*, *b*, *a*, and *EI*.

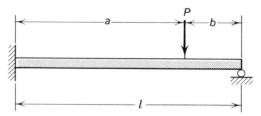

Figure P 5–63

5–64 Determine the vertical reactions and the moment at the fixed support for the beam in Figure P 5–64, in terms of *P*, *l*, *a*, and *EI*.

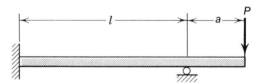

Figure P 5–64

5–65 Determine the vertical reactions and the moment at the fixed support for the beam in Figure P 5–65, in terms of w_0, *l*, and *EI*.

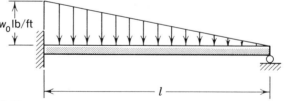

Figure P 5–65

5–66 Determine the vertical reactions for the beam in Figure P 5–66 in terms of M_A, *l*, and *EI*.

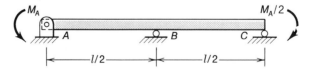

Figure P 5–66

5–67 Determine the vertical reactions and the moment at the fixed support for the beam in Figure P 5–67, in terms of w_0, l, and EI.

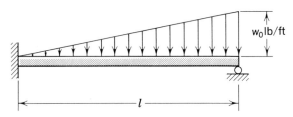

Figure P 5–67

5–68 Determine the vertical reactions and the moment at the fixed support for the beam in Figure P 5–68, in terms of M, l, and EI.

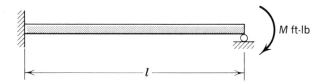

Figure P 5–68

5–69 Determine the vertical reactions and the moment at the fixed support for the beam in Figure P 5–69, in terms of w, a, l, and EI.

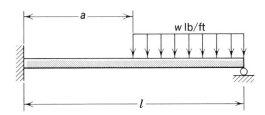

Figure P 5–69

5–70 Determine the vertical reactions and the moment at the fixed support for the beam in Figure P 5–70, in terms of w_0, l, and EI.

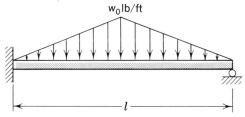

Figure P 5–70

5–71 Determine the vertical reactions and the moment at the fixed support for the beam in Figure P 5–71, in terms of w, l, and EI.

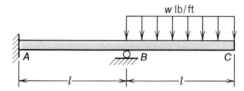

Figure P 5–71

5–72 Determine the vertical reactions and the moment at the fixed support for the beam in Figure P 5–72, in terms of w_0, l, and EI.

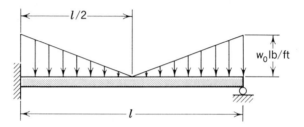

Figure P 5–72

5–73 Determine the vertical reactions and the moment at the fixed support for the beam in Figure P 5–73, in terms of w, l, a, and EI.

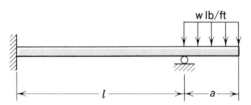

Figure P 5–73

5–74 Determine the deflection of the free end of the beam in Figure P 5–74, in terms of P, w, a, l, and EI.

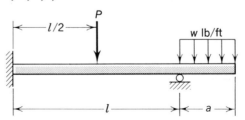

Figure P 5–74

5–75 Determine the reactions (including moments) for the beam in Figure P 5–75 in terms of w, l, and EI.

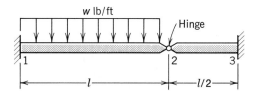

Figure P 5–75

5–76 Determine the vertical reactions and moments at the fixed supports for the beam in Figure P 5–76, in terms of P, b, a, and EI.

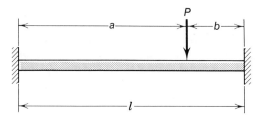

Figure P 5–76

5–77 Determine the vertical reactions and moments at the fixed supports for the beam in Figure P 5–77, in terms of w, l, and EI.

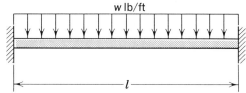

Figure P 5–77

5–78 Determine the vertical reactions and moments at the fixed supports for the beam in Figure P 5–78, in terms of w_0, l, and EI.

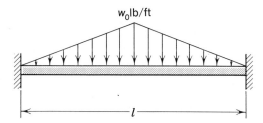

Figure P 5–78

5–79 Determine the vertical reactions and moments at the fixed supports for the beam in Figure P 5–79, in terms of w_0, l, and EI.

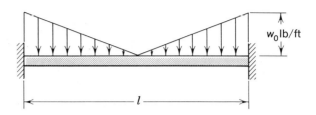

Figure P 5–79

5–80 Determine the vertical reactions and moments at the fixed supports for the beam in Figure P 5–80, in terms of w, a, l, and EI.

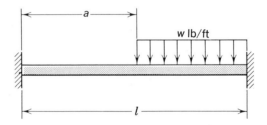

Figure P 5–80

5–81 Determine the vertical reactions and moment at the fixed support for the beam in Figure P 5–81, in terms of w, l, and EI.

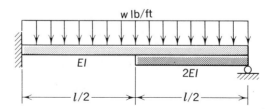

Figure P 5–81

5–82 Determine the vertical reaction and moments at the fixed supports for the beam in Figure P 5–82, in terms of w, l, and EI.

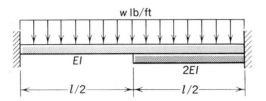

Figure P 5–82

5—5 The Three-Moment Theorem

For one- or two-span indeterminate beams, the methods of analysis presented in Sections 5–2, 5–3, or 5–4 are adequate. For several spans, these methods prove rather tedious and cumbersome, and consequently, they are sources for errors in calculation. The method of analysis to be presented here, the three-moment theorem, is a more desirable approach for such beams.

The basis on which the three-moment theorem is developed was originally established over a century ago by B. P. E. Clapeyron (1799–1864) who observed the continuity requirement over a support: *The slopes of the*

(a)

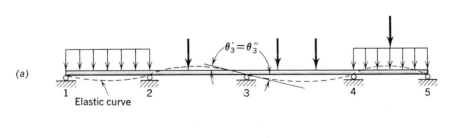

(b)

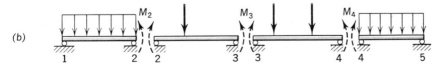

Figure 5–3

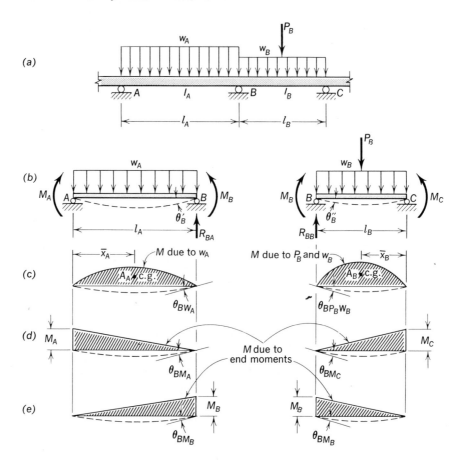

Figure 5–4

elastic curves at a common support of two spans of a continuous beam must be equal. This is illustrated in Figure 5–3a at support 3, the dashed line representing an exaggerated assumed elastic curve. The important thing to note here is that because the beam is continuous, the elastic curve over an interior support is a smooth curve; hence, $\theta_3' = \theta_3''$. Recognizing this fact, we may cut the continuous beam at the supports in order to get a series of simply supported beams which are loaded with the transverse loads and the moments at the supports shown in Figure 5–3b. The magnitude and direction of these moments must be such that we preserve the continuity (smooth elastic curve). Thus, we observe that the moments are the redundant quantities in this approach. If we knew what the magnitudes and orientation were for these moments, we could then determine the reactions and subsequently go on to determine the V, M, or δ diagrams.

To establish the three-moment theorem, let us begin with Figure 5–4a. It represents two adjacent spans arbitrarily selected from a continuous beam. The common support for the two spans is B. Thus, when we cut the beam and load it as shown in Figure 5–4b, we must make the two slopes, $\theta'_B = \theta''_B$. This sets up a relationship between M_A, M_B, M_C, and the transverse loads. By taking two spans at a time and treating them accordingly, we may set up all the equations we need to supplement those of equilibrium for the final analysis. In Figure 5–4b, all the moments are assumed positive, as shown, and, therefore, a minus sign in our results merely means a reversal of direction. Hence, we begin with

$$\theta'_B = \theta''_B \tag{a}$$

and from superposition (see Figures 5–4c, 5–4d, and 5–4e) we have

$$\theta'_B = \theta_{BW_A} + \theta_{BM_A} + \theta_{BM_B}$$

and

$$\theta''_B = \theta_{B_PW_B} + \theta_{BM_C} + \theta_{BM_B}$$

In terms of loads and moments (see Table 4–1), for a constant EI with a span l

$$\theta'_B = \left(\frac{A_A \bar{x}_A}{l_A} + \frac{M_A l_A}{6} + \frac{M_B l_A}{3} \right) \frac{1}{EI_A}$$

and

$$\theta''_B = \left(\frac{A_B \bar{x}_B}{l_B} + \frac{M_C l_B}{6} + \frac{M_B l_B}{3} \right) \frac{1}{EI_B}$$

where A_A and A_B represent the areas under the moment diagrams caused by the transverse loads on the two spans and whose centroids are at $\bar{x}_A$ and $\bar{x}_B$ from supports A and C, as shown in Figure 5–4c. Substituting into Equation (a) we get

$$M_A l_A + 2M_B\left(l_A + \frac{l_A}{l_B} l_B \right) + M_C l_B \left(\frac{l_A}{l_B} \right)$$

$$= -\frac{6A_A}{l_A} \bar{x}_A - \frac{6A_B}{l_B} \bar{x}_B \times \left(\frac{l_A}{l_B} \right) \tag{5-1}$$

which is the three-moment equation for a general case. When I is constant for the entire beam, that is, $I_1 = I_2 = \cdots I_n$,

$$M_A l_A + 2M_B(l_A + l_B) + M_C l_B = -\frac{6A_A \bar{x}_A}{l_A} - \frac{6A_B \bar{x}_B}{l_B} \tag{5-2}$$

For *uniform loads*, Equation (5–2) may be simplified further to

$$M_A l_A + 2M_B(l_A + l_B) + M_C l_B = -\frac{w_A l_A^2}{4} - \frac{w_B l_B^2}{4} \tag{5-2'}$$

As mentioned previously, we take two *adjacent* spans and write a three-moment equation for each pair until we have as many three-moment equations as redundants. Solving these equations yields the moments for which we are looking. Then we may determine the reactions. In this regard, note that the resultant reaction B, for example, is the algebraic sum of the two reactions R_{BA} and R_{BB}, Figure 5–4b.

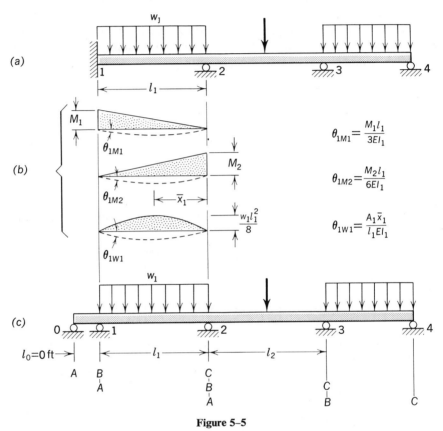

Figure 5–5

If the one end of the beam is fixed, as shown in Figure 5–5a, one additional redundant is encountered, and therefore we need one additional equation. However, because the slope θ_1 at support 1 is zero, the needed equation may be written on that basis:

$$\theta_1 = 0 = \frac{M_1 l_1}{3EI_1} + \frac{M_2 l_1}{6EI_1} + \frac{A_1 \bar{x}_1}{l_1 EI_1}$$

from which

$$2M_1 l_1 + M_2 l_1 = -\frac{6A_1 \bar{x}_1}{l_1} \qquad (5\text{–}3)$$

We detect a certain resemblance of this equation to Equation (5-2). In fact, Equation (5-2) reduces to Equation (5-3) when $L_A = 0$ and $A_A = 0$, and the corresponding subscripts are changed from B and C to 1 and 2, respectively. This leads to an interesting result: The fixed-end beam may be converted into a "modified" beam with an imaginary span $L_0 = 0$, as shown in Figure 5-5c. For this beam the three-moment equation may be applied directly, as explained above. The first equation for this condition will then be the same as Equation (5-3). Example 5-8 illustrates this.

Example 5—7:

Given: The continuous beam shown in Figure *a*. EI = constant.

Find: (*a*) The moments at the supports, and (*b*) the reactions.

Procedure: The beam is statically indeterminate to the second degree. Hence, two equations from the three-moment theorem plus those from equilibrium will suffice for the solution. The first equation is written using the first and second spans, as shown in Figure *b*, where 1, 2, and 3 correspond to *A*, *B*, and *C*, respectively, in the three-moment Equation (5-2). Equation (5-2) is preferable over Equation (5-1) because, in our case, EI is a constant. Thus,

$$M_1 l_1 + 2M_2(l_1 + l_2) + M_3 l_2 = -\frac{6A_1 \bar{x}_1}{l_1} - \frac{6A_2 \bar{x}_2}{l_2} \qquad (a)$$

The second three-moment equation is written using spans l_2 and l_3 as shown in Figure *c*.

$$M_2 l_2 + 2M_3(l_2 + l_3) + M_4 l_3 = -\frac{6A_2(l_2 - \bar{x}_2)}{l_2} - \frac{6A_3 \bar{x}_3}{l_3} \qquad (b)$$

The values for A_1, A_2, and A_3, and for $\bar{x}_1$, $\bar{x}_2$, and $\bar{x}_3$ are given in Figures *b* and *c* directly under the spans. For spans loaded with only uniform load,

$$A = \frac{wl^2}{8} \times \frac{2l}{3} = \frac{wl^3}{12}$$

and $\bar{x} = l/2$. Applied to our case,

$$A_1 \cdot \bar{x}_1 = \frac{w_1 l_1^4}{24} \quad \text{and} \quad A_2 \cdot \bar{x}_2 = \frac{w_2 l_2^4}{24}$$

When substituted into Equation (*a*), these values result in a general three-moment equation that is applicable whenever the spans are loaded with only uniformly distributed loads.

$$M_1 l_1 + 2M_2(l_1 + l_2) + M_3 l_2 = -\frac{w_1 l_1^3}{4} - \frac{w_2 l_2^3}{4} \qquad (a')$$

For spans 2 and 3, Equation (*b*) becomes

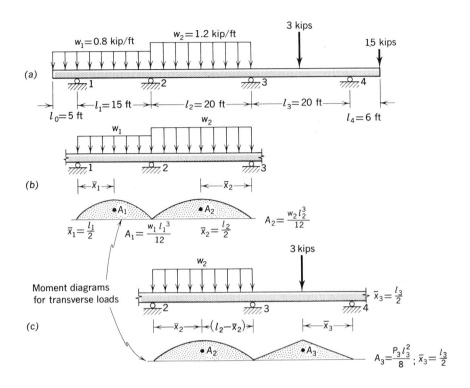

Moment diagrams for transverse loads

(a)
(b)
(c)

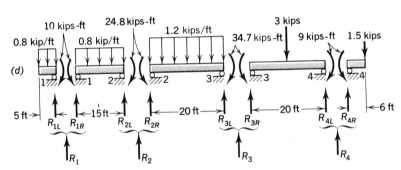

(d)

$$M_2 l_2 + 2M_3(l_2 + l_3) + M_4 l_2 = -\frac{w_2 l_2^3}{4} - \frac{6P_3 l_3^2}{16} \qquad (b')$$

In our case $M_1 = -(0.8 \times 5)(\frac{5}{2}) = -10$ kips-ft, and $M_4 = -1.5 \times 6$ $= -9$ kips-ft. Substituting these values, with those for the w's, l's, and P in Equations (a') and (b'), we get

$$-15(10) + 70M_2 + 20M_3 = \frac{-0.8(15)^3}{4} - \frac{1.2(20)^3}{4} \qquad (a')$$

and

$$-9(20) + 20M_2 + 80M_3 = \frac{-1.2(20)^3}{4} - \frac{6(3)(20)^2}{16} \qquad (b')$$

Solving the two equations simultaneously we have

$$M_2 = -34.7 \text{ kips-ft}$$

and

$$M_3 = -24.8 \text{ kips-ft}$$

The moments are shown in the free-body diagrams in Figure *d*. From here the reactions may be found by statics:

$$R_1 = R_{1L} + R_{1R} = 4 + 5 = 9 \text{ kips (upward)}$$

$$R_2 = R_{2L} + R_{2R} = 7 + 17.5 = 24.5 \text{ kips (upward)}$$

$$R_3 = R_{3L} + R_{3R} = 6.5 + 2.785 = 9.285 \text{ kips (upward)}$$

$$R_4 = R_{4L} + R_{4R} = 0.285 + 1.5 = 1.785 \text{ kips (upward)}$$

As a check, $\Sigma F_y = 0$ shows the solution for the reactions to be correct.

Example 5—8:

Given: The beam shown in Figure *a*. *EI* is uniform along the entire length of the beam.
Find: The *V* and *M* diagrams for this beam.
Procedure: The beam is modified so we may use the three-moment equation [Equation (5–2)] as shown in Figure *b*. The first equation applies to supports 0, 1, and 2; the second equation applies to supports 1, 2, and 3, corresponding to the subscripts *A*, *B*, and *C* in Equation (5–2).

$$M_0(0) + 2M_1(0 + 10) + M_2(10) = 0 \qquad (a)$$

$$M_1(10) + 2M_2(10 + 15) + M_3(15) = -\frac{850(15)^3}{4} \qquad (b)$$

Solving, from (*a*) we have

$$M_1 = -\tfrac{1}{2}M_2$$

Substituting in (*b*) gives

$$M_2 = -15{,}925 \text{ ft-lb}$$

Thus

$$M_1 = (-\tfrac{1}{2})(-15{,}925) = +7963 \text{ ft-lb}$$

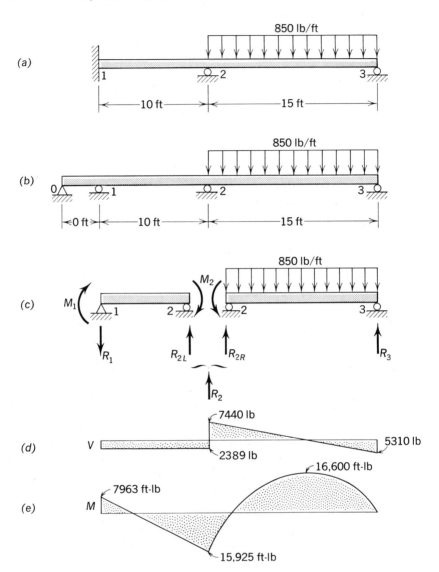

(a)

850 lb/ft

1 ⌐2 3

—10 ft— —15 ft—

(b)

850 lb/ft

0 1 2 3

<0 ft> —10 ft— —15 ft—

(c)

M_1 1 2 M_2 2 3

850 lb/ft

R_1 R_{2L} R_{2R} R_3

R_2

(d)

V

7440 lb

2389 lb

5310 lb

(e)

M

7963 ft-lb

16,600 ft-lb

15,925 ft-lb

The moments are shown acting on the two spans set in equilibrium in Figure c. From $\Sigma M_{R1} = 0$, we get

$$7963 + 15{,}925 = 10R_{2L}$$

$$R_{2L} = 2389 \text{ lb} \qquad \text{(upward as shown)}$$

From $\Sigma F_y = 0$:

$$R_1 = 2389 \text{ lb (downward as shown)}$$

From $\Sigma M_{R3} = 0$:

$$R_{2R} = \frac{(850 \times 15 \times 15/2) + 15{,}925}{15} = 7440 \text{ lb (upward as shown)}$$

Thus

$$R_2 = R_{2L} + R_{2R} = 9829 \text{ lb}$$

From $\Sigma F_y = 0$:

$$R_3 = 5310 \text{ lb (upward as shown)}$$

The shear and moment diagrams are shown in Figures d and e, respectively.

PROBLEMS

For the following problems assume EI is uniform along the length of the entire beam, unless specified otherwise.

5–83 Determine the reactions and moments at the supports for the beam in Figure P 5–83.

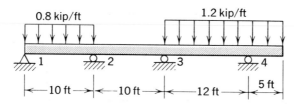

Figure P 5–83

5–84 Determine the reactions and moments at the supports for the beam in Figure P 5–84.

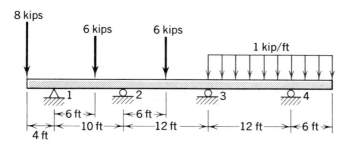

Figure P 5–84

5–85 Determine the reactions and moments at the supports for the beam in Figure P 5–85.

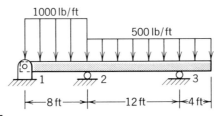

Figure P 5–85

5–86 Determine the reactions and moments at the supports for the beam in Figure P 5–86.

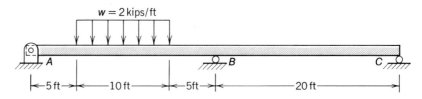

Figure P 5–86

5–87 Determine the reactions and moments at the supports for the beam in Figure P 5–87.

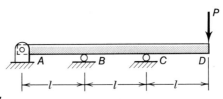

Figure P 5–87

5–88 Determine the reactions and moments at the supports for the beam in Figure P 5–88.

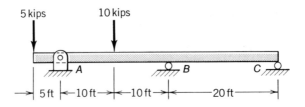

Figure P 5–88

5–89 Determine the reactions and moments at the supports for the beam in Figure P 5–89.

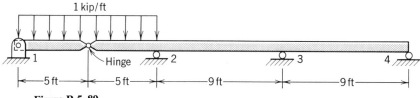

Figure P 5–89

5–90 Determine the reactions and moments at the supports for the beam in Figure P 5–90.

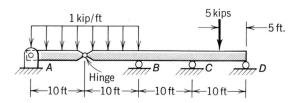

Figure P 5–90

5–91 Determine the reactions and moments at the supports for the beam in Figure P 5–91.

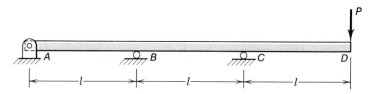

Figure P 5–91

5–92 Determine the reactions and moments at the supports for the beam in Figure P 5–92.

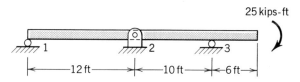

Figure P 5–92

5-93 Determine the reactions and moments at the supports for the beam in Figure P 5-93.

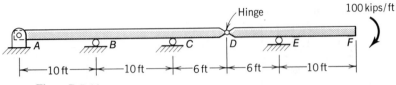

Figure P 5-93

5-94 Determine the reactions and moments at the supports for the beam in Figure P 5-94.

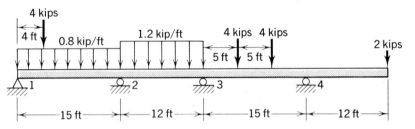

Figure P 5-94

5-95 Determine the reactions and moments at the supports for the beam in Figure P 5-95.

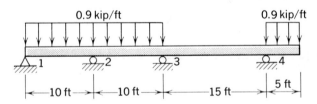

Figure P 5-95

5-96 Determine the reactions and moments at the supports for the beam in Figure P 5-96.

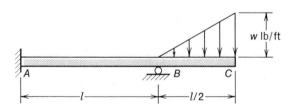

Figure P 5-96

5–97 Determine the reactions and moments at the supports for the beam in Figure P 5–97.

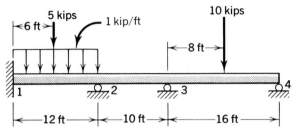

Figure P 5–97

5–98 Determine the reactions and moments at the supports for the beam in Figure P 5–98.

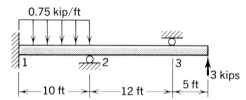

Figure P 5–98

5–99 Determine the reactions and moments at the supports for the beam in Figure P 5–99.

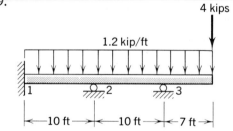

Figure P 5–99

5–100 Determine the reactions and moments at the supports for the beam in Figure P 5–100.

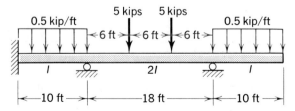

Figure P 5–100

5–101 Determine the reactions and moments at the supports for the frame in Figure P 5–101.

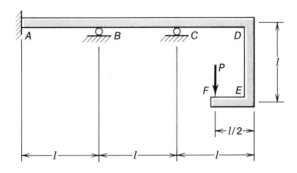

Figure P 5–101

5–102 Determine the reactions and moments at the supports for the beam in Figure P 5–102.

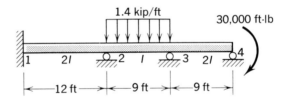

Figure P 5–102

5–103 Determine the reactions and moments at the supports for the beam in Figure P 5–103.

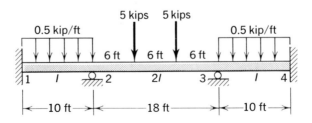

Figure P 5–103

5—6 Plastic Analysis

In Section 3–7 we discussed the behavior of a beam at a section subjected to plastic deformation. The maximum moment, or the related maximum loads that the beam can carry we termed the ultimate moment or ultimate loads, respectively. Assuming perfect elastoplastic behavior of the material (see Figure 3–13a), the ultimate loads for statically determinate beams are comparatively easy to determine. For such cases, the plastic moment occurs at the point of maximum elastic moment of the beam. Subsequently, from the determined plastic moment, we can resolve the ultimate load that the beam can carry (see Section 3–7). For indeterminate beams, however, the procedure for determining the ultimate loads is a little more involved, and only an introduction of the procedure will be presented in this section.[1]

Considering the statically indeterminate beam in Figure 5–6a, let us investigate the ultimate uniform load w_u that the beam can carry. Here, too, we assume an elastoplastic material. The moment diagram by parts, as determined for elastic conditions, is shown in Figure 5–6b. The maximum negative and positive moments occur at points A and C, respectively. Within the elastic limit of the material, the negative moment will be larger than the positive. Hence, if the uniform load is increased steadily, point A would reach the plastic state first, thus becoming a *plastic hinge* at that section. Adding still more load, the beam deflects further, rotating at point A, with a plastic moment M_p (Figure 5–6d) at A, exerting a constant resistance as rotation increases. Eventually the beam reaches a plastic state at the next point of maximum moment — point C as shown in Figure 5–6c. Thus, another plastic hinge is formed at C, and for a uniform section, the internal resisting moment at C is also M_p; the same magnitude as that at A, but acting as shown in Figure 5–6e. When plastic hinges are formed at both A and C, we have what is commonly called a *mechanism*, and the beam continues to collapse under any increase in load. The load causing this state of deformation is known as the *ultimate load*.

With the mechanism formed, we have a statically determinate beam. Thus, the summation of moments about A (Figure 5–6e) gives us

$$R_B = \frac{\frac{1}{2}(w_u l^2) - M_p}{l} = \frac{w_u l}{2} - \frac{M_p}{l} \qquad (a)$$

The summation of moments about C (Figure 5–6e) gives us

$$M_p = R_B d - \frac{w_u d^2}{2} \qquad (b)$$

---

[1] For a more elaborate treatment of plastic analysis, see L. S. Beedle, *Plastic Design of Steel Frames*, John Wiley & Sons, Inc., New York, 1958.

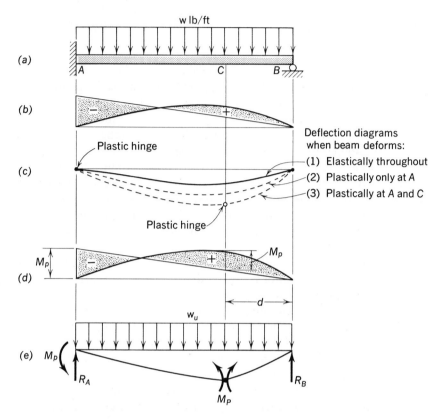

Figure 5–6 (*a*) The statically indeterminate beam. (*b*) *M* diagram of elastic conditions. (*c*) Stages of deflection. (*d*) *M* diagram of plastic conditions. (*e*) Mechanism.

R_B and d can be related by recognizing that the moment is a maximum where the shear is zero. Thus, we have

$$0 = R_B - dw_u$$

from which

$$d = \frac{R_B}{w_u} \qquad (c)$$

Substituting this value for d into Equation (*b*) we get

$$M_p = \frac{1}{2w_u}(R_B)^2$$

which, combined with Equation (*a*), yields

$$M_p = \frac{1}{2w_u}\left[\left(\frac{w_u l}{2}\right)^2 - 2\left(\frac{w_u l}{2}\right)\left(\frac{M_p}{l}\right) + \left(\frac{M_p}{l}\right)^2\right]$$

which simplifies to the quadratic form:

$$M_p^2 - 3w_u l^2 M_p + \tfrac{1}{4}w_u^2 l^4 = 0$$

Solving, gives

$$M_p = w_u l^2 (1.5 \pm 1.414)$$

from which

$$w_u = \frac{M_p}{(1.5 - 1.414)l^2} = \frac{11.62 M_p}{l^2}$$

The value of M_p can be easily obtained in terms of the yield stress of the material and the cross-sectional dimensions for the beam, as explained in Section 3–7. For example, for a *rectangular* section

$$M_p = \sigma_y \frac{bh^2}{4}$$

and thus w_u can be readily determined. Example 5–9 will illustrate this further.

Example 5—9:

Given: The beam shown in Figure *a*. Note that the section is that given in Example 3–8.

Find: The ultimate uniform load w_u that the beam can carry.

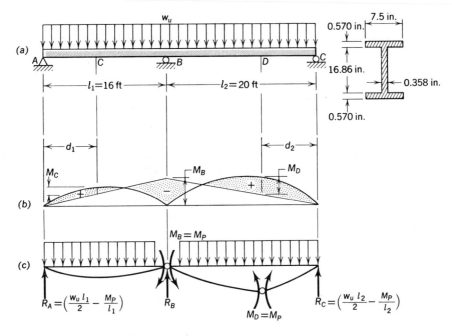

Procedure: From Example 3–8, the ultimate moment this steel section can carry is $M_p = 4 \times 10^6$ in.-lb $= \frac{1}{3} \times 10^6$ ft-lb. Thus, if the two spans were equal, we would have symmetry of deflection, with the slope of the beam over support B equal to zero, and consequently the ultimate load would be the same as for the beam in Figure 5–6. However, we do not have symmetry and, therefore, this result does not apply. To analyze the beam, we will begin by assuming that span B–C forms the mechanism first, with plastic hinges at B and D. The deflected shape for this state is shown in Figure c. Thus, from equilibrium, we have

$$\Sigma \, M_B = 0: \quad M_p = \frac{w_u l_2{}^2}{2} - R_C l_2 \tag{a}$$

and

$$\Sigma \, M_D = 0: \quad M_p = -\frac{w_u d_2{}^2}{2} + R_C d_2 \tag{b}$$

Within span l_2, the maximum moment $M_D = M_p$ occurs when the shear is zero. Thus, we get

$$d_2 = R_C / w_u \tag{c}$$

Solving Equations (a), (b), and (c) simultaneously, we get

$$w_u = \frac{11.62 M_p}{l_2{}^2} \tag{d}$$

which is the same result obtained for the beam in Figure 5–6a. We expected this because the "plastic" conditions of the two beams were the same. The value of w_u may be determined from Equation (d) because we are given that $l_2 = 20$ ft and $M_p = \frac{1}{3} \times 10^6$ ft-lb. Therefore,

$$w_u = \frac{11.62 \times \frac{1}{3} \times 10^6}{(20)^2} = \frac{11.62}{12} \times 10^4 \; \text{lb/ft}$$

Now we must check that our assumption (that span B–C forms the plastic hinges first) is correct. If the moment at C is less than M_p, the assumption is correct, and the correct answer is that $w_u = (11.62/12) \times 10^4$ lb/ft.

The moment at C is a maximum when the shear in span A–B is zero; that is, from Figure c

$$\frac{w_u l_1}{2} - \frac{M_p}{l_1} = w_1 d_1$$

from which

$$d_1 = \frac{l_1}{2} - \frac{M_p}{l_1 w_u} \tag{e}$$

Thus, the moment at C is

$$M_C = \left(\frac{w_u l_1}{2} - \frac{M_p}{l_1}\right) d_1 - \frac{w_u d_1{}^2}{2} \qquad (f)$$

Substituting the value of d_1 from (e) into Equation (f) we get

$$M_C = \frac{w_u l_1{}^2}{8} - \frac{M_p}{2}\left(1 - \frac{M_p}{w_u l_1{}^2}\right) \qquad (g)$$

For $\omega_u = (11.62/12) \times 10^4$ lb, $l_1 = 16$ ft and $M_p = \frac{1}{3} \times 10^6$, M_C becomes

$$M_C = 0.167 \times 10^6 < \tfrac{1}{3} \times 10^6 = M_p$$

Thus, our assumption was correct and the ultimate uniform load the beam can carry is $(11.62/12) \times 10^4$ lb/ft.

PROBLEMS

5–104 Determine the ultimate load P that the beam in Figure P 5–104 can carry, given that $\sigma_y = 2000$ psi.

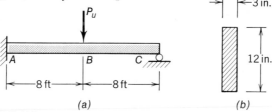

(a)   (b)

Figure P 5–104

5–105 Determine the ultimate load P_u that the beam in Figure 5–105 can carry in terms of M_P.

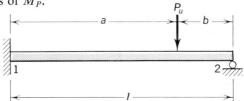

Figure P 5–105

5–106 Determine the ultimate load w_u that the beam in Figure P 5–106 can carry in terms of M_P.

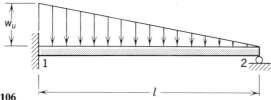

Figure P 5—106

5–107 Determine the ultimate load w_u that the beam in Figure P 5–107 can carry in terms of M_P.

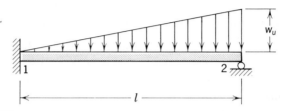

Figure P 5–107

5–108 Determine the ultimate load w_u that the beam in Figure P 5–108 can carry in terms of M_P.

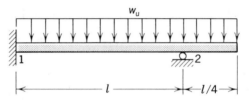

Figure P 5–108

5–109 Determine the ultimate load P_u that the beam in Figure P 5–109 can carry in terms of M_P.

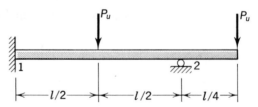

Figure P 5–109

5–110 Determine the ultimate load P_u that the beam in Figure P 5–110 can carry in terms of M_P.

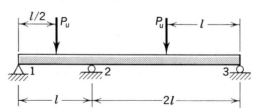

Figure P 5–110

5-111 Determine the ultimate load w_u that the beam in Figure P 5-111 can carry in terms of M_P.

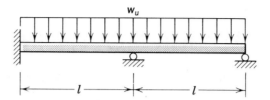

Figure P 5-111

5-112 Determine the ultimate load P_u that the beam in Figure P 5-112 can carry in terms of M_P.

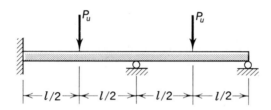

Figure P 5-112

5-113 For the beam in Figure P 5-105 determine the ultimate load P_u, given that the beam is of rectangular section 2 in. × 8 in., $l = 12$ ft, $a = 7$ ft, and $\sigma_y = 1500$ psi.

5-114 For the beam in Figure P 5-111 determine the ultimate load w_u, given that the beam is a standard 8-in. I beam, 24 lb, $l = 16$ ft, and $\sigma_y = 35,000$ psi.

Chapter 6 ◆ ENERGY CONCEPTS

6—1 Introduction

When a body deforms as a consequence of the applied load, external work is done on the body by the external loads. On the other hand, as deformation takes place, *internal work*, more commonly referred to as *strain energy*, is stored within the body as potential energy, completely recoverable, if the proportional limit of the material is not exceeded. In accordance with the principle of conservation of energy, we may assume that the internal work equals the external work, and thus we establish the basis for use in relating deformation to external loads.

The *deformation-load* relationships based on the energy principles to be presented in this chapter shall be limited to linearly elastic systems, although the energy principles, as such, are applicable to any conservative deformable system. These relationships offer a powerful tool for analyzing structures; they are especially useful for statically indeterminate structures and beam deflections. Energies related to axial, bending, torsional, and shear "loads" are considered separately in the next four sections. It is to be understood, however, that total energy of a system subjected to any combination of these loads is the sum of the energies stored in the system by each type of load. We shall denote this total energy by U. It is thus possible for us to determine the total deformation of a member subjected to more than one load, a situation frequently encountered in engineering.

For a linearly elastic system, the load-deformation relationship is a straight line, as shown in Figure 6–1. If P represents a general load and δ represents the deformation of the body caused by and in the direction of the load, then an increase in P by an amount dP results in a corresponding increase in deformation and, consequently, in an increment of strain energy dU stored in the system. The total energy stored in the system is therefore

360

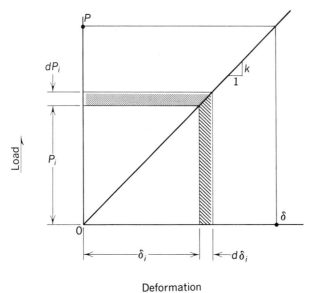

Figure 6–1

$$du_i = Pd\delta_i$$
$$P = k\delta$$
$$\int du_i = U = \int Pd\delta_i$$
$$= k\int \delta d\delta$$
$$= \frac{k\delta^2}{2} = \frac{P\delta}{2}$$

Then, $U = \Sigma_i u_i$

the sum of all such increments between specified limits of load for which linearity holds.

$$U = \tfrac{1}{2}\sum_{i=1}^{n} P_i\delta_i = \tfrac{1}{2}\cdot P\cdot\delta \qquad (6\text{--}1)$$

Equation (6–1) is a general expression which takes a more specific form when the load is specified as an axial or shear force, or moments (bending or torsion), as illustrated in the following sections. Also, note that the equation represents the area under the curve, that is, the area of the triangle of base δ and height P.

6—2 Strain Energy due to Normal Stresses

Let us consider a specific case of strain energy, the energy stored in a bar from an axial force P (tension or compression). Hence, we examine the bar in Figure 6–2, of a uniform cross section A, subjected to the tensile

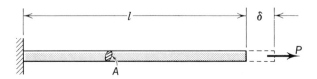

Figure 6–2

load P, as shown. In terms of P, the axial deformation of the bar is $\delta = Pl/AE$. From Equation (6–1), the total energy stored in the bar is

$$U = \frac{P}{2}\left(\frac{Pl}{AE}\right) = \frac{P^2 l}{2AE} \quad \text{(in terms of load)} \qquad (6\text{--}2a)$$

or

$$U = \frac{\delta^2 AE}{2l} \quad \text{(in terms of axial deformation)} \qquad (6\text{--}2b)$$

Both forms are useful for various cases of analysis. In terms of stress and strain ($\sigma = P/A$, $\epsilon = \delta/l$), the two expressions for energy are

$$U = \frac{\sigma^2 Al}{2E} \quad \text{(in terms of stress)} \qquad (6\text{--}3a)$$

$$U = \frac{\epsilon^2 EAl}{2} \quad \text{(in terms of strain)} \qquad (6\text{--}3b)$$

We note that the product Al in the above equation represents the volume. Commonly represented by u, the energy per unit volume is

$$u = \frac{\sigma^2}{2E} \qquad (6\text{--}4a)$$

or

$$u = \frac{\epsilon^2 E}{2} \qquad (6\text{--}4b)$$

When the stress at the proportional limit σ_p is substituted into Equation (6–4a), the unit strain represents the *modulus of resilience*, a quantity that is useful to measure the ability of a material to absorb energy.

The above discussion implied static loads. Actually, the same concept may be extended to impact loads. Consider, for example, the freely falling weight W in Figure 6–3, and assume the weight of the bar to be negligible compared to W. The weight W will be assumed to fall freely for a height h, starting from rest and then coming to rest by means of a head attached to the end of the rod, as shown. As it hits the head, the weight will elongate the bar by an amount δ. The work done by the weight while falling must

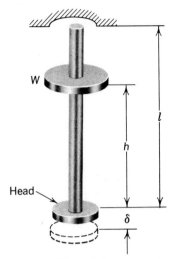

Figure 6–3

equal the energy stored in the bar. Thus, from Equation (6–2)

$$W(h + \delta) = \frac{\delta^2 AE}{2l} \qquad (a)$$

from which

$$\frac{Wl}{AE} = \frac{\delta^2}{2(h + \delta)} \qquad (b)$$

The quantity $Wl/AE = \delta_{st}$, the elongation of the bar due to W under static conditions. Substituting δ_{st} into Equation (b), and solving for δ, the instantaneous deformation, we obtain

$$\delta = \delta_{st} + \sqrt{(\delta_{st})^2 + 2\delta_{st} \cdot h} \qquad (6\text{--}5)$$

From the relationship $\delta = \sigma(l/E)$, Equation (6–5) becomes

$$\sigma = \sigma_{st} + \sqrt{(\sigma_{st})^2 + (2Eh\sigma_{st}/l)} \qquad (6\text{--}6)$$

For both static and dynamic loads, the equations were derived on the assumption that the proportional limit of the material was not exceeded. Consequently, the expressions are not valid for plastic conditions.[1]

Example 6–1:

Given: The structure in Figure *a*. Assume the two truss members to be of the same material and that the cross-sectional area of member *BC* is 3 times that of *AB*.

---

[1] See F. R. Shanley, *Strength of Materials*, McGraw-Hill Book Company, Inc., New York, 1957.

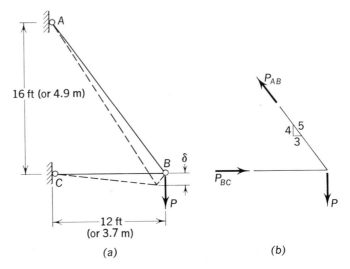

(a)   (b)

Find: The vertical deflection of point *B*.

Procedure: The deflected shape of the truss is shown (exaggerated) by the dashed lines. Thus, point *B*, and with it load *P*, travels vertically *and* horizontally. However, because *P* has no horizontal component, the total work done on the system is the *average* value of the load ($P/2$) times the vertical travel (δ). This external work is equal to the total energy stored in the two members (neglecting deformations in hinges and supports). From Equation (6–2a), we have

$$\frac{P \cdot \delta}{2} = \left(\frac{P^2 l}{2AE}\right)_{AB} + \left(\frac{P^2 l}{2AE}\right)_{BC}$$

From $\Sigma F_y = 0$ and $\Sigma F_x = 0$, we have

$$P_{AB} = \tfrac{5}{4}P \quad \text{and} \quad P_{BC} = \tfrac{3}{4}P$$

Substituting these values and those for lengths and areas, we obtain

$$\frac{P\delta}{2} = \frac{(\tfrac{5}{4}P)^2 \times 20}{2AE} + \frac{(\tfrac{3}{4}P)^2 \times 12}{2(3A)E}$$

from which

$$\delta = \frac{125P}{4AE} + \frac{9P}{4AE} = \frac{67}{2}\frac{P}{AE}$$

Example 6–2:

Given: A 3-ton weight slides down a steel rod (assume there is no friction force between rod and weight), as shown in the figure.

Find: The stress in the rod caused by the weight as it is released from a height *h* at (a) $h = 1$ ft, and (b) $h = 0$ ft.

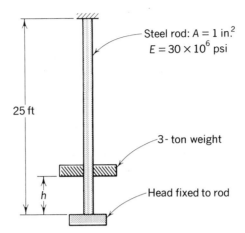

Steel rod: A = 1 in.²
E = 30 × 10⁶ psi

25 ft

3- ton weight

Head fixed to rod

h

Procedure: We may use Equation (6–6)

$$\sigma_{st} = \frac{P}{A} = \frac{6000}{1} = 6000 \text{ psi}$$

Thus, for $h = 1$ ft,

$$\sigma = 6000 + \sqrt{(6000)^2 + \frac{2 \times 30 \times 10^6 \times 1 \times 6000}{25}}$$

$$\sigma = 6000(1 + 20) = 126,000 \text{ psi}$$

For $h = 0$ ft,

$$\sigma = 6000(1 + 1) = 12,000 \text{ psi}$$

Interestingly, we note from the last result, that a suddenly applied load (the weight is merely released) causes a stress in the rod twice as large as if the load was applied statically (gradually).

PROBLEMS

6-1 An aluminum bar, 1 in. diameter and 15 ft long, is stretched by a 6000-lb force. Determine the strain energy stored in the bar. $E_a = 10 \times 10^6$ psi.

6-2 Determine the strain energy stored in a steel (493 lb/ft³) bar that is l ft long, and has a cross-sectional area A sq in., hanging vertically under its own weight.

6-3 A steel bar ($E = 30 \times 10^6$ psi) 27 ft long, 1.5 in. in diameter, hangs vertically and, besides its own weight, supports a 2-ton static load at the lower end. Determine (*a*) the maximum stress in the bar, (*b*) the elongation of the bar, (c) the modulus of resilience, and (d) the total energy stored in the bar.

6-4   If the weight in Problem 6–3 was to drop suddenly a distance of 18 in., determine (a) the maximum stress in the bar, (b) the maximum elongation of the bar, and (c) the total elastic strain energy stored in the bar.

6-5   Determine the maximum height from which the weight in Problem 6–3 may be dropped if the stress in the bar is not to exceed 65,000 psi.

6-6   Determine the total energy stored in truss members in Figure P 6–6 if the members of the truss have the same E and the cross-sectional areas as shown. Neglect the weight of the members.

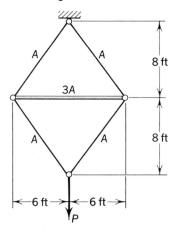

Figure P 6–6

6-7   Determine the vertical travel of the force P in Figure P 6–6 if the following data are given: $A = 1.2$ sq in., $P = 14$ kips, $E = 10 \times 10^6$ psi. (Neglect the weight of the bars.)

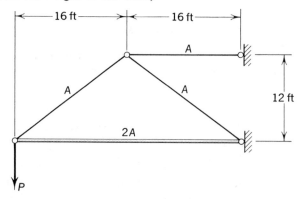

Figure P 6–8

6-8 Determine the total strain energy stored in the truss system shown in Figure P 6–8 if the members have the same E and if the cross-sectional areas are as shown.

6-9 For the truss in Figure P 6–8, determine the total strain energy stored in the system given that: $A = 3$ sq in., $P = 35$ kips, and $E = 30 \times 10^6$ psi. How much will the load P travel vertically?

6-10 Determine the total strain energy stored in the truss system shown in Figure P 6–10 if E and A is the same for all members.

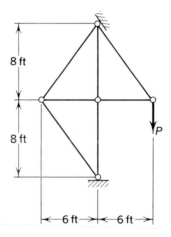

Figure P 6–10

6-11 Find the vertical travel of the load P in Figure P 6–10 given that $P = 20$ kips, $A = 1.5$ sq in., and $E = 30 \times 10^6$ psi.

6-12 Determine the vertical travel of the point C in Figure P 6–12 given that $P = 15$ kips, $A = 1$ sq in. (all members), and $E = 30 \times 10^6$ psi.

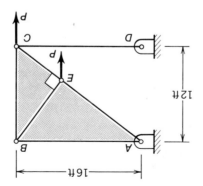

Figure P 6–12

6—3 Strain Energy due to Torsion

The discussion of external work and strain energy presented in Section 6–1 will now be extended to cases of round shafts subjected to pure torque. Here, too, we assume linearity of deformation. Thus, considering the round shaft in Figure 6–4a, the angular deformation is proportional to the torque T as the torque is increased from zero to a final value of T. This torque-deformation relationship is represented generally in Figure 6–4b. The external work done by the torque is, therefore, the average torque $(T/2)$ times the rotation θ. Hence, the energy stored in the shaft is

$$U = (T/2)(\theta) = (T/2)(Tl/JG) = T^2l/2JG \tag{6-7}$$

where $\theta = Tl/JG$, as expressed in Equation (2–4). In terms of angular deformation, the total strain energy stored in the shaft is

$$U = (\theta JG/2l)\theta = \theta^2 JG/2l \tag{6-8}$$

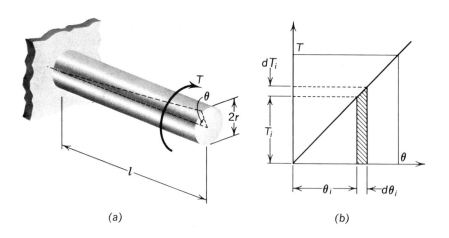

(a) (b)

Figure 6–4

Also, in terms of the maximum shearing stress, from $\tau = Tr/J$, and $\theta = Tl/JG = (\tau J/r)l/JG = \tau l/rG$, the total energy is

$$U = (\tau J/2r)(\tau l/rG) = \tau^2 Jl/2r^2 G \tag{6-9}$$

Example 6—3:

Given: A round shaft made up of two sections (same material) as shown in the figure.

Find: The ratio of the strain energies stored in each section.

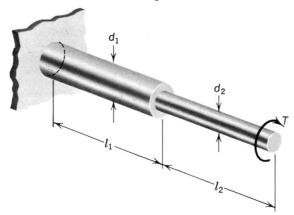

Procedure: Because both sections are subjected to the same torque, we may write [see Equation (6-7)]

$$U_1 = T^2 l_1 / 2 J_1 G \quad \text{and} \quad U_2 = T^2 l_2 / 2 J_2 G$$

from which

$$\frac{U_1}{U_2} = \frac{l_1 J_2}{l_2 J_1} = \left(\frac{l_1}{l_2}\right)\left(\frac{\pi d_2^4/32}{\pi d_1^4/32}\right) = \left(\frac{l_1}{l_2}\right)\left(\frac{d_2}{d_1}\right)^4$$

Example 6—4:

Given: A round bar, of diameter $d = 2r$, is bent to form a circular arc, of radius R, and fixed at one end, as shown in Figure a. A load P perpendicular to the plane in which the curved bar lies, bends and twists the bar.

Find: The strain energy stored in the bar from torsion alone.

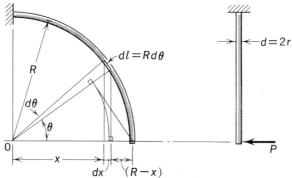

(a) A round bar bent to form a circular arc. (b) Side view.

Procedure: From the figure, we note that at any point x from point 0 (the center of the circle) the torque varies, and thus Equation (6–7) has to be modified accordingly. The expression for torque is

$$T = P(R - x) \qquad (a)$$

In terms of θ and R,

$$x = R \cdot \cos \theta \qquad (b)$$

from which Equation (*a*) becomes

$$T = PR(1 - \cos \theta) \qquad (c)$$

For an increment of length dl, the strain energy in torsion [Equation (6–7)] becomes

$$dU = T^2(dl)/2JG = T^2 R d\theta/2JG \qquad (d)$$

Hence, substituting for T in Equation (*d*), the total energy becomes

$$U = \int_0^{\pi/2} \frac{[PR(1 - \cos \theta)]^2 R \; d\theta}{2JG}$$

or

$$U = \frac{P^2 R^3}{2JG} \int_0^{\pi/2} (1 - 2 \cos \theta + \cos^2 \theta) d\theta$$

which integrates to

$$U = (P^2 R^3/2JG)(\theta - 2 \sin \theta + \theta/2 + \sin 2\theta/4)_0^{\pi/2}$$

Evaluating the integral, and substituting $\pi(2r)^4/32 = J$, we have

$$U = \left(\frac{P^2}{rG}\right)\left(\frac{R}{r}\right)^3\left(\frac{3}{4} - \frac{2}{\pi}\right) = \left(\frac{0.113P^2}{rG}\right)\left(\frac{R}{r}\right)^3$$

PROBLEMS

6–13 A solid shaft fits into another tubular shaft of the same material, as shown in Figure P 6–13. Assuming no friction between the two shafts, determine the ratio of the strain energies absorbed by the inside and outside shafts, in terms of the given dimensions and properties, if they are twisted through the same angle at the free end.

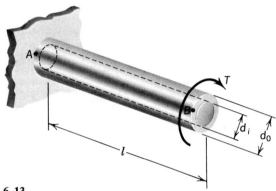

Figure P 6–13

6–14 A round solid shaft is fixed at both ends and has the dimensions shown in Figure P 6–14. Determine the ratios of the energies absorbed by each section if (*a*) both sections are made of the same material, and (*b*) the material in the two sections is different.

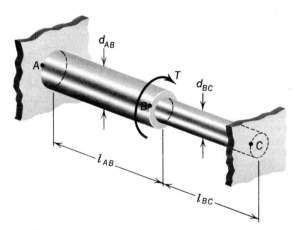

Figure P 6–14

6–15 A round shaft, of uniform diameter, and same material, is subjected to a torque T at point C, and another torque $2T$ at point B, as shown in Figure P 6–15. Determine the ratio l_{AB}/l_{BC} so that both sections will absorb the same amount of strain energy.

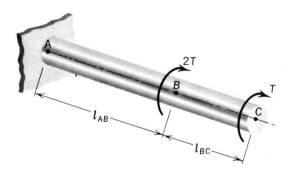

Figure P 6–15

6–16 For the shaft in Figure P 6–15, determine the ratios of the strain energies U_{AB}/U_{BC} when the ratio $l_{AB}/l_{BC} = \frac{1}{3}$.

6–17 A closely coiled helical spring, subjected to an axial load of 150 lb, has the following characteristics: $R = 1$ in., $r = \frac{1}{8}$ in., 15 turns, $G = 12 \times 10^6$ psi. Determine (a) the total strain energy absorbed by the spring, and (b) the deflection of the spring (travel of load P relative to the other end).

6–18 Two closely coiled springs are connected in series and stretched by an axial load of 100 lb, as shown in Figure P 6–18. Each spring is made of a $\frac{1}{4}$-in.-diam wire and has mean diameter of 2 in. If one is steel ($G_{St} = 12 \times 10^6$ psi) and has 20 turns, and the other is phosphor bronze ($G_{Ph} = 6 \times 10^6$ psi) and has 10 turns, determine (a) the total energy absorbed by each spring, and (b) the total travel of force P.

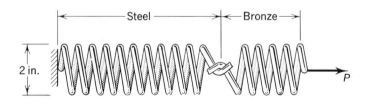

Figure P 6–18

6–19 A semicircular arch, in a horizontal plane and fixed at one end, is subjected to "transverse" force P, as shown in Figure P 6–19. Determine (a) the total energy, due to torsion alone, that the bar absorbs, and (b) the transverse deflection of point B caused by torsional deformation only.

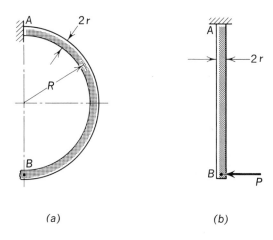

(a)   (b)

Figure P 6–19   (a) A semicircular arch.   (b) Side view.

6—4   Strain Energy due to Bending

Let us consider a small length of a beam subjected to pure bending as shown in Figure 6–5a. As the moment is increased from zero to M, the beam bends, resulting in a curvature $1/\rho = M/EI$ as expressed by Equation (3–5). From $dx \approx dl = \rho d\theta$, we have $d\theta = dx/\rho = (M/EI) dx$ or

$$\theta = \int_0^l \frac{M dx}{EI} = \frac{Ml}{EI} \tag{a}$$

Thus, we see that the angular deformation varies linearly with the moment as shown by Figure 6–5b. The work that the "average" moment (a couple) does as it rotates through an angle θ is equal to the area under the M-θ diagram. This is therefore the strain energy stored in the beam.

$$U = \left(\frac{M}{2}\right)\theta = \int_0^l \frac{M^2 dx}{2EI} = \left(\frac{M}{2}\right)\left(\frac{Ml}{EI}\right) = \frac{M^2 l}{2EI} \tag{6–10}$$

or, in terms of angular displacement, we have

$$U = (\theta EI/2l)\cdot\theta = \theta^2 EI/2l \tag{6–11}$$

Both equations resemble closely Equations (6–7) and (6–8), respectively, which are applicable for pure torsion. In terms of maximum bending stress,

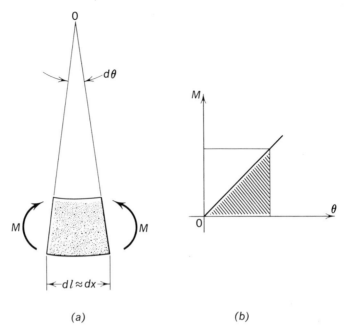

(a)   (b)

Figure 6–5

from $\sigma_{max} = Mc/I$, we have

$$U = \frac{(\sigma_{max}I/c)^2 l}{2EI} = \frac{\sigma_{max}^2 Il}{2 \cdot c^2 \cdot E} \tag{6-12}$$

An interesting comparison of energies absorbed by bars under axial and bending stresses may be easily made for *rectangular sections*.

In tension: $U_{tension} = \dfrac{\sigma_{max}^2 (bd) l}{2E}$ |see Equation (6–3a)|

In bending: $U_{bending} = \dfrac{\sigma_{max}^2 (bd^3/12) l}{2(d/2)^2 E}$ |see Equation (6–12)|

from which

$$\frac{U_{tension}}{U_{bending}} = 3$$

Interestingly, the energy stored in tension may be 3 times as large as that in bending for the same stress.

More frequently than not, a beam in an engineering problem is not subjected to only pure bending; shear forces are also present. The strain energy in shear will be discussed in the following section. For now, we shall assume that the beam is long and relatively shallow and therefore the effect of the shear is negligible. Over an element of length dx, the varying bending moment M_x may be assumed constant. Hence, from Equation (6–10), the total strain energy in a beam of length l is

$$U = \int \frac{M_x{}^2 dx}{2EI} \qquad (6\text{–}13)$$

where M_x is expressed as a function of x (explained in Chapters 3 and 4). The following examples will illustrate some uses for the above expressions.

Example 6—5:

Given: The cantilever beam shown in Figure a.
Find: (a) The strain energy stored in the beam, and (b) the maximum deflection of the beam.

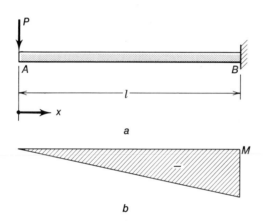

a

b

Procedure: The moment diagram is shown in Figure b. For convenience, the origin of x is assumed to be at point B, as shown in Figure a; hence, we write the equation for moment:

$$M_x = -Px$$

Substituting in Equation (6–13) and integrating, the expression for the total energy becomes

$$U = \int_0^l \frac{(-Px)^2 dx}{2EI} = \frac{P^2 x^3}{6EI} \bigg]_0^l = \frac{P^2 l^3}{6EI}$$

Letting the external work done by the load as the beam deflects be equal to the strain energy, we get the maximum deflection (at point B):

$$\frac{1}{2} \cdot P \cdot \delta_B = \frac{P^2 l^3}{6EI}$$

from which

$$\delta_B = \frac{Pl^3}{3EI}$$

which is the same expression derived by other methods in Chapter 4.

Example 6—6:

Given: The simply supported beam in Figure *a*.
Find: (*a*) The strain energy stored in the beam, and (*b*) the maximum deflection of the beam.

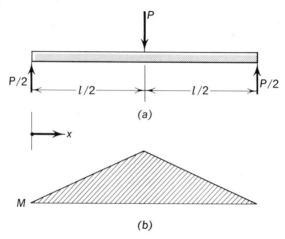

(a)

(b)

Procedure: The moment diagram is shown in Figure *b*. Because of symmetry, it is not necessary to write the equation for moment for the whole beam in order to find the total energy; we merely find the energy for one-half and then double it, that is,

$$M_x = (P/2)\, x \qquad \text{(for } 0 < x < l/2)$$

Thus,

$$U = 2 \int_0^{l/2} \frac{(\frac{1}{2}Px)^2 dx}{2EI} = \frac{P^2 x^3}{12EI}\Bigg]_0^{l/2} = \frac{P^2 l^3}{96EI}$$

The maximum deflection occurs at the center of the span. Hence, letting the external work equal the total strain energy, we obtain the maximum deflection of the beam

$$\frac{1}{2} P \cdot \delta_c = \frac{P^2 l^3}{96EI}$$

from which

$$\delta_c = \frac{Pl^3}{48EI}$$

which agrees with answers for δ_c, determined by other methods in Chapter 4.

PROBLEMS

For the following problems, assume EI is constant along the length of the beam.

6–20 Determine the slope at B of the beam in Figure P 6–20 subjected to the pure moment M using the energy approach.

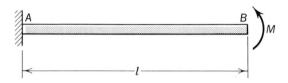

Figure P 6–20

6–21 Compare the energies stored in the two beams shown in Figure P 6–21, assuming that the two beams are identical, and the stresses are within the elastic limit of the material.

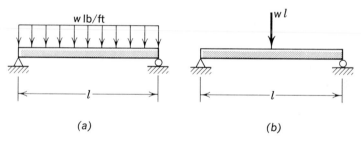

(a)   (b)

Figure P 6–21

6–22 Compare the energies stored in the two beams shown in Figure P 6–22, assuming that the two beams are identical, and the stresses are within the elastic limit of the material.

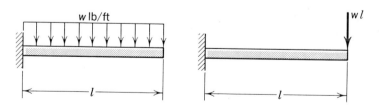

Figure P 6–22

6–23 For the beam in Figure P 6–23, determine the total strain energy stored in the beam.

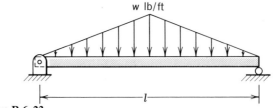

w lb/ft

Figure P 6–23

6–24 A weight W is dropped from a height h onto the free end of a cantilever beam, as shown in Figure P 6–24. If the load were applied gradually, it would cause a deflection δ_{st} at the free end. Determine the instantaneous deflection δ_i in terms of δ_{st} and h. [Hint: $W(h + \delta_i) = \frac{1}{2} W_i \cdot \delta_i$ where $W_i = W(\delta_i/\delta_{st})$, and $\delta_i = W_i l^3/3EI$.]

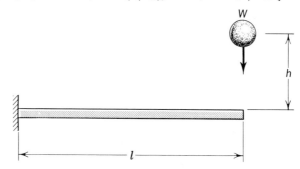

W

h

l

Figure P 6–24

6–25 A weight W is dropped from a height h onto the center of a simply supported beam, as shown in Figure P 6–25. If the load were applied gradually, it would cause a deflection δ_{st} at the center. Determine the instantaneous deflection δ_i in terms of δ_{st} and h. [Hint: $W(h + \delta_i) = \frac{1}{2} W_i \delta_i$ where $W_i = W(\delta_i/\delta_{st})$, and $\delta_i = W_i l^3/48 \, EI$.]

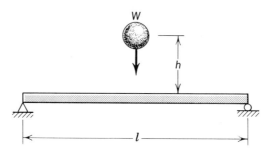

W

h

l

Figure P 6–25

6–26 For Problem 6–24 determine the instantaneous bending stress σ_i in terms of the static stress σ_{st}, h, l, E, and I.

6–27 For Problem 6–25 determine the instantaneous bending stress σ_i in terms of the static stress σ_{st}, h, l, E, and I.

6–28 Determine the "bending" strain energy absorbed by beam A–B in Figure P 6–28. Consider joints B and C as rigid.

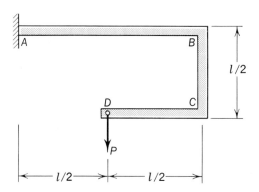

Figure P 6–28

6—5 Strain Energy due to Shear

Let us consider a small element subjected to pure shear as shown in Figure 6–6. The shear stresses on the four faces must be numerically equal, as shown in Section 1–11; here the stresses are merely represented by τ.

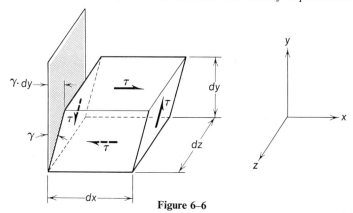

Figure 6–6

On the upper face the *shear force* is $V = \tau \cdot dx \cdot dz$. As the average force travels an amount $\gamma \cdot dy$, it does work which is equal to $\frac{1}{2}(\tau dx \cdot dz)\gamma \cdot dy = \frac{1}{2} \cdot V \cdot \gamma dy$. Because the shear forces on the other faces do no work, the following expression for work represents the total energy stored in the element; that is,

$$U = \iiint_{\text{volume}} \frac{\tau \cdot \gamma \cdot dy \cdot dz \cdot dx}{2} \qquad (a)$$

Assuming a constant cross-sectional area A, $\gamma = \tau/G = V/AG$; Equation (a) becomes

$$U = \iiint_{\text{volume}} \frac{\tau^2 \cdot dy \cdot dz \cdot dx}{2G} = \frac{V^2}{2GA} \int_l dy \qquad (6\text{--}14)$$

The strain energy per unit volume is merely

$$u = \frac{\tau^2}{2G} \qquad (6\text{--}15)$$

Example 6–7:

Given: A rectangular section shown in Figure a, subjected to a uniform shear force V over a length l.

Find: The total strain energy due to shear in the beam.

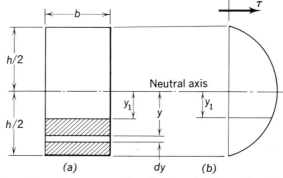

Procedure: The shear-stress distribution is shown in Figure b. Once we determine an expression for the shear stress, we may use Equation (6–14) to determine the total strain energy. Thus,

$$\tau = \frac{V}{Ib} \int_{y_1}^{h/2} b \cdot y \cdot dy = \frac{V}{2I}\left(\frac{h^2}{4} - y^2\right) \quad \text{[see Equation(3–9)]}$$

Substituting in Equation (6–14), we obtain

$$U = \iiint_{\text{volume}} \frac{V^2}{8I^2G}\left(\frac{h^2}{4} - y^2\right)^2 \cdot dy \cdot dz \cdot dx$$

The quantities dy, dz, and dx range between $-h/2$ to $+h/2$, 0 to b, and 0 to l, respectively. Thus,

$$U = \frac{V^2}{8I^2G} \int_0^l \int_0^b \int_{-h/2}^{h/2}\left(\frac{h^2}{4} - y^2\right)^2 dy \cdot dz \cdot dx$$

Integrating,[2] we have

$$U = \frac{V^2 h^5 bl}{240 \cdot I^2 G}$$

_____

[2] The integration is left as an exercise for the student.

6—6   Principle of Virtual Work

The principle of virtual work is probably the most direct and versatile method for computing deflections. It is indeed a powerful tool of analysis, applicable not only to deflections but also to the analysis of reactions for both elastic as well as plastic conditions. Stevinus (1548–1620) suggested the method and Galileo (1564–1642) made use of it, as a result of their investigations on pulleys and inclined planes. But it was Johann Bernoulli (1667–1748) who, in a letter to Varignon in 1717, gave a formal statement of the principle (he called it the *principle of virtual velocities*). As used here, the word *virtual* will be taken to mean an imaginary or hypothetical force, or deformation, either finite or infinitesimal. Hence, the resulting work is only imaginary or hypothetical in nature.

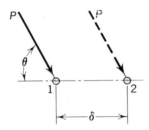

Figure 6–7

As a prelude to the above principle, let us consider the action of a single force P acting as shown in Figure 6–7. The work that the force does as it travels from point 1 to 2 is equal to the horizontal distance between the points, times the horizontal component of the force; that is,

$$U = (P \cos \theta) \cdot \delta = P(\delta \cos \theta) \tag{a}$$

Equation (*a*) may be interpreted as: (1) the work is the product of displacement and component of force in the direction of displacement, or (2) the work is the product of the force and the component of the displacement in the direction of the force.

Now, let us consider a perfectly rigid body subjected to a system of forces which places the body in equilibrium, as shown in Figure 6–8. Because the body is perfectly *rigid*, no internal deformation of the body takes place, and consequently, the internal work on a rigid body must be zero. Thus, as the rigid body is translated by some other cause unrelated to the forces shown, so that a point on the body travels an amount δ from a position 1 to 2, all the work that is done is external. This is a consequence of the components of the forces acting on the body as they travel through the

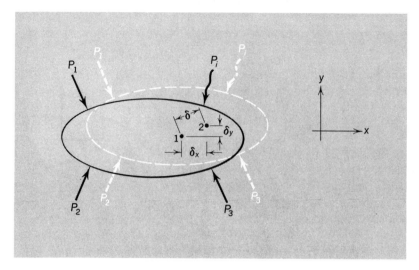

Figure 6–8

distance that the body is translated. The total work done by all the forces during this translation is

$$U = \delta_x \cdot \Sigma\, P_{ix} + \delta_y \cdot \Sigma\, P_{iy} = \delta_r \cdot P_r \qquad (b)$$

where P_r represents the resultant force of the force system. But when the system is in equilibrium, $P_r = 0$, making Equation (b) zero regardless of the direction of δ_r. Rewriting Equation (b), we obtain

$$\delta_x\, \Sigma\, P_{ix} + \delta_y\, \Sigma\, P_{iy} = 0 \qquad (c)$$

A comparative result is obtained if, instead of the forces and translation, we considered moments and rotation, or a combination thereof. Subsequently, we arrive at:

> BERNOULLI PRINCIPLE OF VIRTUAL DISPLACEMENTS: If a system of forces and/or moments (couples) acting on a rigid body is in equilibrium and remains in equilibrium as the body is given a small virtual displacement, the total virtual work done by this system of forces and/or moments is equal to zero.

For a rigid system, the principle is used to impose the condition of equilibrium that is compatible with the restraints of the system. Example 6–8 illustrates a typical use of this principle in structural analysis.

Let us now consider a deformable system. Assume Figure 6–9a represents such a system, a deformable body subjected to a system of forces which keep the body in equilibrium for the duration of their action. The body may be a truss, a beam, a frame, or merely an axially loaded member; the external effects may be either forces or moments. We will classify these

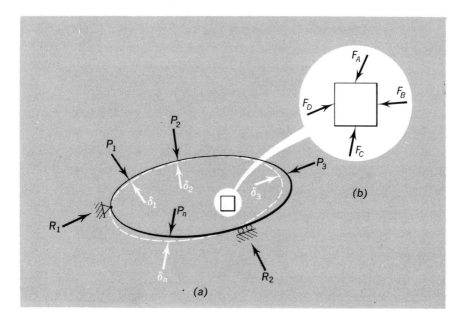

Figure 6–9

as real external forces, denoted by P, and the stresses they cause on an element within the body as real internal forces, denoted by F, as shown in Figure b. Now we impose a *virtual deformation*, not necessarily related to the system of forces. The body as a whole might deform as shown by the dashed line in Figure a. This is consistent with the restraints of the system. Internally, each of say n elements which constitute the body may be deformed and displaced (translated and/or rotated) as a rigid element. The total work resulting from only the displacement of rigid elements (of all the elements) is zero as shown by Equation (c). Hence, the total work in the system is equal to the sum of the products of the real forces F and virtual deformation ϵ_i (strain) of each element; that is, $W_i = \Sigma F \cdot \epsilon_i$. Externally, the P forces[3] also do work during the virtual displacement, associated with the virtual deformation, and the total external work is $W_e = \Sigma P \cdot \delta$. According to the principle of the conservation of energy, the external work must equal the internal work; that is,

$$W_e = W_i \qquad\qquad (6\text{–}16)$$

or

$$W_e - W_i = 0 \qquad\qquad (6\text{–}16a)$$

---

[3] The P forces are assumed to be of constant magnitude during the virtual deformation.

which may be summarized as follows:

> PRINCIPLE OF VIRTUAL WORK: If a deformable body is in equilibrium under a system of external forces and remains in equilibrium while being subjected to a virtual deformation which is compatible with the conditions of restraints, the external virtual work done by the external forces (P forces) is equal to the internal virtual work done by the internal forces (F forces).

As mentioned in the Introduction, the concept of virtual work is widely used in computing deflections of structures (trusses, frames, beams, or torsion members). To get to more specific conditions, let us consider the beam in Figure 6–10. If we were interested in the vertical deflection at point

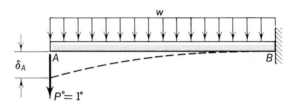

Figure 6–10

A, it is convenient to apply a unit virtual load at A, just before the application of the real load w, which causes the deflection δ_A. Hence, as the beam deflects, the unit load travels down, resulting in external virtual work equal to $1° \cdot \delta_A$. The internal real deformations in the beam caused by the real load w times the respective internal virtual forces caused by the unit virtual loads result in the internal work in the beam. For an element in the beam of length dx and an area dA, located at y distance from the neutral axis of the beam, the *real* deformation is $(My/EI)\,dx$. The virtual force on this element is $(my/I)dA$. Hence, the total virtual internal work (virtual strain energy) is

$$U_i = \iiint_V \left(\frac{my}{I}\right)dA\left(\frac{My}{EI}\right)dx = \int_l \frac{mM dx}{EI}, \text{ since } \iint y^2 dA = I$$

However, because the internal work equals the external work, we obtain the expression for the deflection desired.

For pure bending

$$1° \cdot \delta_A = \int_l \frac{m \cdot M \cdot dx}{EI} \tag{6–17}$$

where m is the moment at any point on the beam caused by the unit load, and M is the moment at any point on the beam caused by the real loads.

In a similar fashion, we can obtain the following expressions.[4] For pure torsion

$$1° \cdot \delta = \int_l \frac{t \cdot T \cdot dx}{GJ} \qquad (6\text{–}18)$$

where t is the virtual torque and T the real torque. For trusses

$$1° \cdot \delta = \Sigma f \cdot \left(\frac{Fl}{AE}\right) \qquad (6\text{–}19)$$

where f is the virtual force in each member of the truss and the term in parenthesis represents the real axial deformation of the truss members caused by the real loads (or perhaps temperature changes).

This method of approach, a special presentation of the virtual work concept, is frequently referred to as the *unit load method*, sometimes credited to James Maxwell (1830–1879) and Otto Mohr (1835–1918) who formulated an independent development of the method. Examples 6–8 through 6–13 should place the presentation of the method in a better, more specific perspective.

Example 6—8:

Given: The beam in the figure.
Find: The value of reaction R_B, in terms of P and M_A by the method of virtual displacement.

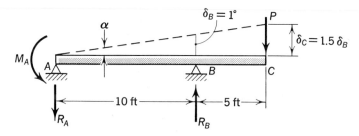

Procedure: Assuming the beam is a rigid body, we give it a virtual displacement equal to unity at point B. By proportion, we find the virtual displacement at C. The work done by R_B and M_A may be assumed positive, and by P negative. According to the principle of virtual displacement, the total energy for the rigid system is

$$U = M \cdot \alpha + R_B \cdot \delta_B - 1.5 \cdot \delta_B \cdot P = 0$$

But

$$\delta_B = 1° \text{ (unity)}; \quad \alpha = \frac{\delta_B}{l} = \frac{1°}{10}$$

---

[4] For derivations and more elaborate treatment of the expression, the student is referred to texts on theory of structures.

Thus, we have

$$M_A \cdot (1°/10) + R_B \cdot 1° - 1.5 \cdot 1°P = 0$$

Solving for R_B,

$$R_B = 1.5 \cdot P - \left(\frac{M_A}{10}\right)$$

The same expression may be obtained from equilibrium. That is, by summing moments about point A, we have

$$R_B = \frac{1.5 \cdot P - M_A}{10} = 1.5P - \frac{M_A}{10}$$

Example 6–9:

Given: The statically indeterminate beam in Figure *a*.
Find: The ultimate load P_u that the beam can carry using virtual work.
Procedure: The plastic analysis approach for statically indeterminate beams was discussed in Section 5–7, to which the student is referred for

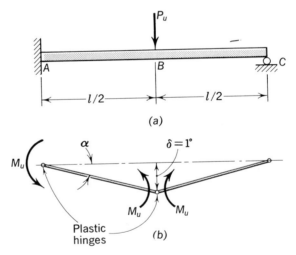

(a)

(b)

any questions on plastic hinges and ultimate moments M_u. Assuming a uniform EI along the length of the beams, the ultimate moments will be of the same magnitude (M_u) at points A and B, oriented and located at the plastic hinges, as shown in Figure *b*. Thus, we may apply a virtual displacement (say $\delta = 1°$) at point B. The total virtual work may be considered as positive by the load P_u and negative by the three moments M_u, once the orientation of the moments is either known or assumed. Hence,

$$P_u \cdot 1° - M_u \alpha - M_u \alpha - M_u \alpha = 0$$

Because

$$\alpha = \frac{1°}{l/2} = \frac{2 \cdot 1°}{l}$$

the above expression becomes

$$P_u - \frac{2}{l} M_u - \frac{2M_u}{l} - \frac{2M_u}{l} = 0$$

from which

$$P_u = 6M_u/l$$

Example 6—10:

Given: The beam in Figure *a*.

Find: (*a*) The deflection at the free end of the beam, and (*b*) the slope of the elastic line at the free end. Use the concept of virtual work (unit load).

Procedure: To determine the deflection [part (*a*)], we may apply a virtual (unit) load at the free end; to determine the slope [part (*b*)], we apply a virtual rotation (unit) moment at the free end. The virtual loads are shown in Figures *b* and *c*, respectively. The moment equations for the respective virtual loadings are shown adjacent to the figures. Thus, we can make use of Equation (6–17) for our results. The

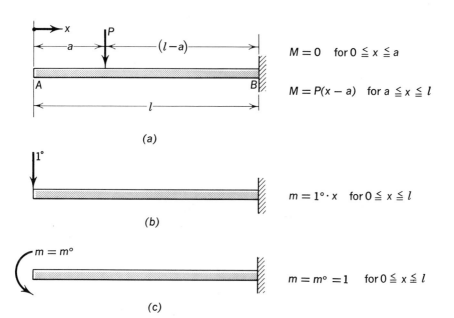

$M = 0 \quad$ for $0 \leqq x \leqq a$

$M = P(x - a) \quad$ for $a \leqq x \leqq l$

(a)

$m = 1° \cdot x \quad$ for $0 \leqq x \leqq l$

(b)

$m = m° = 1 \quad$ for $0 \leqq x \leqq l$

(c)

deflection at the free end becomes

$$1° \cdot \delta_A = \int \frac{mMdx}{EI} = \frac{1}{EI} \int_a^l (1°x)[P(x - a)]dx$$

$$\delta_A = \frac{P}{EI}\left[\left(\frac{x^3}{3} - \frac{x^2a}{2}\right)\right]_a^l = \frac{P}{EI}\left(\frac{l^3}{3} - \frac{l^2a}{2} + \frac{a^3}{6}\right)$$

Now for the slope θ_A, from Figure c, we have

$$m°\theta_A = \int \frac{m° Mdx}{EI} = \frac{1}{EI} \int_a^l (m°)P(x - a)dx$$

$$\theta_A = \frac{P}{EI}\left[\frac{x^2}{2} - ax\right]_a^l = \frac{P}{EI}\left(\frac{l^2}{2} - al + \frac{a^2}{2}\right)$$

Example 6–11:

Given: The beam in Figure a.
Find: The vertical deflection of the beam at midspan.

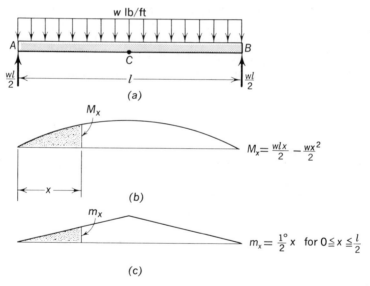

(a)   A beam.   (b) M diagram from w load.   (c) M diagram from unit load.

Procedure: We apply a virtual unit load at midspan, resulting in the virtual moment diagram shown in Figure c. The equations for the respective moments are given adjacent to the diagrams. The deflection δ_c thus becomes

$$\delta_c = \int \frac{m_x M_x dx}{EI} = \frac{2}{EI} \int_0^{l/2} \left(\frac{x}{2}\right)\left(\frac{wlx}{2} - \frac{wx^2}{2}\right)dx$$

Integrating, we have

$$\delta_c = \frac{2w}{EI}\left[\frac{x^3 l}{12} - \frac{x^4}{16}\right]_0^{l/2}$$

Hence,

$$\delta_c = \frac{5wl^4}{384EI}$$

Example 6—12:

Given: The cantilever truss shown in Figure *a*. A concentrated load at point *B* causes the axial deformation shown in the parentheses. Plus indicates elongation, minus indicates contraction.

Find: The vertical and horizontal deflections of point *B*. [*Note:* This is the same problem solved in Example 1-6 by a graphical approach. Compare answers.]

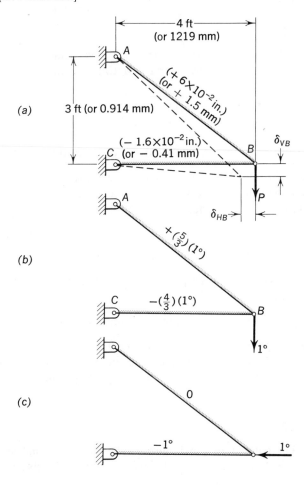

Procedure: For the vertical deflection at B we apply a virtual unit load at B, as shown in Figure b. The direction of the load is assumed downward, indicating a correct assumption if the final deflection is positive, and indicating a wrong assumption if the result is negative. By applying the equations of equilibrium, we determine the virtual forces in the two members. That is,

$$\Sigma F_y = 0: \qquad F_{AB} = +\tfrac{5}{3}(1°) = +\tfrac{5}{3} \text{ (tension)}$$

and

$$\Sigma F_x = 0: \qquad F_{BC} = -\tfrac{1}{3}(1°) = -\tfrac{1}{3} \text{ (compression)}$$

The results are tabulated on the diagram (Figure b). Thus, the total internal work is

$$W_i = (+\tfrac{5}{3})(+ 6 \times 10^{-2}) + (-\tfrac{1}{3})(-1.6 \times 10^{-2})$$
$$= 10 \times 10^{-2} + 2.13 \times 10^{-2}$$
$$W_i = 12.13 \times 10^{-2}$$

The virtual load does work as it travels the vertical distance δ_{VB}. Considering this as *external work* and equating it with the internal work, we obtain δ_{VB}; that is,

$$W_i = W_e = 1° \cdot \delta_{VB} = 12.13 \times 10^{-2} \text{ in.} \qquad \text{[see Equation (6–19)]}$$

or

$$\delta_{VB} = 12.13 \times 10^{-2} \text{ in. (or 3.08 mm)} \downarrow \text{ downward (correct assumption)}$$

For the horizontal deflection at B, δ_{HB}, we apply the unit load horizontally, assumed oriented to the left, as shown in Figure c. By applying the equations of static equilibrium we have

$$\Sigma F_y = 0: \qquad \text{therefore, } F_{AB} = 0$$

$$\Sigma F_x = 0: \qquad F_{BC} = -1° \text{ (compression)}$$

Thus, the work (internal) is $(-1°) \times (-1.65 \times 10^{-2} \text{ in.})$. The horizontal deflection is

$$\delta_{HB} = 1.65 \times 10^{-2} \text{ in. (or 0.42 mm) (correct assumption)}$$

Comparing the results with those from the graphical approach of Example 1–7, we note a slight error in the graphical approach. By selecting a larger scale, the result from the graphical approach can be decidedly improved.

Example 6–13:

Given: The truss shown in Figure a. The cross-sectional area of truss members is 3 in.^{2} [$E_s = 30 \times 10^6$ psi (or 200×10^6 kN/m^2)].

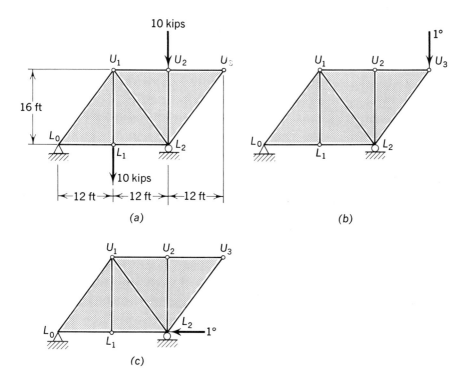

(a)

(b)

(c)

Find: (*a*) The vertical movement of joint U_3 (Figure *b*), and (*b*) the horizontal travel of the roller at support L_2 relative to support L_0. *Procedure:* The forces in the truss members resulting from the respective loads (real loads in Figure *a* and virtual unit loads in Figures *b* and *c*) are tabulated in Table 6–1; plus indicating tension, minus compression.

TABLE 6–1   Forces in the Truss Members

Truss members	L ft	F_p kips	$f_{1°\downarrow}$ kips	$f_{1°\leftarrow}$ kips	$F_p \cdot (f_{1°\downarrow})$	$F_p \cdot (f_{1°\leftarrow})$
$U_1\ U_2$	12	0	+0.75	0	0	0
$U_1\ U_2$	12	0	+0.75	0	0	0
$L_0\ L_1$	12	+3.75	−0.375	−1.0	−1.406 × 12	−3.75 × 12
$L_1\ L_2$	12	+3.75	−0.375	−1.0	−1.406 × 12	−3.75 × 12
$U_1\ L_1$	16	+10	0	0	0	0
$U_2\ L_2$	16	−10	0	0	0	0
$U_1\ L_0$	20	−6.25	+0.625	0	−3.906 × 12	0
$U_1\ L_2$	20	−6.25	−0.625	0	+3.906 × 12	0
$U_3\ L_2$	20	0	−1.25	0	0	0
Σ					−2.812 × 12	−7.5 × 12

From Equation (6–19) we have

$$1°\delta = \Sigma f \cdot (Fl/AE)$$

Because $EA = 30 \times 10^3$ ksi $\times$ 3 in.^{2} = 90×10^3 kips, the vertical deflection at U_3 is

$$1 \text{ kip} \cdot \delta_{U_3} = \frac{1}{EA} \Sigma f F \cdot l$$

$$= \left(\frac{1}{90 \times 10^3 \text{ kips}}\right)(-2.812 \text{ kips}^2 \text{ ft})\left(\frac{12 \text{ in.}}{\text{ft}}\right) \times 12$$

$\delta_{U_3} = -0.375 \times 10^{-3} \times 12 = \mathbf{-4.5 \times 10^{-3}}$ in. (or 1.14×10^{-1} mm) (*upward* because the result is *negative*)

Similarly, the horizontal deflection at L_2 is

$$\delta_{L_2} = \frac{1}{90 \times 10^3} (-7.5)(12)(12) = \mathbf{-12 \times 10^{-3}} \text{ in. (or } 3.05 \times 10^{-1} \text{ mm)}$$

(to the right)

PROBLEMS

Solve the following problems using *virtual work*.

6–29 Determine the reactions for the beam shown in Figure P 6–29.

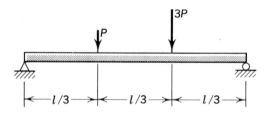

Figure P 6–29

6–30 Determine the value of k and of the left reaction for the beam in Figure P 6–30.

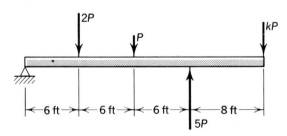

Figure P 6–30

6–31 Determine the ultimate load P_u for the beam in Figure P 6–31 in terms of M_P.

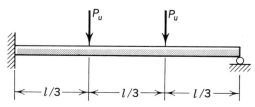

Figure P 6–31

6–32 Determine the ultimate load P_u for the beam in Figure P 6–32 in terms of M_P.

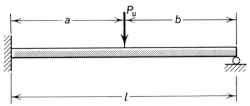

Figure P 6–32

6–33 Determine the maximum deflection and the slope at the left end of the beam in Figure P 6–33, in terms of EI, P, and l.

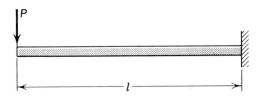

Figure P 6–33

6–34 Determine the maximum deflection and the slope at the left end of the beam in Figure P 6–34, in terms of EI, w, and l.

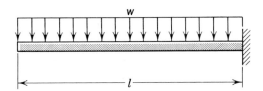

Figure P 6–34

6–35 Determine the maximum deflection and the slope at the left end of the beam in Figure P 6–35, in terms of EI, w_0, and l.

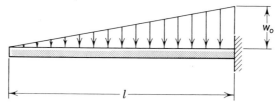

Figure P 6–35

6–36 Determine the maximum deflection and the slope at the left end of the beam in Figure P 6–36, in terms of EI, M, and l.

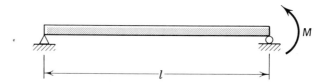

Figure P 6–36

6–37 Determine the maximum deflection and the slope at the left end of the beam in Figure P 6–37, in terms of EI, w_0, and l.

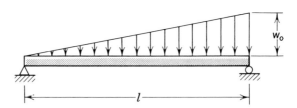

Figure P 6–37

6–38 Determine the maximum deflection of the beam in Figure P 6–38 in terms of EI, w, and l.

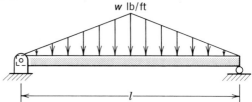

Figure P 6–38

6–39 Determine the maximum deflection and the slppe at the left end of the beam in Figure P 6–39, in terms of *EI*, *P*, and *l*.

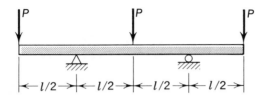

Figure P 6–39

6–40 Determine the deflection at the center of the span of the steel beam shown in Figure P 6–40.

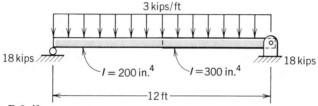

Figure P 6–40

6–41 Determine the maximum deflection and the slope at the left end of the beam in Figure P 6–41, in terms of *EI*, *w*, and *l*.

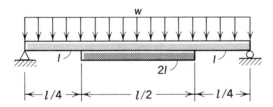

Figure P 6–41

6–42 Determine the slope and deflection at the overhanging end of the beam in Figure P 6–42, in terms of *EI*, *P*, and *l*.

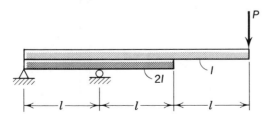

Figure P 6–42

6–43 Determine the deflection and slope at point C for the frame in Figure
P 6–43, in terms of EI, w, and l (the horizontal and vertical members
have the same EI).

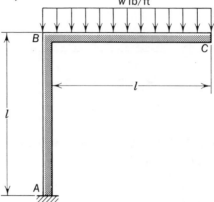

Figure P 6–43

6–44 Determine the deflection and slope at point D for the frame in Figure
P 6–44, in terms of EI, w, and l (the horizontal and vertical members
have the same EI).

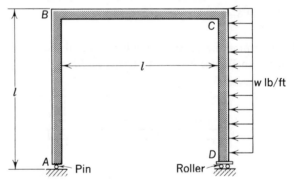

Figure P 6–44

6–45 Determine the vertical and horizontal deflection at point 2 shown in
Figure P 6–45 in terms of AE (both members have the same AE).

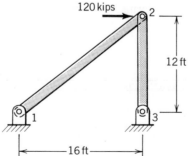

Figure P 6–45

6–46 Determine the vertical deflection at point *B* in Figure P 6–46 if *AE* = 2 × 10⁷ lb for each member.

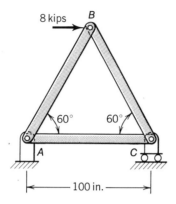

Figure P 6–46

6–47 Determine the horizontal deflection at point *C* in Figure P 6–47 in terms of *AE* (both *AC* and *CB* have the same *AE*).

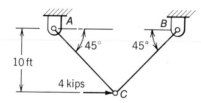

Figure P 6–47

6–48 Determine the horizontal and vertical deflection of point *C* in Figure P 6–48 if *AE* = 90 × 10⁶ lb.

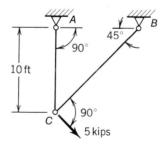

Figure P 6–48

6–49 Determine the horizontal and vertical travel of point B in Figure P 6–49 in terms of P, A, l, E, if members AB and BC have the same AE.

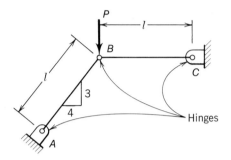

Figure P 6–49

6–50 In Figure P 6–50 determine the horizontal deflection of point B in terms of P, l, and AE.

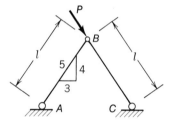

Figure P 6–50

6–51 Determine the horizontal and vertical movement of point B in Figure P 6–51 if $W = 7000$ lb and $AE = 60 \times 10^6$ lb for both members.

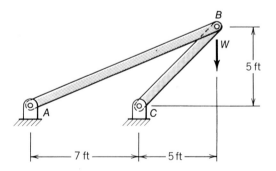

Figure P 6–51

6-52 Determine the vertical travel of the 3.5 kip load in Figure P 6-52. All members are steel.

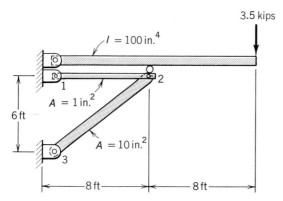

Figure P 6–52

6-53 Determine the horizontal and vertical movement of point B in Figure P 6–53 if $AE = 24 \times 10^7$ lb for both members.

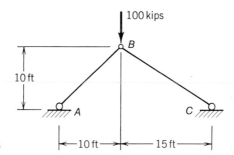

Figure P 6–53

6-54 Determine the horizontal deflection of points B and C in Figure P 6–54 if $AE = 40 \times 10^6$ lb for all members.

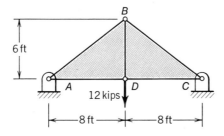

Figure P 6–54

6-55 Determine the vertical movement of point C in Figure P 6–55, in terms of AE (the same for all members).

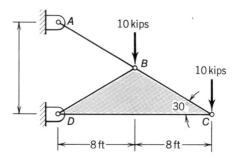

Figure P 6–55

6–56 Determine the horizontal and vertical movement of point C in Figure P 6–56 in terms of AE (the same for all members).

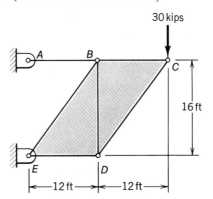

Figure P 6–56

6–57 Determine the vertical and horizontal deflections at point C of the steel truss in Figure P 6–57 if the cross-sectional area of each truss member is 5 in.^{2}

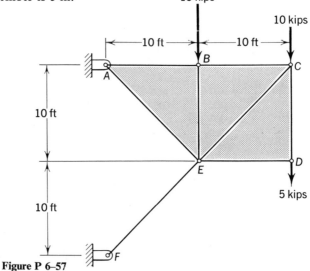

Figure P 6–57

6–58 Determine the vertical and horizontal deflections at point D of the steel truss in Figure P 6–57 if the cross-sectional area of each truss member is 5 in.^{2}

6–59 Determine the vertical and horizontal deflections at point L_2 of the steel truss in Figure P 6–59 if the cross-sectional area of each truss member is 8 in.^{2}

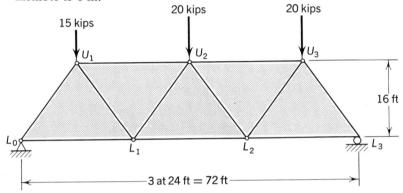

Figure P 6–59

6–60 Determine the horizontal travel of the roller support at L_3 of the truss in Figure P 6–59.

6—7 Castigliano's Theorem

After considerable research on statically indeterminate structures, in 1879 Alberto Castigliano (1847–1884) unveiled his theorem. He called it the *theorem of the differential coefficients of the internal work*. Although the theorem consists of two parts, the first part is seldom used and, as a result, almost forgotten. The second part, a corollary of the first, is commonly known as Castigliano's second theorem, a useful tool for calculating deflections and, after careful interpretation, useful for solving statically indeterminate structures. Slightly expanded from the original statement, *Castigliano's second theorem* may be expressed as follows:

For any elastic system at a constant temperature and on unyielding supports, and which obeys the law of superposition,[5] *the first partial derivative of the strain energy with respect to any one of the external forces of the system, is equal to the displacement of the point of application of that force, and in the direction of that force.* The terms *force* and *displacement* may represent a true force, in which case the displacement would be a linear deflection in the direction of the force. It may be a couple; hence the displacement would then represent an angle change.

---

[5] See Section 1-10.

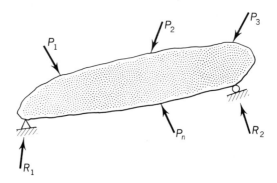

Figure 6–11

To evaluate this theorem, let us consider the body in Figure 6–11, assumed linearly elastic and at a constant temperature, supported by unyielding supports, and subjected to a system of loads, P_1, P_2, P_3,$\cdots$, P_n. As these loads are applied gradually, the body deforms. The loads do work and subsequently, energy U is stored in the body. Now assume that after all the loads have been applied to the body, any of the loads P_n is increased by an amount dP_n. This will result in a small additional work as dP_n travels an increment of distance $d\delta_n$ in the direction of dP_n. The corresponding increase in energy is $(\partial U/\partial P_n)\cdot dP_n$. Hence, the total energy is

$$U + (\partial U/\partial P_n)\cdot dP_n \qquad (a)$$

The final strain energy, however, will be the same regardless of the order in which the loads were applied. Therefore, assume dP_n is applied first and $P_1, P_2, \cdots, P_n$ last. Before the loads $P_1, P_2, \cdots, P_n$ are applied, dP_n produces a deformation $d\delta_n$ which results in an increment of strain energy equal to $\frac{1}{2}(dP_n)(d\delta_n)$. Thus, with dP_n on the body when the loads P_1, P_2, $\cdots$, P_n are applied, it rides through a displacement of δ_n produced by the loads. Consequently, there is produced an additional amount of strain energy $(dP_n)\cdot\delta_n$. Therefore, for this sequence of load, the total strain energy is

$$U + (dP_n)\cdot\delta_n + \frac{1}{2}(dP_n)\cdot(d\delta_n) \qquad (b)$$

where the last term, being an infinitesimal amount of the second order, may be neglected. Hence, we may equate Equations (a) and (b) to obtain

$$U + (\partial U/\partial P_n)\cdot dP_n = U + (dP_n)\delta_n \qquad (c)$$

from which we obtain

$$\partial U/\partial P_n = \delta_n \qquad (6-20)$$

Castigliano's Theorem in Mathematical Terms. There is a marked similarity between Castigliano's theorem and the virtual work (unit load) method presented in Section 6–6. Consider, for instance, the expression for energy for various types of loads and structures as they apply in Equations (6–17), (6–18), and (6–19). The comparison may start with:

Bending:

    Virtual work:

$$\delta = \int_l (m)\left(\frac{M\,dx}{EI}\right)$$

    Castigliano's theorem[6]:

$$\delta = \frac{\partial U}{\partial P_i} = \frac{\partial}{\partial P_i}\left\{\int_{l_2} \frac{M^2 dx}{EI}\right\} = \int_l \frac{\partial}{\partial P_i}\left\{\frac{M^2}{2EI}\right\}dx = \int_l \frac{M}{EI}\frac{\partial M}{\partial P_i}\,dx$$

Torsion:

    Virtual work:

$$\delta = \int_l (t)\left(\frac{T\,dx}{GJ}\right)$$

    Castigliano's theorem:

$$\delta = \frac{\partial U}{\partial P_i} = \int_l \left(\frac{\partial T}{\partial P_i}\right)\left(\frac{T\,dx}{GJ}\right)$$

Trusses:

    Virtual work:

$$\delta = \Sigma\,(f)\cdot\left(\frac{Fl}{AE}\right)$$

    Castigliano's theorem:

$$\delta = \frac{\partial U}{\partial P_i} = \Sigma\left(\frac{\partial f}{\partial P_i}\right)\left(\frac{Fl}{AE}\right)$$

The differentiation inside the integral sign is permissible when P is not a function of x. Hence, the comparison yields the equivalence of $\partial M/\partial P_i = m$, $\partial T/\partial P_i = t$, $\partial F/\partial P_i = f$, and so forth. When there is no load P_i where the deflection is desired, we place an imaginary load P_i at the point and in the direction of the desired deflection. We set the imaginary load equal to zero after we differentiate and before integration takes place. Examples 6–14 through 6–16 further illustrate this. When applied to statically indeterminate structures, Castigliano's theorem provides the means for determining the redundant reactions. (As mentioned previously, the reactions must be unyielding.) Briefly, the procedure consists of (*a*) expressing the strain energy of the system in terms of the applied loads and resulting reactions, (*b*) applying Castigliano's equation, $\partial U/\partial R_i = 0$ (where $i = 1, 2, 3, \cdots, n$) as many times as the number n of redundants. This yields n additional

---

[6] We can differentiate under the integral sign as though no integrals were considered since the limits of the integrals are constants.

equations to supplement those from static equilibrium, and (c) solving the equations for the redundants and subsequently solving the problem for stress, deflection, and so forth. Examples 6–17 and 6–18 will illustrate this approach.

Example 6–14:

Given: The cantilever beam shown in Figure a.
Find: The deflection at (a) the free end, and (b) the center of beam (point B).

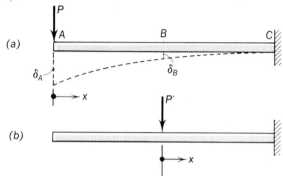

Procedure:

$$\delta_A = \frac{\partial U}{\partial P} = \int_0^l \frac{\partial M}{\partial P} \cdot \frac{M dx}{EI}$$

$$M = -Px \qquad 0 < x < l/2$$

$$\frac{\partial M}{\partial P} = -x$$

Hence,

$$\delta_A = \int_0^l (-x)(-Px)(dx/EI) = Pl^3/3EI \text{ (downward as shown)}$$

For the deflection at B, we place an imaginary load $P' = 0$ at B, as shown in Figure b, because there is no real load applied there. The moment from B to C is

$$M = -Px - P'\left(x - \frac{l}{2}\right) \qquad \frac{l}{2} < x < l$$

and

$$\frac{\partial M}{\partial P'} = -\left(x - \frac{l}{2}\right)$$

from which

$$\delta_B = \frac{\partial U}{\partial P} \int_{l/2}^l \left\{ -\left(x - \frac{l}{2}\right)\left[-Px - \overset{0}{R'}\left(x - \frac{l}{2}\right)\right]\right\} \frac{dx}{EI}$$

$$= \int_{l/2}^l \left(Px^2 - P\frac{l}{2}x\right)dx = P\left[\frac{x^3}{3} - l\frac{x^2}{4}\right]_{l/2}^l = \frac{5Pl^3}{48EI}$$

Example 6–15:

Given: A round bar, of diameter $d = 2r$, is bent to form a circular arc, of radius R, and fixed at one end, as shown in Figure *a*. A load P, perpendicular to the plane in which the curved bar lies, bends and twists the bar. [*Note:* This is the same problem as given in Example 6–4.]

Find: The horizontal deflection of the bar at the free end.

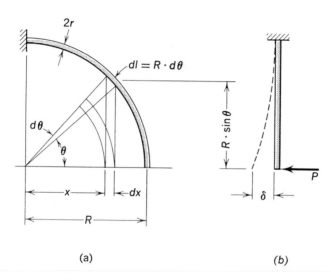

(a)   (b)

(*a*) A round bar bent to form a circular arc. (*b*) Side view.

Procedure: Letting T and M represent torque and moment, respectively, we have

$$T = P(R - x) = PR(1 - \cos \theta); \quad \partial T/\partial P = R(1 - \cos \theta)$$

$$M = P(R \cdot \sin \theta); \quad \partial M/\partial P = R \cdot \sin \theta$$

Thus, the total deflection is the sum of the two effects; that is,

$$\delta = \frac{\partial U}{\partial P} = \int \left(\frac{\partial M}{\partial P}\right)\left(\frac{M}{EI}\right)dl + \int \left(\frac{\partial T}{\partial P}\right)\left(\frac{T}{JG}\right)dl$$

$$\delta = \frac{1}{EI}\int_0^{\pi/2} (R \cdot \sin \theta)(P \cdot R \cdot \sin \theta) \cdot Rd\theta$$

$$+ \left(\frac{1}{JG}\right)\int_0^{\pi/2} R(1 - \cos \theta) \cdot PR(1 - \cos \theta)Rd\theta$$

which yields

$$\delta = (\pi PR^3/4)[1/EI + (3\pi - 8)/GJ]$$

Example 6–16:

Given: The cantilever truss shown in Figure *a*. [*Note:* It is the same structure as presented in Examples 1–7 and 6–12.]

Find: The vertical and horizontal deflection at *B*.

Procedure: In terms of *P*, the forces in the two bars are as shown in Figure *b*. Thus, the total energy in the whole system is

$$U = \Sigma \frac{P^2 l}{2AE} = \left[\left(\frac{1}{2}\right)\frac{(+\frac{5}{3}P)^2 l}{AE}\right]_{AB} + \left[\frac{1}{2}\frac{(-\frac{4}{3}P)^2 l}{AE}\right]_{BC}$$

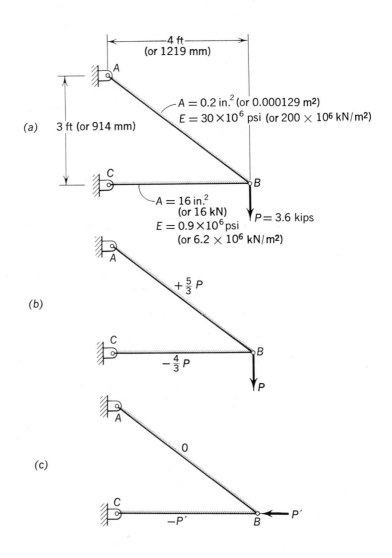

Therefore, the vertical deflection at B is

$$\delta_{B\downarrow} = \frac{\partial U}{\partial P} = \left(\frac{P(5/3)^2(60)}{0.2 \times 30 \times 10^6}\right) + \frac{P(4/3)^2(48)}{16 \times 0.9 \times 10^6}$$

$$\delta_{B\downarrow} = 33.6P/10^6 = 33.6 \times 3.6 \times 10^3/10^6$$

$$\delta_{B\downarrow} = \mathbf{12.12 \times 10^{-2}\ in.\ (or\ 3.08\ mm)}$$

which compares with the answers found in Examples 1–6 and 6–12. To determine the horizontal movement at B, we place an imaginary load P', as shown in Figure c. Thus, the horizontal deflection becomes

$$\delta_{B\leftarrow} = \partial U/\partial P' = \Sigma\ (\partial F/\partial P')(Fl/AE)$$

where $\partial F/\partial P' = 0$ for AB and $\partial F/\partial P' = -1$ for BC. Hence,

$$\delta_B = \left[(0)\left(\frac{+\tfrac{5}{3}Pl}{AE}\right)\right]_{AB} + \left[\frac{(-1)(-\tfrac{4}{3}Pl)}{AE}\right]_{BC} = \frac{4Pl}{3AE}\bigg]_{BC}$$

When substituting $P = 3600$ lb, $l = 48$ in., $A = 16$ in.², and $E = 0.9 \times 10^6$ psi, the deflection becomes

$$\delta_B = .1.6 \times 10^{-2}\ \mathbf{in.\ (or\ 0.41\ mm)}$$

Example 6—17:

Given: The beam shown in the figure. [*Note:* See also Example 5–3.]
Find: The reactions at the supports.

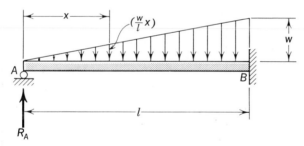

Procedure: The beam is statically indeterminate to the first degree. Thus, one equation based on Castigliano's theorem may be used to solve for one of the redundants; hence, we reduce the structure to one that is statically determinate. From there, the equations of static equilibrium would be sufficient to complete the problem. The moment equation at any point x is, by superposition,

$$M_x = R_A \cdot x - (wx/l) \cdot x/2 \cdot x/3 = R_A x - wx^3/6l$$

Thus, the deflection expression, for zero deflection at support A, becomes

$$\delta_A = \partial U/\partial R_A = \int_0^l (\partial M_x/\partial R_A)(M_x/dxEI) = 0$$

and

$$\partial M_x/\partial R_A = x$$

which yields

$$\frac{1}{EI}\int_0^l (x)\left[R_A x - \left(\frac{wx^3}{6l}\right)\right]dx = \frac{1}{EI}\left[\left(\frac{R_A l^3}{3}\right) - \left(\frac{wl^4}{30}\right)\right] = 0$$

from which

$$R_A = wl/10 \uparrow \text{ (as assumed)}$$

and from $\Sigma F_y = 0$ and $\Sigma M_A = 0$, we get

$$R_B = 4/10(wl)\uparrow \quad \text{and} \quad M_B = - wl^2/15 \;\curvearrowright$$

Example 6–18:

Given: Three wires of equal length and of the same cross-sectional area, support a load P, as shown in Figure a.
Find: The axial load, in terms of P, that each wire carries.

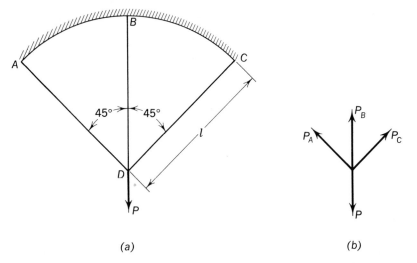

(a) (b)

Procedure: We shall assume that P_B is our redundant member. Also note that $P_A = P_C$, assumed to act as shown in Figure b. Thus,

$$P_A = P_C = (P - P_B)/\sqrt{2}$$

The total energy in the system is

$$U = \frac{1}{2}\left[2\frac{((P - P_B)/\sqrt{2})^2 l}{AE} + \frac{P_B^2 l}{AE}\right] = \frac{l}{2AE}[(P - P_B)^2 + P_B^2]$$

and because the deflection at $B = 0$, we get

$$\delta_B = \frac{\partial U}{\partial P_B} = \frac{l}{2AE}[-2(P - P_B) + 2P_B] = 0$$

Solving for P_B, we obtain

$$P_B = \tfrac{1}{2}P \quad \text{and therefore,} \quad P_A = P_{Bc} = P/2\sqrt{2} = \mathbf{0.35P}$$

PROBLEMS

Solve the following problems by using Castigliano's theorem.

6–61 Determine the slope at point B of the beam shown in Figure P 6–61, in terms of M, l, and EI.

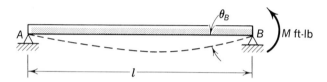

Figure P 6–61

6–62 Determine the deflection and slope of the free end of the cantilever beam in Figure P 6–62, in terms of w, l, and EI.

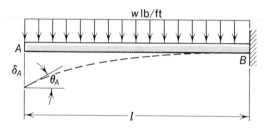

Figure P 6–62

6–63 Determine the deflection and slope of the free end of the cantilever beam in Figure 6–63, in terms of w_0, l, and EI.

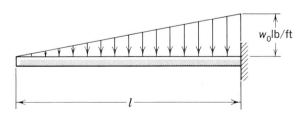

Figure P 6–63

6-64 Determine the deflection and slope at point C for the cantilever frame in Figure P 6-64, in terms of EI, w, and l (the horizontal and vertical members have the same EI).

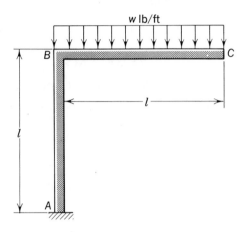

Figure P 6-64

6-65 Determine the deflection and slope at point D for the frame in Figure P 6-65, in terms of EI, w, and l (the horizontal and vertical members have the same EI).

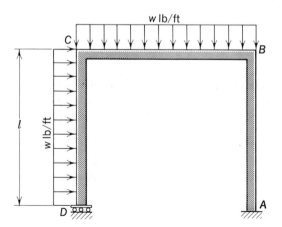

Figure P 6-65

6–66 Determine the horizontal deflection of point *C* in Figure P 6–66 in terms of *AE* (same *AE* for both members).

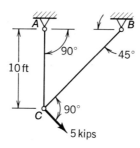

Figure P 6–66

6–67 Determine the horizontal deflection of point *B* in the frame in Figure P 6–67 in terms of *P*, *l*, and *AE* (same for both members).

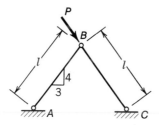

Figure P 6–67

6–68 Determine the vertical and horizontal deflections at point *C* of the steel truss in Figure P 6–68 if the cross-sectional area of each truss member is 1 in.²

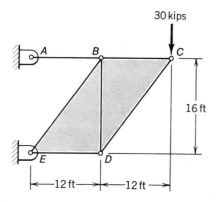

Figure P 6–68

6-69 Determine the vertical travel of point C of the steel truss in Figure P 6-69 if the cross-sectional area of each member is 5 in.^{2} and if $P = 15$ kips.

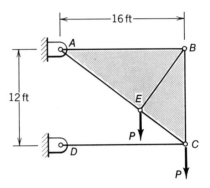

Figure P 6-69

6-70 Determine the vertical and horizontal deflections at point C of the steel truss in Figure P 6-70 if the cross-sectional area of each truss member is 5 in.^{2}

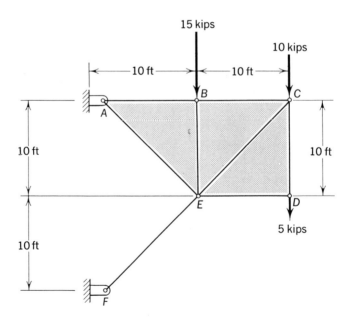

Figure P 6-70

6–71 Three steel wires of equal diameter support a load as shown in Figure P 6–71. Determine the diameter of a wire if the stress in the wire is not to exceed 20,000 psi.

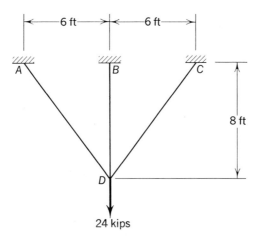

Figure P 6–71

6–72 Determine the reaction at A in terms of M and l for the beam in Figure P 6–72.

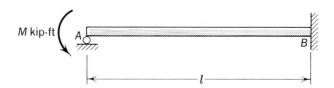

Figure P 6–72

6–73 Determine the reaction at point B in terms of P for the beam in Figure P 6–73.

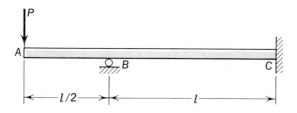

Figure P 6–73

6–74 Determine the vertical reactions at the two supports, in terms of P for the beams shown in Figure P 6–74.

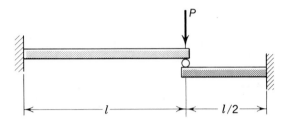

Figure P 6–74

6–75 Determine the moment at point A of the system in Figure P 6–75.

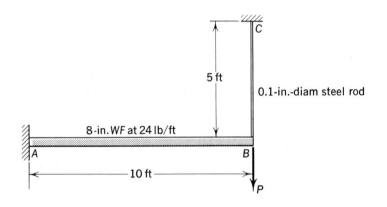

Figure P 6–75

Chapter 7 • COLUMNS

7—1 Introduction

Columns as considered in this chapter may have several utilitarian functions and may fall into various groups of structural components. The word *column* is perhaps most frequently associated with vertical members that support a building or a bridge, or other structures. More broadly, however, and especially from the point of analysis, we must include in the term column any slender compression member which may buckle at loads appreciably smaller than those which would result in a crushing failure. Hence, most generally, we may define columns as *slender structural members primarily subjected to axial loads and which may tend to buckle under such loads.*

There are but few cases in which the column load is perfectly axial; in most instances, columns are subjected to both compression and bending. Figure 7–1 shows some different loadings to which a column may be subjected. The column may support a load, such as shown in Figure 7–1*a*, that is transmitted from a beam or some other structural member; or the column may be part of a "frame" which transmits a moment and a vertical force to the column, as shown in Figure 7–1*b*. Axial force and a transverse thrust, Figure 7–1*c*, may be still another loading combination.

For purposes of discussion, it is convenient to categorize compression members into three groups: *short columns or struts, intermediate columns,* and *long columns*. The criteria separating these groups are not well established. Perhaps the most accepted basis for this purpose is the *slenderness ratio*, defined as the ratio of length to radius of gyration of the column section in the direction of impending bending. Short members (small slenderness ratio) usually fail by crushing when subjected to axial or slightly eccentric loads, and thus may be analyzed as discussed in Chapter 1. As the slenderness ratio increases, however, and we reach the intermediate and long column level, we approach a condition of buckling failure. Material imperfections or lateral disturbances may cause long columns to deflect

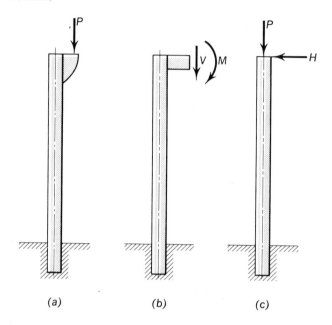

Figure 7–1 Columns subjected to axial and bending loads.

and subsequently buckle even for axially applied loads. Eccentrically applied loads or a combination of axial load and a bending moment (as shown in Figure 7–1) yield similar consequences.

7—2 Eccentrically Loaded Short Columns

For short columns subjected to eccentric compressive loads, we may safely assume only negligible *sway* of one end relative to the other. Hence, only the initial eccentricity of the load need be considered in the bending analysis. Consider a short column supporting an eccentric load, as shown in Figure 7–2a. The column is assumed to possess an axis of symmetry in the plane of bending. We may then superimpose the effects of compression and bending to obtain the net stress σ_n over the section, as shown in Figure 7–2b. Consistent with the conditions set forth for elastic bending of beams, the bending stress varies linearly with the distance from the centroidal axis of the column. The maximum and minimum (absolute values) compression stresses occur at the extreme fibers, and are

maximum:
$$\sigma_n = P/A + (Pe)c_1/I = P/A(1 + A \cdot e \cdot c_1/I) \qquad (7\text{–}1)$$

minimum:
$$\sigma_n = P/A - (Pe)c_2/I = P/A(1 - Aec_2/I) \qquad (7\text{–}2)$$

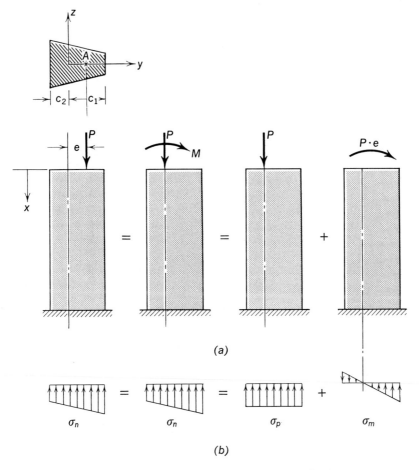

Figure 7–2 (*a*) External loading. (*b*) Internal effects.

where A represents the cross-sectional area, and I represents the moment of inertia of the section about the z axis.

Equation (7–1) above may be rearranged to focus attention on the maximum load P that a column may carry, given a certain allowable stress.

$$P_{max} = \frac{\sigma_{max} \cdot A}{1 + Aec_1/I} \qquad (7\text{--}3)$$

As a corollary to this thinking, we might want to investigate the maximum eccentricity e, for which there can be no tension on the left face of the column in Figure 7–2. This means that the minimum stress expressed by Equation (7–2) must be equal to zero; that is,

$$0 = P/A(1 - A \cdot e \cdot c_2/I) \qquad (a)$$

from which we get

$$e_{max} = I/Ac_2 \tag{7-4}$$

Note that e is independent of the magnitude of P. For a rectangular section, as shown in Figure 7-3a, the e for zero stress on a b face is $e = (bd^3/12)/(bd)(d/2) = d/6$ on either side of the z axis; for a circular section of diameter D, the respective e is $D/8$. For a rectangular section with the load eccen-

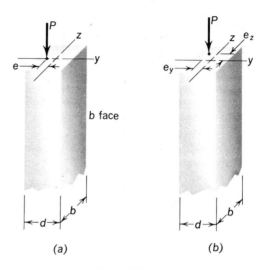

(a)   (b)

Figure 7-3

tric in two directions, as shown in Figure 7-3b, the stress at any point may be determined by

$$\sigma = \frac{P}{A} \pm \frac{(P \cdot e_y)(d/2)}{I_z} \pm \frac{(P \cdot e_z)(b/2)}{I_y} \tag{7-5}$$

For $\sigma = 0$, from Equation (7-5), we obtain

$$1 = (6/d) \cdot e_y + (6/b) \cdot e_z \tag{b}$$

From this result we obtain four points, commonly known as *kern points*, two on each axis at a distance of ⅙ of the respective dimensions. Because the expression is linear, the four points may be connected by straight lines to give a region within which the load may be applied anywhere without creating tension stresses on any portion of the column. This area, known as the *kern region*, is shown in the shaded area of Figure 7-4a for a rectangular section, and in Figure 7-4b for a circular section. For reinforced concrete columns, the kern region is only slightly greater.[1]

---

[1] See John N. Cernica, *Fundamentals of Reinforced Concrete*, Addison-Wesley Publishing Company, Reading, Massachusetts, 1964.

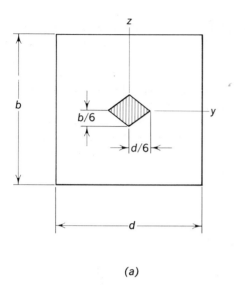
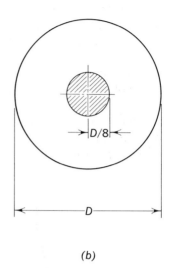

(a)   (b)

Figure 7–4

7—3 Euler's Formula — Long Columns

It was pointed out in Section 7–1 that even for axially loaded columns, we may experience buckling failures of long columns at loads appreciably lower than those which would cause short-column failures. The slightest lateral disturbance might be a sufficient cause for lateral deflection and subsequent buckling failure. Hence, the lateral deflection for long columns may no longer be considered negligible as we did for short columns, and therefore we must modify our analysis accordingly.

Columns present a structural problem which received much attention in the past and which is still not completely solved. Various methods of column analysis are available; perhaps the most famous of these is the Euler formula. It is named after the Swiss mathematician Leonhard Euler (1707–1783) who derived it in a paper written in 1757 and published in 1759. Although Euler's formula is limited to long, slender columns (as we shall see a little later in this section), it marked the significant beginning of column research, both theoretical and experimental. Euler recognized over two centuries ago a version of the formula $EI \cdot (d^2y/dx^2) = \pm M$ and proceeded to develop his equations. It is beyond the purpose here to give an elaborate treatise of Euler's work; we will be mainly concerned with a brief and general coverage of concepts, derivation, and limitations of Euler's formula.[2]

---

[2] See J. A. Van den Broek, "On the Strength of Columns," *Am. J. Phys.*, vol. 15, No. 4, pp. 309–318, July 1947.

Consider a long slender column with pinned (hinged) ends subjected to an axial load P, as shown in Figure 7-5. The material is assumed to be homogeneous and follows Hooke's law within the range of load. For relatively small loads P, an induced lateral deflection by an external lateral force will disappear with the removal of the force; hence, we have a laterally stable condition. However, if we gradually increase the P load, we reach a point where the load will preserve the induced deflection even after the lateral force is removed. At this point, we reach what is known as the *critical load* P_{cr} of the column. The procedure for determining such a load will be the purpose of this section.

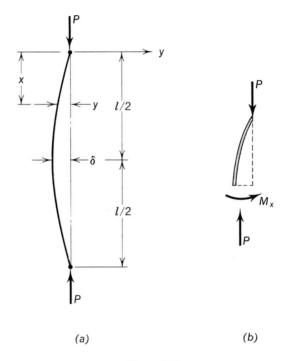

(a)   (b)

Figure 7-5

The bending moment at any section of a deflected column (Figure 7-5) is

$$M_x = -Py \qquad (a)$$

For small lateral deflection, the differential equation of the elastic curve is applicable. For our case, assuming a uniform bending stiffness, we may write

$$EI(d^2y/dx^2) = -Py \qquad (b)$$

M is considered negative with y assumed, as shown in Figure 7–5 (that is, we have a negative change in slope with increasing x). Then, letting $k = \sqrt{P/EI}$, Equation (b) becomes

$$d^2y/dx^2 + k^2y = 0 \qquad (c)$$

This is a linear, second-order, homogeneous differential equation with constant coefficients.[3] Its solution is

$$y = C_1 \sin (kx) + C_2 \cos(kx) \qquad (d)$$

where C_1 and C_2 are arbitrary constants to be evaluated from conditions consistent with the restraints and deformations of the column. From Figure 7–5, we observe that

at $x = 0$: $y = 0$ at $x = l$: $y = 0$

at $x = l/2$: $dy/dx = 0, y = \delta$

from which we can proceed to evaluate the constants C_1 and C_2. Thus, from Equation (d) we have, at $x = 0$,

$$y = 0 = C_1 \sin k(0) + C_2 \cos k(0)$$

from which $C_2 = 0$. At $x = l/2$ $y = \delta$, $dy/dx = 0$, hence, we obtain

$$y = \delta = C_1 \cdot \sin k(l/2) \qquad (e)$$

and

$$dy/dx = 0 = C_1 \cdot k \cdot \cos k(l/2) \qquad (f)$$

Because $k \neq 0$, from Equation (f), we have

$$0 = C_1 \cdot \cos k(l/2) \qquad (g)$$

and solving, from Equations (e) and (g), we get

$$C_1 = \delta$$

The solution [Equation (d)] becomes

$$y = \delta \cdot \sin(kx) \qquad (h)$$

From the condition that $y = 0$ at $x = l$, we have

$$0 = \delta \cdot \sin(kl)$$

from which

$$\sin kl = 0, \text{ because } \delta \neq 0 \text{ when } P = P_{cr}$$

Thus, the relationship

$$kl = n\pi, \text{ where } n = 1, 2, 3 \qquad (i)$$

---

[3] See any text on differential equations.

must hold true. For a laterally unsupported column, P_{cr} is obtained when $n = 1$. Then, substituting $k = \sqrt{P_{cr}/EI}$ into Equation (i), we get

$$\left(\sqrt{P_{cr}/EI}\right) \cdot l = n\pi = 1\pi$$

from which

$$P_{cr} = \pi^2 EI/l^2 \qquad (7\text{--}6)$$

Equation (7–6) is Euler's column formula for a pinned-end column shown in Figure 7–5. The deflection equation for this column becomes

$$y = \delta \cdot \sin (\pi/l)x \qquad (h')$$

For columns having other end conditions, the same procedure may be employed to determine the respective critical loads and deflection equations.

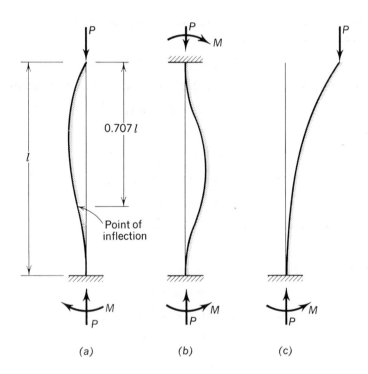

(a)   (b)   (c)

Figure 7–6

However, if any segment of the length of any such column corresponds to that of our pinned-end column, shown in Figure 7–5, we may obtain the critical loads by merely modifying our l in Equation (7–6). For example, the columns shown in Figure 7–6 may be assumed to have points of inflections as shown. Hence, from Equation (7–6), we obtain the critical loads:

$$P_{cr} = \frac{\pi^2 EI}{(2l)^2} = \frac{\pi^2 EI}{4l^2} \qquad \text{(Figure 7–6c)} \qquad (7–7)$$

$$P_{cr} = \frac{\pi^2 EI}{(l/2)^2} = \frac{4\pi^2 EI}{l^2} \qquad \text{(Figure 7–6b)} \qquad (7–8)$$

$$P_{cr} = \frac{\pi^2 EI}{(0.707l)^2} = \frac{2\pi^2 EI}{l^2} \qquad \text{(Figure 7–6a)} \qquad (7–9)$$

Interestingly, we note that the critical load is proportional to the flexural rigidity EI, and is inversely proportional to the square of the length l of the column. It is independent of the strength of the material. Hence, the obvious inference is that for maximum stability, the material is to be distributed as far as possible from the neutral axis so that I becomes a maximum. Less obvious, however, is the extent to which this can be carried out before the thickness becomes critically small, resulting in localized "wrinkling" failure. Experimental findings coupled with some sound engineering judgment, is perhaps the one reliable answer to such a question.

Associated with the critical load P_{cr} is the critical stress σ_{cr}. Dividing both sides of Equation (7–6) by A, the cross-sectional area of the column, we get the *average* critical stress

$$\sigma_{cr} = \frac{P_{cr}}{A} = \frac{\pi^2 E}{l^2}\left(\frac{I}{A}\right) = \left[\frac{\pi^2 E}{(l/r)^2}\right] \qquad (7–10)$$

where $r = $ radius of gyration $= \sqrt{I/A}$. For any given material, Equation (7–10) may more conveniently be written as

$$\sigma_{cr} = K\left[\frac{1}{(l/r)^2}\right] \qquad (7–10a)$$

and plotted as shown in Figure 7–7. The graphical depiction is known as Euler's curve, which is a useful presentation of the relationship obtained by Euler.

Several observations are necessary at this time: We note that the curve in Figure 7–7 represents a general case of Equation (7–10). For every material for which the E differs, a different, yet specific, curve is obtained ($K = \pi^2 E$). Also, the derivation of Euler's formula and the subsequent plotting of Euler's equation was based on the stated or implied assumptions that (a) the material is perfectly elastic, and (b) the material contains no imperfections. This leads us to stipulate two limitations: The *first* is the maximum stress σ_i for which the curve is theoretically applicable. Commonly, σ_i is taken to be the proportional stress σ_p of the material, although this choice is subject to discussion. We note that the stress used to plot Euler's curve represents an *average* stress over the cross section. Actually, from bending superposed with compression, some fibers experience stress higher (some lower) than the average. The *second* limitation is related to the material

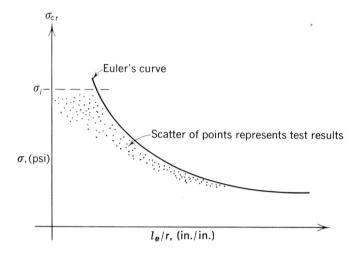

Figure 7–7

consistency. Experimental investigations revealed that because no column is free of all imperfections (material quality, workmanship, usage, and so forth), buckling occurs at values of P_{cr} lower than those predicted by Euler's curve. Although for long columns, test results show only insignificant variations from Euler's curve, for the shorter to intermediate-length range, the deviation is quite significant. Experimental results are generally the criteria used for determining the *slenderness ratio* which establishes the boundary between short, intermediate, and long ranges. For example, for steel, commonly accepted values of these ranges are[4]

Short	Intermediate	Long
$0 < l_e/r < 60$	$60 < l_e/r < 120$	$120 < l_e/r < 300$

Hence, we conclude that Euler's formula is satisfactory for long, slender columns, and not for intermediate or short ones. The analysis of intermediate columns will be introduced later.

Example 7—1:

Given: A rectangular steel tube 3 in. × 5 in. (or 76 mm × 127 mm) outside dimensions and 2½ in. × 4½ in. (or 64 mm × 114 mm) inside dimensions, as shown in the figure, serves as a column 14 ft (or 4.3 m) long, pinned at each end. Assume $E_{St} = 29 \times 10^6$ psi (or 200×10^6 kN/m²).
Find: (a) The critical axial load P_{cr} of the column, and (b) the critical stress in the column.

---

[4] See, for example, *The Alcoa Structural Handbook* for information on aluminum alloys.

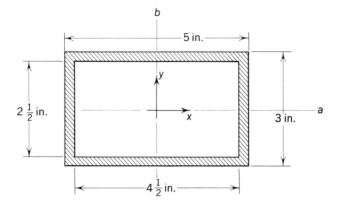

Procedure: The least radius of gyration is the one yielding the largest (l_e/r) which subsequently results in the critical load and stress [see Equation (7–10)]. Thus, we have

$$I_a = \frac{5(3)^3}{12} - \frac{4.5(2.5)^3}{12} = 5.40 \text{ in.}^4 \text{ (minimum } I)$$

$$A = 5 \times 3 - 2.5 \times 4.5 = 3.75 \text{ in.}^2$$

$$r_a = \sqrt{I/A} = \sqrt{5.40/3.75} = 1.2 \text{ in.}$$

$$\frac{l_e}{r} = \frac{(14 \text{ ft})(12 \text{ in.}/\text{ft})}{1.2} = 140$$

This slenderness ratio is within the range for which Euler's formula may be assumed reliable. From Equation (7–6) we obtain

$$P_{cr} = \frac{\pi^2 EI}{l^2}$$

$$P_{cr} = \frac{\pi^2(29 \times 10^6)(5.4)}{(14 \times 12)^2} = \textbf{54,700 (or 243 kN)}$$

$$\sigma_{cr} = \frac{P_{cr}}{A} = \frac{54,700}{3.75} = \textbf{14,586 psi (or 100.57 MN/m}^2)$$

σ_{cr} may also be obtained directly from Equation (7–10); that is,

$$\sigma_{cr} - \frac{\pi^2 E}{(l_e/r)^2} = \frac{\pi^2(29 \times 10^6)}{(140)^2} = \textbf{14,586 psi (or 100.57 MN/m}^2)$$

Example 7—2:

Given: A round steel column fixed at one end and free at the other (see Figure 7–6a), 6 in. (or 152 mm) outside diameter and 5½ in. (or 140 mm) inside diameter, is to support an axial load of 10 kips (or 44.5 kN) with a safety factor of 2.0. Assume $E = 30 \times 10^6$ psi (or 200×10^6 kN/m²). *Find:* The maximum length that the column may have.

Procedure: Euler's formula will be assumed reliable for the resulting r, then it will be checked. The critical load in our case becomes

$$P_{cr} = (10,000)(2.0) = 20,000 \text{ lb}$$

Using Equation (7–7) we have

$$P_{cr} = 20,000 = \pi^2 EI/4l^2$$

For a circular section,

$$I = \frac{\pi}{64}(d_o{}^4 - d_i{}^4) = \frac{\pi}{64}(6^4 - 5.5^4) = 18.7 \text{ in.}^4$$

Therefore,

$$l = \left(\frac{\pi^2 EI}{4 \times 20,000}\right)^{1/2} = \left(\frac{\pi^2 \times 30 \times 10^6 \times 18.7}{8 \times 10^4}\right)^{1/2} = 264 \text{ in.}$$
$$(\text{or } 6706 \text{ mm})$$

We now check (l_e/r). If $l_e/r > 120$, we shall assume that our answer is satisfactory; that is, Euler's formula would be assumed applicable.

$$r = \sqrt{\frac{I}{A}} = \sqrt{\frac{18.7}{4.52}} = 2.04 \text{ in. (or 52 mm)}$$

Therefore,

$$\frac{l_e}{r} = \frac{264}{2.04} = 129 > 120$$

The assumption, therefore, holds.

PROBLEMS

Unless otherwise specified, assume there is an axial load (no eccentricity) and $E_s = 30 \times 10^6$ psi.

7–1   A solid square steel bar, 1.5 in. × 1.5 in. and 7 ft long, fixed at both ends, supports a load $P = P_{cr}/3$. Determine (a) the load P, and (b) the critical stress of the column.

7–2   Solve Problem 7–1 if the column is free at one end and fixed at the other end.

7–3   An American Standard I beam, 3 in. × 2⅜ in. at 5.7 lb, is to serve as a column 8 ft long, assumed to be pinned at both ends. Determine (a) the critical load for the column, and (b) the critical stress.

7–4   Determine the maximum length that the column in Problem 7–3 may be if it is to support a load of 5 kips with a safety factor of 3.

7–5   A square steel tube, 4 in. × 4 in. (outside dimensions) × 3/16-in. wall thickness, is to be used as a column 16 ft long, fixed at both ends. Determine the critical load for the column.

7-6   A rectangular steel tube, 4 in. × 2 in. (outside dimensions) × 3/16-in. wall thickness, is to be used as a column 10 ft long, free at one end and fixed at the other. Determine the critical load and stress for the column.

7-7   Two steel angles, 3½ in. × 2½ in., at 7.6 lb/ft, fastened together with the long legs back to back, Figure P 7-7, serve as a pinned-end column 12 ft long. Determine the critical stress of the column. The two angles are assumed to act as a unit.

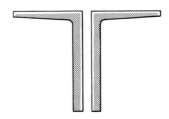

Figure P 7-7

7-8   Solve Problem 7-7 assuming the column is fixed at both ends.

7—4   The Secant Formula

Euler's formula derived in Section 7-3, assumed, among other things, that the load was axially applied; that is, $e = 0$. However, we pointed out that this is a rather unrealistic assumption because of material and manufacturing imperfections, and errors from assumptions regarding end conditions or the locations of loads. The secant formula, although somewhat cumbersome to apply, does take eccentricity into account.

Consider the fundamental case of a long slender column with pinned ends, which is subjected to an eccentric load (Figure 7-8). As the load P is gradually increased, the column will start to bend, forming the shape shown. Thus, the bending moment at any section is

$$M = Py \qquad (a)$$

where $y = e +$ bending deflection. Thus, we have

$$EI(d^2y/dx^2) = -Py \qquad (b)$$

and letting $k = \sqrt{P/EI}$, Equation (b) becomes

$$d^2y/dx^2 + k^2y = 0 \qquad (c)$$

The general solution for this equation is

$$y = C_1 \cdot \sin(kx) + C_2 \cdot \cos(kx) \qquad (d)$$

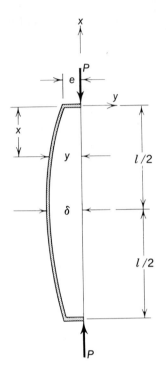

Figure 7–8

To evaluate the constants, we assume the *boundary* conditions to be as follows:

$$\text{at } x = 0: \ y = e \qquad \text{at } x = l: \ y = e$$
$$\text{at } x = l/2: \ dy/dx = 0, \ y = \delta$$

and we find from Equation (*d*)

$$e = C_1(0) + C_2 \cdot 1$$

from which $C_2 = e$. Substituting $C_2 = e$ into Equation (*d*), we have

$$y = C_1 \cdot \sin(kx) + e \cdot \cos(kx) \tag{e}$$

At $x = l/2$, $dy/dx = 0$. Hence, from Equation (*e*) we obtain

$$dy/dx = 0 = C_1 \cdot k \cdot \cos k(l/2) - e \cdot k \cdot \sin k(l/2)$$

from which

$$C_1 = \frac{e \cdot \sin k(l/2)}{\cos k(l/2)}$$

Hence, Equation (*e*) becomes

$$y = e\left[\frac{\sin k(l/2)}{\cos l(k/2)} \sin kx + \cos kx\right] \tag{7–11}$$

The maximum deflection ($y = \delta$) occurs when $x = l/2$. From Equation (7–11) we have

$$\delta = e\left[\frac{\sin^2 k(l/2)}{\cos k(l/2)} + \cos k\left(\frac{l}{2}\right)\right] = e \cdot \sec\left(k\,\frac{l}{2}\right) \qquad (f)$$

or

$$\delta = e \cdot \sec\sqrt{\frac{P}{EI}}\left(\frac{l}{2}\right) \qquad (g)$$

The maximum compressive stress may, within the proportional limit, be obtained by superposition of the axial and bending stresses at the middle of the column, where the deflection is a maximum and consequently the moment is also a maximum. Hence, we have

$$\sigma_{\max} = \frac{P}{A} + \frac{M_{\max} \cdot c}{I} = \frac{P}{A} + \frac{(P \cdot \delta)c}{I}$$

or from $I = Ar^2$

$$\sigma_{\max} = \frac{P}{A}\left[1 + \frac{ec}{r^2}\sec\sqrt{\frac{P}{EA}}\left(\frac{l}{2r}\right)\right] \qquad (h)$$

In terms of the average compressive stress P/A, Equation (h) yields

$$\frac{P}{A} = \frac{\sigma_{\max}}{1 + (ec/r^2)\sec\left((l/2r) \cdot \sqrt{P/EA}\right)} \qquad (7\text{--}12)$$

Equation (7–12) is known as the *Secant formula* for eccentrically loaded slender columns. We note that the relation between $\sigma_{\max}$ and P is not linear; there is no proportionality between P and the deflection δ and the related σ that it produces. Hence, stresses caused by different loads cannot be superposed; the forces, however, can be superposed, and then the stresses can be calculated.

The quantity $\sigma_{\max}$ is the proportional limit of the material. For steel, however, it is convenient, and acceptably accurate, to assume $\sigma_{\max} = \sigma_{y.p.}$, if the stress-strain diagram of the material is reasonably straight up to $\sigma_{y.p.}$. For structural steels $\sigma_{y.p.} = 40,000$ psi is perhaps acceptable. The quantity (ec/r^2) is called the *eccentricity ratio;* (l/r) has already been defined as the *slenderness ratio.*

For any given $\sigma_{\max}$ and a selected eccentricity ratio, we may solve Equation (7–12) by trial and error, and plot a curve of (P/A) versus (l/r). If we hold $\sigma_{\max}$ fixed, and vary the eccentricity ratio, we get a family of curves (a curve for each eccentricity ratio) as shown only schematically in Figure 7–9. For long columns (that is, large l/r ratios) the curves approach the Euler curve; the departure is significant for short and intermediate columns.

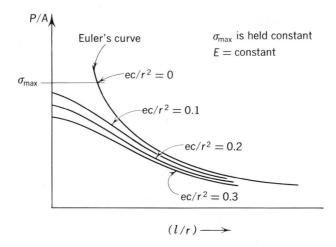

Figure 7–9

We must recognize that the value of P/A found is one which, theoretically, results in impending yielding and subsequent failure. Hence, for design purposes, the safe load is necessarily only a fraction of the "yielding" load; that is, the design must incorporate a safety factor $n = P_{\text{yielding}}/P_{\text{safe}}$.

Inherent in the theoretically correct approach are some unanswered practical questions. For example, how reasonable are our assumptions regarding end conditions, manufacturing and/or erection accuracies, material homogeneity (E for concrete-steel composites, for example) and eccentricity e? The answers to these questions are not easily obtained; they vary with the large range of cases (loads, materials, uses, and so forth) encountered. The one answer that may be extracted from all this, however, is that these "imperfections" do induce bending. Hence, we may express the bending moment as a centrally applied load times an eccentricity e. But how large is e, and, therefore, how large is the eccentricity ratio ec/r^2, a parameter we fixed (along with σ_{max}) for any of the curves shown in Figure 7–9? It is virtually impossible to guess an eccentricity ratio which would encompass accurately the various effects on eccentricity. Consequently, our recourse is the results of tests of columns and/or common sense and experience. A committee[5] of the American Society of Civil Engineers made a rather comprehensive study of test results of axially loaded, pin-ended steel (structural grade) columns commonly used in structural work. They found that an $ec/r^2 = 0.25$ used in the secant formula yields results which are in close agreement with those of tests.

---

[5] "Final Report of Special Committee on Steel Column Research," *Trans. Am. Soc. of Civil Engineers*, vol. 98, 1933.

Example 7—3:

Given: A pin-ended steel column is to be 20 ft long, and is to carry an axial load of 80 kips and a bending moment of 280 in.-kips (in only one direction) with a safety factor of $n = 2.5$. Assume $\sigma_{max} = \sigma_{y.p.} = 40$ ksi, and $E_s = 30 \times 10^3$ ksi.
Find: A suitable size of the section.

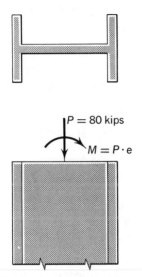

$P = 80$ kips

$M = P \cdot e$

Procedure: We may use the secant formula. From Equation (*h*), we have

$$\sigma_{max} = \frac{P}{A}\left[1 + \frac{ec}{r^2}\sec\sqrt{\frac{P}{AE}\left(\frac{l}{2r}\right)}\right]$$

The bending moment and axial load can be converted into an equivalent eccentric load of 80 kips, eccentric $e = 280/80 = 3.5$ in. Hence, substituting for the quantities σ_{max}, e, E, l, and for $P = n(80) = 2.5(80) = 200$ kips, the above equation becomes

$$40 = \frac{200}{A}\left[1 + \frac{3.5c}{r^2}\sec\sqrt{\frac{200}{30 \times 10^3 A}\left(\frac{20 \times 12}{2r}\right)}\right]$$

or

$$\frac{200}{A}\left(1 + \frac{3.5c}{r^2}\sec\frac{9.8}{r}\sqrt{\frac{1}{A}}\right) - 40 = 0$$

Now it is a matter of selecting various structural sizes until the equation is satisfied. It is obvious, however, that $200/A$ must be less than 40 in order for the equation to equal zero. Thus, $A > 5$ in.^{2}. As a start, let us assume a standard I section. From Table B–2, Appendix B, we select

10 in. at 25.4 lb; $A = 7.38$ in.²; $c = 5$ in.; $r = 4.07$ in.

Thus,

$$\frac{200}{7.37}\left(1 + \frac{3.5 \times 5}{(4.07)^2} \sec \frac{9.8}{4.07}\sqrt{\frac{1}{7.38}}\right) - 40 = 0$$

$$27.1[1 + 1.05 \sec (0.88)] - 40 = 31.8 \neq 0$$

Let us try a larger section:

12 in. at 40.8 lb; $A = 11.84$ in.², $c = 6$ in.; $r = 4.77$ in.

$$16.9\left(1 + 0.92 \sec \frac{9.8}{4.77}\sqrt{\frac{1}{11.84}}\right) - 40 = 0$$

$$16.9(1 + 1.12) - 40 = -4.1 \neq 0, \text{ but close}$$

Thus, our design calls for a standard I section 12 in. deep, 40.8 lb/ft, placed as shown in the figure.

PROBLEMS

For the following problems, assume $E = 30 \times 10^6$ psi, and $\sigma_{y.p.} = 40,000$ psi. Assume that the eccentric load or bending moment is resisted by the *largest* section modulus.

7–9 An 8-in. I section at 23 lb/ft is to serve as a pin-ended column 14 ft long, in order to support an eccentric load of 70 kips with a safety factor of 2.5. Using the secant formula, determine the maximum eccentricity at which the load may be safely placed.

7–10 Determine the maximum load, eccentric 2 in. from the center line of the column, that the column of Problem 7–9 can safely carry with the safety factor of 2.5.

7–11 Determine the safety factor that is inherent in the column of Problem 7–9 if the 70-kip load is eccentric 2.5 in.

7–12 A 12-in. WF-65-lb section (steel) is to serve as a pin-ended column 22 ft long and is to support an axial load of 280 kips with a safety factor of 2.5. Using the secant formula, determine the maximum bending moment (in addition to the axial load) that the column may safely carry.

7–13 Determine the maximum load, eccentric 5 in. from the center line of the column, that the column of Problem 7–12 can safely carry with the safety factor of 2.5.

7–14 Determine the safety factor that is inherent in the column of Problem 7–12 if the 280-kip load is eccentric 3½ in.

7–15 A standard steel pipe, 10.75 in. outside diameter and 0.365-in.-thick wall, is to serve as a pin-ended column 20 ft long and is to support an eccentric load of 120 kips with a safety factor of 2.5. Using the secant formula, determine the maximum eccentricity at which the load may be safely placed.

7–16 Determine the maximum load, eccentric 1.75 in. from the center line of the column, that the column of Problem 7–15 can safely carry with a safety factor of 2.0.

7–17 Determine the safety factor that is inherent in the column of Problem 7–15 if the 120-kip load is eccentric 2.5 in.

7–18 Determine the safety factor that is inherent in Problem 7–9 if the load is 35 kips eccentric 5 in. Compare answers with Problem 7–11.

7–19 Determine the safety factor that is inherent in the Problem 7–12 if the load is 140 kips eccentric 7 in. Compare answers with Problem 7–14.

7–20 Determine the safety factor that is inherent in Problem 7–15 if the load is 60 kips eccentric 5 in. Compare answers with Problem 7–17.

7—5 Empirical Column Formulas

By now, it has become apparent that both the Euler and the secant equations are limited. Euler's equation is limited to long columns loaded only axially. It is inaccurate for short or intermediate-length columns, and not even applicable to eccentrically loaded columns. On the other hand, the secant formula is applicable, with acceptable results, to all columns, axially or eccentrically loaded. But the eccentricity e (or ec/r^2) has to be either known, measurable, or assumed. The assumed value may be subject to question, and hence, the formula is questionably reliable. Furthermore, even when applicable, the labor involved makes the use of the secant formula rather unpopular.

In view of the above so-called limitations, and in an attempt to simplify the calculations, many empirical column formulas have been proposed. Most, however, would fit into the realm encompassed by the *straight line*, the *parabolic*, and *Gordon-Rankine* formulas.

Straight Line Formula. First proposed by W. H. Burr in 1882, the formula reads

$$\left(\frac{P}{A}\right)_w = \sigma_w - \alpha\left(\frac{l}{r}\right) \tag{7–13}$$

where σ_w is the allowable working stress for the material in question, and α is a coefficient selected to fit experimental results ($\sigma_w = 16,000$ psi and $\alpha =$

70 are values frequently suggested for structural grade steels and for the $30 < l/r < 120$ length range).

Parabolic Formula. This formula was proposed by J. B. Johnson in 1893, and has the general form

$$\left(\frac{P}{A}\right)_w = \sigma_w - \beta\left(\frac{l}{r}\right)^2 \qquad (7\text{-}14)$$

where again σ_w represents the allowable working stress and β an experimental coefficient (values for $\sigma_w = 17{,}000$ psi and $\beta = 0.485$ are commonly used for structural steel columns for which $0 < l/r < 120$ applies).

Gordon-Rankine Formula. Proposed during the middle of the nineteenth century by two English engineers, Gordon and Rankine, the formula is of the general form

$$\left(\frac{P}{A}\right)_w = \frac{\sigma_w}{1 + \varphi(l/r)^2} \qquad (7\text{-}15)$$

where, as previously, σ_w and φ represent the working stress and an experimental coefficient, respectively. (Values for $\sigma_w = 18{,}000$ psi and $\varphi = 1/18{,}000$ are commonly used for structural steel column for which $0 < l/r < 120$ applies.)

In addition to these formulas, the American Institute of Steel Construction (AISC, 1970) recommends formulas for eccentrically loaded columns of structural steel,

$$\frac{f_a}{F_a} + \frac{f_{bx}}{F_{bx}} + \frac{f_{by}}{F_{by}} \le 1 \qquad (7\text{-}16)$$

and for $f_a/F_a > 0.15$,

$$\frac{f_a}{F_a} + \frac{C_{mx}f_{bx}}{(1 - f_a/F_{ex}')F_{bx}} + \frac{C_{my}f_{by}}{(1 - f_a/F_{ey}')F_{by}} \le 1 \qquad (7\text{-}17)$$

where the various quantities are specific and restricted in the AISC code.[6] Thus, it is advisable that one who uses these formulas refer to the code.

It is beyond the scope here to present a detailed explanation of the various formulas; it is merely the intention to acquaint the reader with the general forms and expressions found in the column-design field. One interesting observation, however, may be added: The results from the various formulas are comparative within the restrictions governing the applicability of the various equations; the variation of these results is no greater than those of results from tests. Figure 7–10 is a schematic, general comparison of the formulas presented.

---

[6] For a more detailed treatment, see *Manual of Steel Construction*, American Institute of Steel Construction, 7th ed., 1970, pp. 5–22, 5–23.

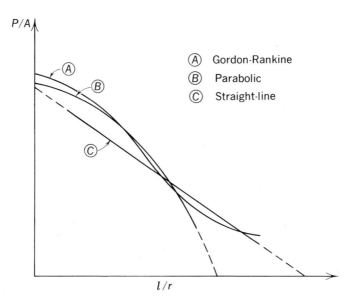

Figure 7–10

7—6 Inelastic Action of Columns

In the preceding sections we have shown that *long* columns buckle at a stress level (average stress over the cross section) below the proportional limit. *Short* struts fail only by compression (no appreciable buckling) at stresses well above the proportional limit. *Intermediate* columns, however, may fail by buckling, rather than compression, but at stresses much higher than those related to long-column failures, and above the proportional limit. This inelastic failure is the topic to be discussed here.

In 1889, F. R. Engesser proposed a method for analyzing columns in the inelastic range. He suggested that (in Euler's equation) E be replaced by E_t, the tangent modulus (defined as the slope of the stress-strain diagram at *any* stress). This approach then became known as the *tangent modulus theory*. Engesser assumed that the column remained straight up to buckling failure and that the tangent modulus of elasticity (see Figure 7–11) remained constant for the whole cross section. On this basis, he predicted the critical stress (Engesser stress) by modifying Euler's formula (for a *pin-ended* column) to

$$\sigma_{cr} = \frac{P_{cr}}{A} = \frac{\pi^2 E_t}{(l/r)^2} \qquad (a)$$

where E was merely replaced by E_t. Then he realized that, as the column bends, the fibers on the convex side of the column would undergo a decreasing compression stress with increased bending, and subsequently the

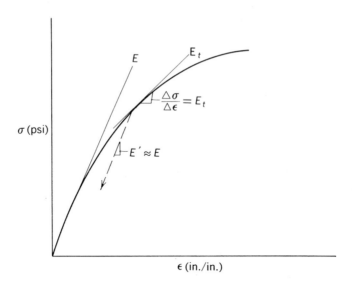

Figure 7–11

modulus E' for those fibers would be about the same E as for a straight column. That is, from Figure 7–11,

convex-side fibers: $E \approx E'$ *concave-side fibers:* $E > E_t$

Hence, Engesser replaced the tangent modulus theory by the *reduced modulus theory.* Again, for a pin-ended column, the critical stress may be obtained by substituting E_r(reduced modulus) for E in Euler's formula.

$$\sigma_{cr} = \frac{P_{cr}}{A} = \frac{\pi^2 E_r}{(l/r)^2} \tag{b}$$

where $E_r > E_t$ and is dependent on both E and E_t, and, to some extent, on the shape of the cross section. For example, for a rectangular shape[7]

$$E_r = \frac{4E_t E}{(E_t^{1/2} + E^{1/2})^2}$$

    Experimental results fall in between those predicted by the tangent modulus and the reduced modulus theories. Convincing evidence was presented to this effect in 1947 by Professor F. R. Shanley,[8] and since then by various researchers. Figure 7–12 generalizes the column behavior when tested, and compares test results with those predicted by the tangent modu-

---

[7] E_r may be derived from the equilibrium of forces acting on a section of a bent column. Such derivations are beyond the scope of this book.

[8] F. R. Shanley, "Inelastic Column Theory," *J. Aero Sci.,* vol. 14, No. 5, May 1947.

lus and reduced modulus theories. The column buckles when the load reaches P_t, the tangent modulus load. The ultimate load P_u that the column may support is, however, generally greater than P_t and less than P_r, the reduced-modulus load.

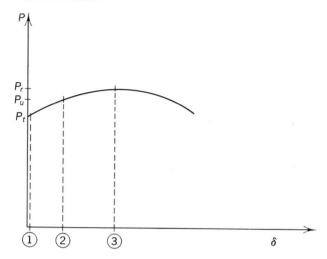

Figure 7–12

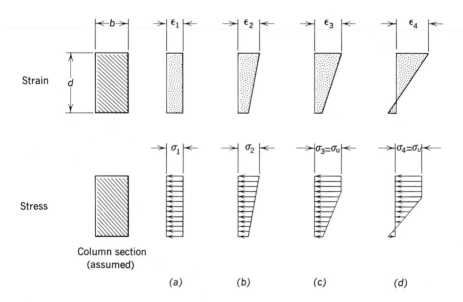

Figure 7–13 Approximate strain and stress distribution over column section with increasing bending. (Actually, the distribution may be a curved line.)

The qualitative analysis of the deformation and stresses over a critical section (usually at midlength of a pin-ended column) may be partially correlated to the analysis given in Section 3-7 for beams. Referring to Figure 7-12, for a small δ (say, point 1), the bending is small and the corresponding strain and stress (rectangular cross section of column assumed) distribution over the depth of the column section is almost rectangular in shape, as shown in Figure 7-13a. The effects of increasing the load may be followed in Figures 7-13b, 7-13c, and 7-13d in relation to Figure 7-12. Actually the E is not constant over the cross section after yielding occurs. (In Section 3-7, we idealized the stress-strain relationship.) The E changes, as indicated in Figure 7-11, which is a fact intentionally not taken into account for the schematic analysis in Figure 7-13 because of the complications encountered from such considerations.

Chapter 8 ◈ RIVETED AND WELDED JOINTS

8—1 Introduction

In the preceding discussions we disregarded, generally speaking, considerations related to the manufacture or construction of structural components. For example, we analyzed pressure vessels but disregarded questions on how one is manufactured; we discussed trusses but intentionally neglected their manufacture; that is, the connecting of one member to another to obtain the truss unit. For us it was convenient and expedient to disregard methods and details of manufacture. Nevertheless, to the designer this is indeed a problem to be solved, and hence one to which he attaches significant importance. In this realm falls the design of *connections*.

There are various ways of connecting one member to another: rivets, welds, bolts and pins, hooks, or threaded couplings. Perhaps more than one of these may serve a particular purpose. In this chapter, however, we will limit our discussion to welded and riveted (or bolted) connections. Some such connections are shown in Figures 8–1a through 8–1e. Both types provide a relatively tight and a relatively stiff joint. However, because of improved welding techniques, the welded joint is rapidly becoming the predominant method of connecting.

Most riveted or welded joints fall into two general categories: *lap* and *butt* joints. The *lap joint* is formed by overlapping two plates and then joining them by either rivets or welding as shown in Figures 8–1a and 8–1c. The *butt joint* is obtained by butting together the two ends of the plates to be connected and then joining them by means of cover plates or butt welding, as shown in Figures 8–1b, 8–1d, or 8–1e. The number of rows of rivets used for each main plate is the criteria which identify the joint as single riveted (Figure 8–1c), double riveted (Figure 8–1d), triple riveted, and so forth. When only some of the rows of rivets engage both cover plates (Figure 8–1e),

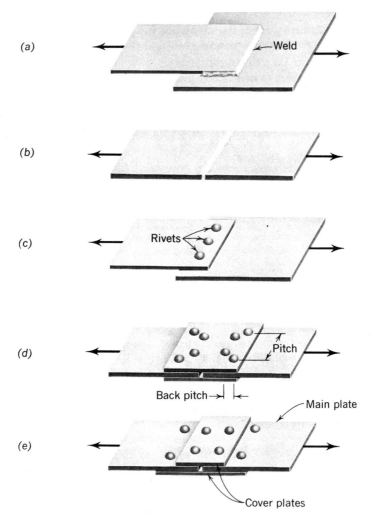

Figure 8–1 (*a*) Lap joint. (*b*) Butt-weld joint. (*c*) Single-riveted lap joint. (*b*) Double-riveted butt joint. (*e*) Double-riveted butt-joint pressure type.

the joint is known as a *pressure joint*. The spacing between the rivets in a row is known as the *pitch;* the distance between rows is known as *back pitch* (Figure 8–1*d*). The general design practice calls for a constant pitch for any given row; subsequently, the joint will consist of a repeating pattern of rivets. Hence, it is convenient to analyze merely a typical repeating section rather than the whole joint.

Riveting is generally performed by heating the rivet having the shape shown in Figure 8–2*a* (although cold riveting is common, especially in the aluminum industry), then flattening one head (by pneumatic hammers, for example) to form the rivet shown in Figure 8–2b. The diameter of the rivet holes are larger than the rivet diameters, usually by 1/16 in. to account for the expanded rivet (when heated) and to provide a reasonable toler- ance for insertion. When the head is flattened, however, the rivet diameter does increase somewhat and thus, partially at least, "filling" the hole. Hence, for analysis purposes, the diameter of the rivet may be assumed equal to the diameter of the hole. As the rivet cools, it shortens (and decreases in diame- ter) and subsequently squeezes the plates together. Thus, it contributes to the rigidity of the joint, which provides impending frictional resistance under the action of load.

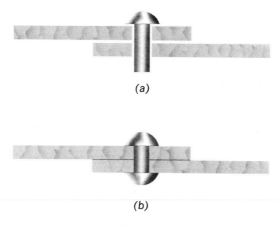

(a)

(b)

Figure 8–2

8—2 Analysis of Riveted Joints

As a preliminary to a more simplified assumption regarding the analy- sis or designing of riveted joints, we may view a simple joint under load from an intuitive approach (Figure 8–3). As the load P is increased, the joint may deform, as shown somewhat exaggerated in Figure 8–3b. We may safely surmise that the resisting forces include friction forces between the plates; also shear, bending, and some tension in the rivet, as well as on the plates. A much more risky conjecture, however, would be the prediction of the exact interrelationship of these forces to each other; also, what and how they contribute to the elastic strength of the joint. Needless to say, the problem becomes much more complex with increasing number of rows of rivets, or with eccentrically loaded joints. Hence, to simplify the problem, it is common practice to neglect the friction completely, to disregard any bending in the plates and in the rivets, and to assume that the shear stress is uniformly distributed over the cross-sectional area of the rivets. With the

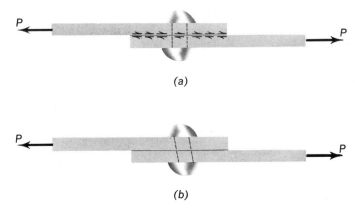

(a)

(b)

Figure 8–3

adoption of a suitable factor of safety, this approach has been reliably and satisfactorily used for most rivet designs.

Based on this simplified approach, we may establish various possible modes of failures of riveted joints: The rivets may fail in shear, as shown in Figure 8–4a. In such a case, we may increase the strength of the joint by increasing the area in shear, that is, by increasing the diameter of the rivets, or by adding more rivets. A second possible type of failure is shown in Figure 8–4b. It is a compression failure of the plate behind the rivet. We may increase the compression strength of the joint by increasing the compression (or bearing) area; that is, by increasing the diameter of the rivet or by increasing the thickness of the plate, or both. A third possible mode of failure is tearing of the plate, as shown in Figure 8–4c. We can increase the tensile strength of the joint by either increasing the plate's thickness or width, or both. Still another possible, although not very probable, mode of failure is shearing of the plate behind the rivet, as shown in Figure 8–4d. By providing sufficient plate length behind the rivet, say a length 2 to 3 times the diameter of the rivet, such shear failures are unlikely. Other types of failure are possible but not common; hence, we shall limit our discussion to those mentioned here.

The strength of a riveted joint is obviously a function of the size and type of material of which it is made. Steel, being by far the most common material used, will be discussed here. The ASME Boiler Construction Code for Unfired Pressure Vessels, 1971 and AISC 1970 Code serve as good guides in the design of boiler joints and steel-building-construction joints, respectively.

Table 8-1 presents a rather limited summary of allowable stresses related to riveted (or bolted) joint design. For a more detailed treatment and explanation, one is encouraged to review these Codes and several others (American Society of Civil Engineers for Aluminum Alloys; American As-

sociation of State Highway Officials for Steel Bridges; American Railway
Engineers Association).

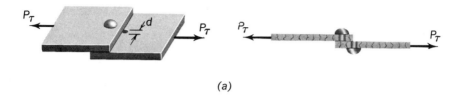

(a)

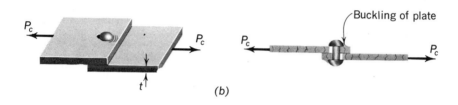

Buckling of plate

(b)

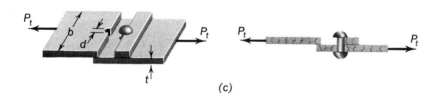

(c)

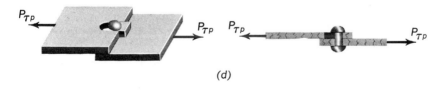

(d)

Figure 8–4 (a) Shearing of rivets. (b) Compression or bearing of
plate. (c) Tearing of plate. (d) Shearing of plate behind rivet.

Table 8-1   Allowable Stresses—Riveted Joints

Code Type	American Institute of Steel Construction for Structural Steel Buildings, 1970 Ed.		A.S.M.E. Boiler Construction Code for Unfired Pressuere Vessesls, 1971
		Yield Stress—F_y (ksi) 36.0 42.0 45.0	13.7
Tension (ksi)	Tension on the net section, $F_t = 0.60F_y = 0.50F_{TS}$ where F_{TS} = minimum tensile strength	22.0 25.2 27.0	for SA-S15 Plate
Bearing (ksi)	Bearing on projected area of bolts in bearing-type connections and on rivets: $F_p = 1.35F_y$	48.6 56.7 60.8	18.0

Shear (ksi)	Power Driven Shop and Field Rivets
	
	   ASTM
	   Designation
		A502-1 A502-2	

	Shear F_v, ksi	15.0 20.0	9.0

	High Strength Bolts in Friction Type Connections and in Bearing Type Connections with Threads in Shear Planes
	   b ASTM
	   Designation
		A325-F
		A325-N A490-F
			for SA-31A rivets

	Shear F_v, ksi	15.0 20.0	

a For threaded parts of material other than $F_y = 36$ ksi steel, use $F_v = 0.30\,F_y$.

b The letter suffixes following the ASTM Designations A325 and A490 represent the following:

   F: Friction type connection
   N: Bearing type connection with threads included in shear plane

Because the holes reduce the area of the plate, it is obvious that, regardless of the strength of the joint in shear or crushing, the strength of the joint is always less than the strength of the solid plate. Hence, we have a new term

$$\textit{Efficiency of the joint} = \frac{\text{allowable load on joint}}{\text{allowable load on unriveted plate}}$$

Example 8—1:

Given: The double-riveted butt joint of Figure *a*. This represents a typical rivet pattern which repeats itself every 6 in.

Find: (*a*) The allowable strength of the joint, and (*b*) the efficiency of the joint. Assume ASME Boiler Code to govern.

Procedure: Because the cross-sectional area of both cover plates is larger than the area of a main plate, only the main plates will be investigated for bearing. They will, however, be investigated for tensile strength. Also, because the distance behind the rivet (row 2) is sufficiently large (greater than 2 diameters), the shearing of the plate will not be investigated. Hence, we investigate strengths for shear of rivets, crushing of plate, tension of plate, or a combination thereof.

Shearing of all 5 rivets (Figure *b*): The rivets are assumed to completely fill the hole. Thus, the area of one rivet is $\pi/4 \, (\%)^2 = 0.307$ in.^{2} The total force P_r that the joint can take in shear of all 5 rivets is (assuming that the load is distributed equally to each rivet)

$$P_r = 5(0.307 \times 9000) \times 2 \text{ shear surfaces /rivet}$$

$$P_r = 5(2763) \times 2 = 27,630 \text{ lb}$$

Bearing of plate behind rivets (Figure *c*):

$$P_c = (18,000 \cdot t \cdot d)5 \text{ rivets} = 18,000 \times \frac{1}{2} \times \% \times 5$$

$$P_c = 28,125 \text{ lb}$$

Tension of main plate (Figures *d* and *e*):

Row 1: The area in tension is $(6 - 2 \times \%) \frac{1}{2} = 2.375$ in.^{2} Therefore,

$$P_{t1} = 13,700 \times 2.375 = 32,538 \text{ lb}$$

Row 2: If tension failure in row 2 occurs, row 1 will first fail either in shear or crushing, whichever results in the smaller force. The shearing of two rivets in row 1 requires a force of $\frac{2}{5} (27,630) = 11,052 < \frac{2}{5} \times (28,125)$. Therefore, shearing of the two rivets will have to take place simultaneously with the tearing of the plate in row 2.

Area of plate, row 2, $= (6 - 3 \times \%) \cdot \frac{1}{2} = 2.06$ in.^{2}

The allowable force in tension across row 2 then becomes

$$P_{t2} = (2.06)(13,700) + 11,052 = 39,274 \text{ lb}$$

Obviously, tearing across row 2 is not possible because the joint would fail much before P_{t2} is reached. (Shear strength is 27,630 lb.)

Tearing of *cover* plates (Figure *f*): The least cross-sectional area of the cover plates is along row 2. This area is

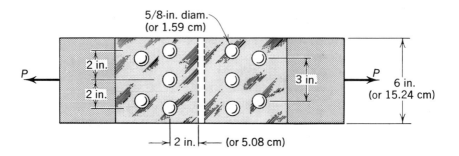

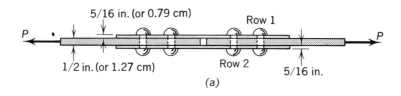

(a)

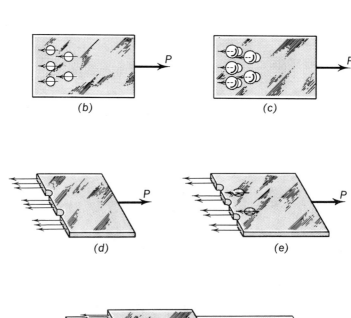

(b)   (c)

(d)   (e)

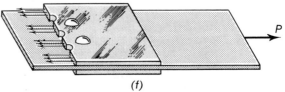

(f)

$$A_2 = (2 \text{ plates})(\tfrac{5}{16})(6 - 3 \times \tfrac{5}{8}) = 2.57 \text{ in.}^2$$

Therefore,

$$P_{t2} = (2.57)(13,700) = 37,675 \text{ lb}$$

Hence, the ultimate strength of the joint is the *smallest* force calculated; that is, $P = 27,630$ is the ultimate strength. Thus, the efficiency of the joint becomes

$$\text{Efficiency} = \frac{27,630}{6 \times \tfrac{1}{2} \times 13,700} = \mathbf{67 \ percent}$$

PROBLEMS

Solve the following problems using the following allowable stresses: $\tau = 15,000$ psi, $\sigma_c = 48,600$ psi, and $\sigma_t = 22,000$ psi. Neglect shearing of the plate.

8–1 Determine the allowable strength and the efficiency of the joint in Figure P 8–1 if the rivets are ¾ in. diameter, the plates are each ½ in. thick and 5 in. wide. The pitch of the rivets is 2 in.

Figure P 8–1

8–2 For Problem 8–1, adjust the width of the plates such that the strength in tension is equal to that of the rivets in shear.

8–3 In Figure P 8–1, determine the diameter to the nearest ⅛ in. of the rivets, if the two plates are each ½ in. thick and 5 in. wide, and that the strength of the joint by shearing of the rivets is approximately the same as that by tension failure in the plate.

8–4 Determine the minimum plate thicknesses of the joint in Figure P 8–1 so that the plates, 5 in. wide, will resist about the same force as necessary to fail the ¾-in. diam rivets in shear.

8–5 Determine the allowable strength and the efficiency of the joint in Figure P 8–5 if the rivets are ⅞ in. diameter, the mainplate thickness t is ¾ in. thick and 9 in. wide. The pitch of the rivets is 3 in.

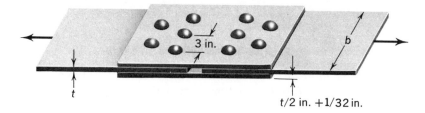

Figure P 8–5

8–6 For Problem 8–5, adjust the width of the plates so that the strength in tension is equal to that of the rivets in shear.

8–7 If the plates in Figure P 8–5 are 9 in. wide and the main plates are ¾ in. thick, determine the diameter to the nearest ⅛ in. of the rivets, so that the strength of the joint by shearing of the rivets is approximately the same as that by tension in the plate.

8–8 Determine the minimum main-plate thickness t of the joint in Figure P 8–5 so that the plates, 9 in. wide, will resist about the same force as ⅞-in. diam rivets in shear.

8–9 Determine the maximum strength and the efficiency of the joint in Figure P 8–9 if the rivets are ⅞ in. diameter, the main-plate thickness t is ¾ in. thick and 9 in. wide. The pitch of the rivets is 3 in.

Figure P 8–9

8–10 For Problem 8–9, adjust the width of the plates so that the strength in tension is equal to that of the rivets in shear.

8–11 If the plates in Figure P 8–9 are 9 in. wide and the main plates are ¾ in. thick, determine the diameter to the nearest ⅛ in. of the rivets, so that the strength of the joint by shearing of the rivets is approximately the same as that by tension in the plate.

8–12 Determine the allowable strength and the efficiency of the joint in Figure P 8–12 if the rivets (single shear) are ¾ in. diameter and the plate is ½ in. thick and 8 in. wide.

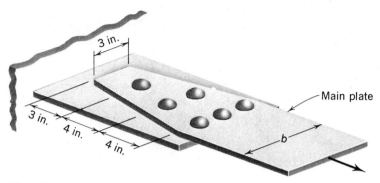

Figure P 8–12

8–13 For Problem 8–12, adjust the width of the plate (keeping the left end 3 in.) so that the strength in tension is equal to that of the rivets in shear.

8–14 If the plate in Figure P 8–12 is ½ in. thick and 8 in. wide, determine the diameter, to the nearest ⅛ in., of the rivets so that the strength of the joint by shearing of the rivets is approximately the same as that by tension in the plate.

8–15 Determine the minimum plate thickness of the joint in Figure P 8–12 so that the plate, as shown, will resist about the same force as the ¾-in. diam rivets in shear.

8—3 Eccentric Loads on Riveted Joints

The preceding discussion was limited to riveted joints subjected to a load passing through the centroid of the group of rivets. Subsequently, we assumed that for such a case the rivets were equally stressed and that the resisting forces of the rivets were oriented in the same direction, but opposite to the applied force. Although these assumptions are acceptably reasonable and useful for many analysis problems, modifications must be made when the load does not pass through the centroid of the rivet group.

Consider a group of rivets under the action of a load P, eccentric a distance e from 0, the centroid of the group, as shown in Figure 8–5a. The effect of this force is equivalent to that of a force through 0 and a clockwise couple, as shown in Figures 8–5b and 8–5c. Hence, the rivet forces must resist the downward translation and clockwise rotation, and must therefore be oriented, as shown in Figure 8–6b and 8–6c. With the load P evenly distributed to n rivets, the magnitude of the resisting forces in Figure 8–6b is merely P/n, oriented as shown.

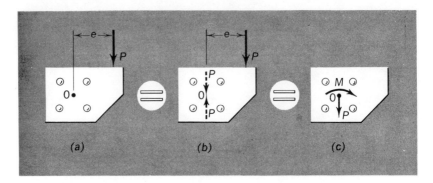

Figure 8–5

The evaluation of magnitudes and directions of the forces in Figure 8–6c is based on certain assumptions, and needs a little more detailed viewing: The external couple will rotate the plate clockwise about some point, say point O, as shown in Figure 8–7. If we assume that straight lines on the plate remain straight after the couple is applied, we may then safely assume that the average strain in a rivet is proportional to that rivet's distance from point O; further, if elastic action prevails, the average shear stress in a rivet is also proportional to that radial distance. For any rivet i, we may write

$$\tau_i = kr_i \tag{a}$$

from which

$$F_i = kr_iA_i \tag{b}$$

where k is a proportionality constant, and F_i is the resisting force of the rivet, perpendicular to the radius r_i.

The resultant of the resisting forces F_1, F_2,$\cdots$, F_i must necessarily be a couple equal and opposite in direction to $(P \cdot e)$. Hence, $\Sigma F_x = 0$,

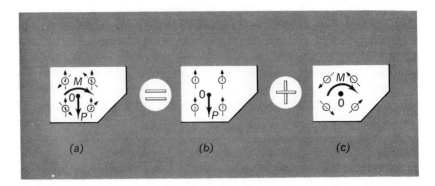

Figure 8–6

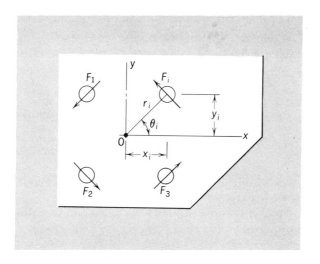

Figure 8–7

$\Sigma F_y = 0$. Thus, applying $\Sigma F_x = 0$, we have

$$\Sigma F_x = F_1 \sin \theta_1 + F_2 \sin \theta_2 + \cdots = \sum_{i=1}^{n} F_i \sin \theta_i = 0$$

and from Equation (b), we obtain

$$\Sigma F_x = \sum_{i=1}^{n} (kr_i A_i)\sin \theta_i = k \sum_{i=1}^{n} y_i A_i = k\bar{y}_i A_A = 0$$

where A_A = area of all rivets. Because $kA_A \neq 0$, $\bar{y}_i = 0$. A similar analysis shows that $\bar{x}_i = 0$. Hence, the center of rotation and centroid of areas of the rivets is the same point, O.

The forces in the rivets, because of rotation only, may be determined using $\Sigma M_0 = 0$. Thus, we have

$$P \cdot e = \Sigma r_i F_i = \Sigma r_i(kr_i A_i) = k \, \Sigma \, r_i^2 A_i$$

For equal-size rivets of area A, from Equation (b), $F_i = kAr_i$. Thus, we have

$$P \cdot e = kA \, \Sigma \, r_i^2 = \left(\frac{F_i}{r_i}\right) \Sigma \, r_i^2$$

where $r_i = \sqrt{x_i^2 + y_i^2}$

$$P \cdot e = \frac{F_i \, \Sigma \, (x^2 + y^2)}{\sqrt{x_i^2 + y_i^2}}$$

Solving for F_i, we get

$$F_i = \frac{P \cdot e \cdot \sqrt{x_i^2 + y_i^2}}{\Sigma \, x^2 + \Sigma \, y^2} \tag{8-1}$$

It must be kept in mind that F_i in Equation (8–1) is only the force in the rivet caused by *rotation*. The translation force was resolved as P/n. Hence, the resultant force R_i in a rivet is the vector sum of these components; that is,

$$R_i = \bar{F}_i + \bar{P}/n \qquad (8\text{–}2)$$

The following example will illustrate the procedure of analysis presented above.

Example 8–2:

Given: The riveted joint shown in Figure *a*. The rivets are ½ in. in diameter (single shear).

Find: (*a*) The shear stress in the rivet carrying the greatest load, and (*b*) the shear stress in the rivet carrying the least load.

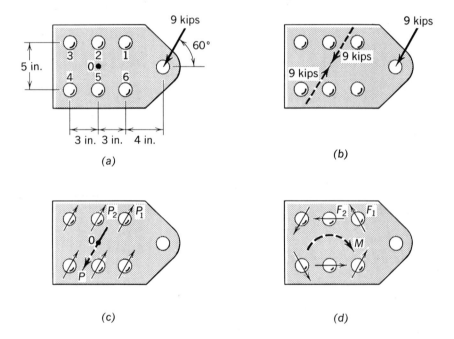

(a)

(b)

(c)

(d)

Procedure: The force may be replaced by a force through 0 and a couple, as shown in Figure *b*. The translational effect is resisted by the forces shown in Figure *c*. Their magnitude (same for each) is

$$\text{Translation: } P_1 = P_2 = \cdots P_6 = \tfrac{9}{6} = 1.5 \text{ kips}$$

The value of the couple is $9 \cdot \text{kips} \cdot \sin 60° \cdot 7 \text{ in.} = 54.56$ kips-in. This is the equivalent of $P \cdot e$ in Equation (8–1). Also, in Equation (8–1), we have

$$\Sigma\, x^2 = 2(3)^2 + 2(0)^2 + 2(-3)^2 = 36$$

$$\Sigma\, y^2 = 3(2.5)^2 + 3(-2.5)^2 = 37.5$$

Hence, for rivets 1, 3, 4, 6: $r = \sqrt{(3)^2 + (2.5)^2} = 3.9$ in. For rivets 2 and 5: $r = 2.5$ in.

From Equation (8–1),

$$F_1 = F_3 = F_4 = F_6 = \frac{54.56 \times 3.9}{36 + 37.5} = 2.9 \text{ kips}$$

$$F_2 = F_5 = \frac{54.56 \times 2.5}{36 + 37.5} = 1.86 \text{ kips}$$

From the orientation of the forces, Figures c and d, we note that rivet 6 may well carry the largest load, while rivet 3 carries the least load. Although this is frequently evident, it is a good practice to investigate all rivets when not sure. In our case, we shall assume that rivets 6 and 3 carry the maximum and minimum loads, respectively; we will then check our assumptions. A diagram of the forces (magnitude and directions) acting on the respective rivets is shown below. From these we

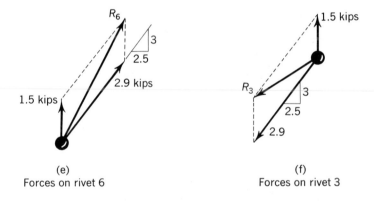

(e)
Forces on rivet 6

(f)
Forces on rivet 3

note that the vector sum of the components of the two forces in the x and y directions must be equal to the respective components of the resultant; that is, for the x component of rivet 6:

$$
\begin{array}{l}
\cos 60^\circ\,(1.5) = 0.75 \rightarrow \\
2.5/3.9\,(2.9) = 1.86 \rightarrow \\
\hline
\qquad R_{6x} \qquad = 2.61 \rightarrow
\end{array}
$$

for the y component of rivet 6:

$$
\begin{array}{l}
\sin 60^\circ\,(1.5) = 1.3 \uparrow \\
3/3.9\,(2.9) \quad = 2.25 \uparrow \\
\hline
\qquad R_{6y} \qquad = 3.55 \uparrow
\end{array}
$$

$$R_6 = \sqrt{R_x{}^2 + R_y{}^2} = \sqrt{2.61^2 + 3.55^2} = 4.41 \text{ kips}$$

For the x component of rivet 3:

$$\begin{array}{l} \cos 60° \,(1.5) = 0.75 \longrightarrow \\ 3/3.9 \,(2.9) \quad\; = 1.86 \longleftarrow \\ \hline \qquad\qquad R_{3x} = 1.11 \longleftarrow \end{array}$$

and for the y component of rivet 3:

$$\begin{array}{l} \sin 60° \,(1.5) = 1.30 \uparrow \\ 2.5/3.9 \,(2.9) = 2.25 \downarrow \\ \hline \qquad\quad R_{3y} = 0.95 \downarrow \end{array}$$

$$R_3 = \sqrt{(1.11)^2 + (0.95)^2} = 1.46 \text{ kips}$$

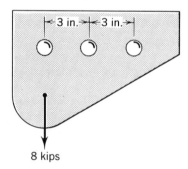

The relative magnitudes of F_1, F_2, F_4, and F_5 can perhaps be made evident by merely sketching; thus, they provide a satisfactory proof that F_6 and F_3 are indeed the largest and smallest forces, respectively, in the group. Hence, the stresses in rivets 6 and 3 are

$$\sigma_6 = \frac{R_6}{A} = \frac{4.41}{\pi/4(0.5)^2} = \textbf{22.5 ksi}$$

$$\sigma_3 = \frac{R_3}{6} = \frac{1.46}{\pi/4(0.5)^2} = \textbf{7.45 ksi}$$

PROBLEMS

8–16 Determine the magnitude and directions of the forces in each of the three rivets in Figure P 8–16.

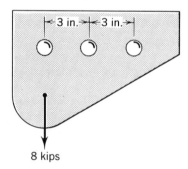

Figure P 8–16

8–17 Determine the magnitudes and direction of the forces in each of the rivets in Figure P 8–17.

8–18 Determine the minimum diameter of the rivets in Figure P 8–17 (same diameter for all rivets) if the shear stress in any of the rivets is not to exceed 10,000 psi.

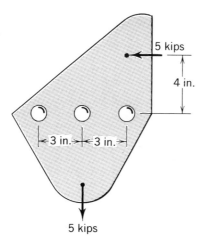

Figure P 8–17

8–19 For the joint in Figure P 8–19, determine the average shear stress across the rivets carrying the greatest and least loads. All rivets are ¾ in. diameter. Assume the rivets to be in single shear.

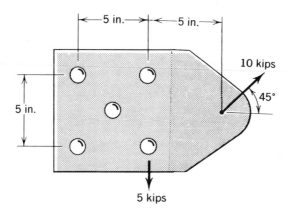

Figure P 8–19

8–20 Determine the least diameter that the rivets in Figure P 8–19 may be if the stress in any of the rivets is not to exceed 14,000 psi.

8–21 Determine the average shear stress across each of the rivets of the joint in Figure P 8–21. The rivets are ⅞ in. diameter, in single shear.

8–22 What is the maximum eccentricity that the 12-kip load in Figure 8–21 may have if the allowable stress (single shear) in any of the rivets in Figure P 8–21 is 16,000 psi?

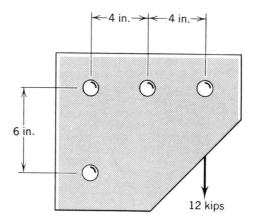

Figure P 8–21

8–23 For the joint in Figure P 8–23, determine the average shear stresses in the rivets carrying the largest and least loads. All the rivets are ⅞ in. diameter, in single shear.

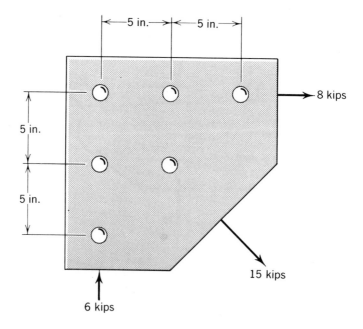

Figure P 8–23

8–24 Determine the least diameter the rivets in Figure P 8–23 may be if the stress in any of the rivets is not to exceed 10,000 psi.

8—4 Welded Joints

Welding in construction has already been mentioned briefly in Section 8-1. In most instances, welding is accomplished through fusion although it can also be done by heating the base metal under high localized pressure, by brazing, or by soldering. As mentioned in Section 8-1, the welded joint may be a butt or a fillet joint (by lapping). Here, we limit our discussion to fusion welding, by far the most common method in general structural fabrication.

Fusion welding consists of melting (by means of an electric arc, gas flame, and so forth) the parts in the vicinity of the juncture, then permitting cooling and subsequent union of the two metals. Additional welding material is usually deposited between the two parts of a butt joint, or to form the fillet in the case of the lap joints.

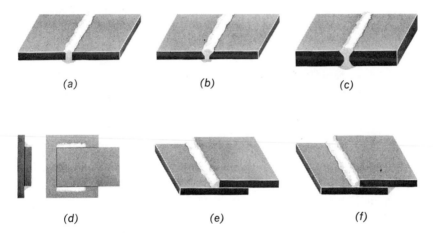

(a)   (b)   (c)

(d)   (e)   (f)

Figure 8–8 (a) Square-groove weld. (b) Single-vee groove butt joint. (c) Double-vee groove butt joint. (d)–(f) Fillet welds.

Figure 8–8 shows two types of welded joints formed by fusion: *butt* welds and *fillet* welds. For fairly thin materials, say less than ⅜ in. (or 9.5 mm) thick, the square groove joint in Figure 8–8a is satisfactory. For plates up to ¾ in. (or 19 mm) thick, the single-vee-groove joint in Figure 8–8b is acceptable; the double-vee in Figure 8–8c is acceptable for thicker materials. The fillet welds shown in Figures 8–8d, 8–8e, and 8–8f are used to join two lapped pieces.

The strength of the fillets (sides and ends) is assumed equal to the resistance which is provided by the shearing of the weld along the throat (through a–b, Figure 8–9), using a working shearing stress of 13,600 psi (or 93.77 mn /m²) (steel) for shielded arc welds. In the process, the coated welding rod (electrode) produces an inert gas which envelops the arc stream and

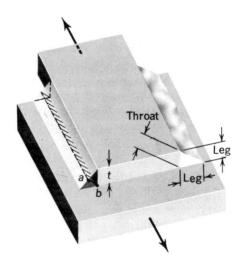

Figure 8–9

thus prevents rapid and excessive oxidation of the molten metal. This proves to be a superior joint to one produced by an unshielded arc. For a 45-degree fillet, the leg is equal to t, thus making the throat equal to $0.707 \cdot (t)$. From this, the working shearing resistance per unit length of fillet weld becomes

$$F/\text{unit length} = (13,600)(0.707t) = 9600t \text{ lb/in.}$$

For butt welds, the allowable stress in tension or compression is 20,000 psi.

Although in our analysis the weld is assumed to slope at 45 degrees, as shown in Figure 8–9, it is desirable to make the fillet slightly concave (Figure 8–8e) in order to reduce stress concentration.

Example 8–3:

Given: A 6-in. × 4-in. × ½-in. angle section is welded to a plate with the 6-in. side parallel to the plate. It carries a load of 96,000 lb, applied along the centroidal axis of the angle, as shown in the figure. Assume the legs of the fillets to be equal to ½ in., and a working shearing stress in the fillets to be 13,600 psi.

Find: The lengths and arrangement of the weld fillets if (a) only the sides of the angle are welded, and (b) the sides as well as the ends are welded.

Procedure: The resistance force provided by 1-in. length of weld is

$$F = 13,600(0.5 \times 0.707) = 4800 \text{ lb}$$

Hence, the total length of fillet weld necessary to carry the load is $96,000/4800 = 20$ in.

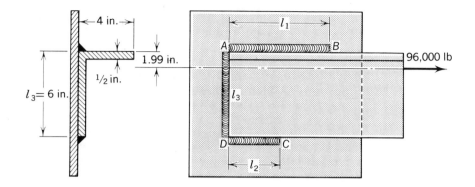

(*a*) For side welds only:

$$l_1 + l_2 = 20 \text{ in.} \qquad (a)$$

From equilibrium we may determine the ratio of l_1/l_2; that is, from $\Sigma M_D = 0$, we have

$$(4800 l_1) \cdot 6 = 4.01 \times 96,000$$

$$l_1 = \mathbf{13.4 \text{ in.}}$$

From Equation (*a*) we obtain

$$l_2 = 20 - 13.4 = 6.6 \text{ in.}$$

(*b*) For side and end welds:

$$l_1 + l_2 + l_3 = l_1 + l_2 + 6 \text{ in.} = 20 \text{ in.}$$

Hence,

$$l_1 + l_2 = 14 \text{ in.} \qquad (b)$$

Again, using $\Sigma M_D = 0$, we obtain

$$(4800\, l_1)(6) + 4800 \times l_3 \times 3 = 4.01 \times 96,000$$

$$l_1 = \frac{4.01 \times 96,000 - 4800 \times 6 \times 3}{4800 \times 6} = 10.4 \text{ in.}$$

From Equation (*b*), we get

$$l_2 = 14 - 10.4 = 3.6 \text{ in.}$$

PROBLEMS

For the following problems, assume there is a working shearing stress in the fillet welds of 13,600-psi and 20,000-psi tensile stress for the base metal. Also, the leg of the weld assumed to be equal to the thickness of the plate.

8-25 A plate, 5 in. wide and ½ in. thick, is loaded axially with a 30,000-lb load. Determine the length of weld on each side of the plate to carry this load.

8-26 For the plate in Problem 8–25, what should be the length of the side welds if the plate is to develop its full working strength?

8-27 A 5-in. × 3-in. × ⅜-in. angle section is welded to a plate with the 5-in. side parallel to the plate. It carries a load of 60,000 lb applied along the centroidal axis of the angle. Determine the minimum lengths of the welds if (a) only the sides are welded, and (b) the weld is along the sides and at the end (as in the figure of Example 8–3).

8-28 For the angle of Problem 8–27, determine the lengths of the side welds if the plate is to develop its full working strength.

8-29 Rework Problem 8–27 with the 3-in. side against the plate.

8-30 A cylindrical tank, 3 ft in diameter and 6 ft long, is constructed of a ⅜-in. plate butt welded along a 45-degree spiral seam. If the allowable working stress in the weld is equal to that of the plate, determine the safe pressure that the tank can withstand.

8-31 Rework Problem 8–30 if the tank is constructed by lapping the plates as shown in Figure P 8–31. Assume that both welds resist the pressure equally. The allowable fillet shear stress in 13,600 psi.

Figure P 8–31

8-32 An 8-in. × 4-in. × 7/16-in. angle is stressed to an axial stress of 16,000 psi. Determine the length of side fillet welds when (a) the 8-in. side of the angle rests against the base plate, and (b) the 4-in. side rests against the plate.

APPENDIX A

461

TABLE A–1 Average Physical Properties of Some Common Metals*

Metal	Density lb per cu in.	Modulus elasticity Psi E Tension	Modulus elasticity Psi G Shear	Elastic stress† Psi Tension	Elastic stress† Psi Shear	Ultimate strength Psi Tension	Ultimate strength Psi Compression	Ultimate strength Psi Shear	Temperature coefficient of linear expansion per °F	Percent elongation in 2 in.
Steel										
hot rolled, 0.2%	0.283	29×10^6	11.6×10^6	38,000	24,000	65,000		46,000	6.6×10^{-6}	30
cold rolled, 0.2%	0.283	29×10^6	11.6×10^6	60,000	35,000	85,000		60,000	6.6×10^{-6}	20
hot rolled, 0.4%	0.283	29×10^6	11.6×10^6	53,000	30,000	82,000		60,000	6.6×10^{-6}	30
hot rolled, 0.8%	0.283	29×10^6	11.6×10^6	75,000	40,000	120,000		100,000	6.6×10^{-6}	10
stainless (18-8)										
annealed	0.285	28×10^6	12.5×10^6	36,000		85,000			9.6×10^{-6}	45
cold rolled	0.285	28×10^6	12.5×10^6	150,000		180,000		60,000	9.6×10^{-6}	10
Aluminum										
cast (99% Al)	0.100	10×10^6	4×10^6	10,000		15,000		11,000	12.8×10^{-6}	20
hard drawn	0.100	10×10^6	4×10^6	22,000		33,000		18,000	12.8×10^{-6}	5
Alloy and temper										
2014-T6	0.100	10.6×10^6	4×10^6	55,000	32,000	63,000	63,000	39,000	12.8×10^{-6}	7
2024-T3	0.100	10.6×10^6	4×10^6	42,000	24,000	64,000	64,000	40,000	12.9×10^{-6}	
6061-T6	0.100	10.4×10^6	3.9×10^6	58,000	34,000	68,000	68,000	41,000	13.1×10^{-6}	5
7075-T6	0.100	10.0×10^6	3.75×10^6	35,000	20,000	42,000	42,000	27,000	13.0×10^{-6}	

Cast iron										
gray	0.261	14×10^6	6×10^6	35,000		25,000	90,000	30,000	6×10^{-6}	1
malleable	0.264	25×10^6	12.5×10^6		25,000	55,000		48,000	6.6×10^{-6}	20
nodular	0.264	25×10^6	12.5×10^6	65,000	45,000	95,000		80,000	6.6×10^{-6}	4
Magnesium										
extrusion	0.064	6.5×10^6	2.4×10^6	20,000		30,000		18,000	14.5×10^{-6}	10
sand cast (AZ63-HT)	0.064	6.5×10^6	2.4×10^6	14,000		40,000		20,000	14.5×10^{-6}	10
Brass — Red 80% Cu										
hard	0.310	15×10^6	5.5×10^6	65,000		85,000			10.4×10^{-6}	4
soft	0.310	15×10^6	5.5×10^6	18,000		43,000			10.4×10^{-6}	50
Bronze cast — 90% Cu, 10% Sn	0.320	15×10^6	6.5×10^6	25,000		65,000			10×10^{-6}	10
Titanium alloy annealed	0.165	17×10^6	6.5×10^6	140,000		165,000		100,000	6.0×10^{-6}	15

*These properties may vary greatly with variation in composition, heat treatment, or mechanical working.

†For more ductile material the elastic strength was based on the 0.2 percent offset.

TABLE A–2 Average Physical Properties of Some Common Woods*

| Species† | Density lb per cu in. | Strength parallel to grain (Psi) | | | | Shearing | |
| | | Compression | | | | | |
		Prop. limit	Ultimate strength	S‡ Working stress	E Modulus of elasticity	Ultimate strength	Working‡ stress
Red cedar	24	4300	5000	600	1.2×10^6	800	150
Douglas fir	30	6500	7400	1200	1.9×10^6	1100	250
Western hemlock	24	5300	6200	900	1.5×10^6	1100	250
White oak	50	4300	7000	900	1.6×10^6	1900	350
Southern pine	25	4800	7100	900	1.9×10^6	1400	200
Redwood	23	4500	6200	1000	1.3×10^6	1000	100
Spruce	24	4100	5600	800	1.3×10^6	1100	200
Birch	43	6200	8300	1500	2.1×10^6	2000	300

*The properties may vary greatly with variation in quality, geographical source, seasoning, water content, and so forth.

†The values assume reasonably dry wood, say, less than 12 percent water content, and reasonably seasoned.

‡These values are suggested based on a reasonable safety factor, and may vary with conditions and designers.

APPENDIX B

Reprinted from *The Manual of Steel Construction*, 7th edition, American Institute of Steel Construction, New York, 1970.

465

PROPERTIES OF GEOMETRIC SECTIONS

SQUARE
Axis of moments through center

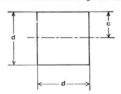

$$A = d^2$$

$$c = \frac{d}{2}$$

$$I = \frac{d^4}{12}$$

$$S = \frac{d^3}{6}$$

$$r = \frac{d}{\sqrt{12}} = .288675\ d$$

$$Z = \frac{d^3}{4}$$

SQUARE
Axis of moments on base

$$A = d^2$$

$$c = d$$

$$I = \frac{d^4}{3}$$

$$S = \frac{d^3}{3}$$

$$r = \frac{d}{\sqrt{3}} = .577350\ d$$

SQUARE
Axis of moments on diagonal

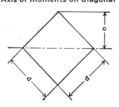

$$A = d^2$$

$$c = \frac{d}{\sqrt{2}} = .707107\ d$$

$$I = \frac{d^4}{12}$$

$$S = \frac{d^3}{6\sqrt{2}} = .117851\ d^3$$

$$r = \frac{d}{\sqrt{12}} = .288675\ d$$

$$Z = \frac{2c^3}{3} = \frac{d^3}{3\sqrt{2}} = .235702d^3$$

RECTANGLE
Axis of moments through center

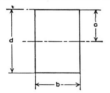

$$A = bd$$

$$c = \frac{d}{2}$$

$$I = \frac{bd^3}{12}$$

$$S = \frac{bd^2}{6}$$

$$r = \frac{d}{\sqrt{12}} = .288675\ d$$

$$Z = \frac{bd^2}{4}$$

PROPERTIES OF GEOMETRIC SECTIONS

RECTANGLE
Axis of moments on base

$A = bd$

$c = d$

$I = \dfrac{bd^3}{3}$

$S = \dfrac{bd^2}{3}$

$r = \dfrac{d}{\sqrt{3}} = .577350\,d$

RECTANGLE
Axis of moments on diagonal

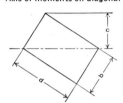

$A = bd$

$c = \dfrac{bd}{\sqrt{b^2 + d^2}}$

$I = \dfrac{b^3 d^3}{6(b^2 + d^2)}$

$S = \dfrac{b^2 d^2}{6\sqrt{b^2 + d^2}}$

$r = \dfrac{bd}{\sqrt{6(b^2 + d^2)}}$

RECTANGLE
Axis of moments any line
through center of gravity

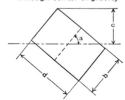

$A = bd$

$c = \dfrac{b \sin a + d \cos a}{2}$

$I = \dfrac{bd(b^2 \sin^2 a + d^2 \cos^2 a)}{12}$

$S = \dfrac{bd(b^2 \sin^2 a + d^2 \cos^2 a)}{6(b \sin a + d \cos a)}$

$r = \sqrt{\dfrac{b^2 \sin^2 a + d^2 \cos^2 a}{12}}$

HOLLOW RECTANGLE
Axis of moments through center

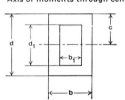

$A = bd - b_1 d_1$

$c = \dfrac{d}{2}$

$I = \dfrac{bd^3 - b_1 d_1^3}{12}$

$S = \dfrac{bd^3 - b_1 d_1^3}{6d}$

$r = \sqrt{\dfrac{bd^3 - b_1 d_1^3}{12\,A}}$

$z = \dfrac{bd^2}{4} - \dfrac{b_1 d_1^2}{4}$

PROPERTIES OF GEOMETRIC SECTIONS

EQUAL RECTANGLES

Axis of moments through center of gravity

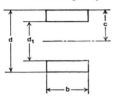

$$A = b(d - d_1)$$

$$c = \frac{d}{2}$$

$$I = \frac{b(d^3 - d_1^3)}{12}$$

$$S = \frac{b(d^3 - d_1^3)}{6d}$$

$$r = \sqrt{\frac{d^3 - d_1^3}{12(d - d_1)}}$$

$$Z = \frac{b}{4}(d^2 - d_1^2)$$

UNEQUAL RECTANGLES

Axis of moments through center of gravity

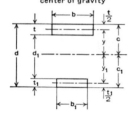

$$A = bt + b_1t_1$$

$$c = \frac{\frac{1}{2}bt^2 + b_1t_1(d - \frac{1}{2}t_1)}{A}$$

$$I = \frac{bt^3}{12} + bty^2 + \frac{b_1t_1^3}{12} + b_1t_1y_1^2$$

$$S = \frac{I}{c} \qquad S_1 = \frac{I}{c_1}$$

$$= \sqrt{\frac{I}{A}}$$

$$Z = \frac{A}{2}\left[d - \left(\frac{t + t_1}{2}\right)\right]$$

TRIANGLE

Axis of moments through center of gravity

$$A = \frac{bd}{2}$$

$$c = \frac{2d}{3}$$

$$I = \frac{bd^3}{36}$$

$$S = \frac{bd^2}{24}$$

$$r = \frac{d}{\sqrt{18}} = .235702\,d$$

TRIANGLE

Axis of moments on base

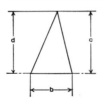

$$A = \frac{bd}{2}$$

$$c = d$$

$$I = \frac{bd^3}{12}$$

$$S = \frac{bd^2}{12}$$

$$r = \frac{d}{\sqrt{6}} = .408248\,d$$

PROPERTIES OF GEOMETRIC SECTIONS

TRAPEZOID

Axis of moments through
center of gravity

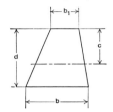

$$A = \frac{d(b + b_1)}{2}$$

$$c = \frac{d(2b + b_1)}{3(b + b_1)}$$

$$I = \frac{d^3 (b^2 + 4 bb_1 + b_1{}^2)}{36 (b + b_1)}$$

$$S = \frac{d^2 (b^2 + 4 bb_1 + b_1{}^2)}{12 (2b + b_1)}$$

$$r = \frac{d}{6(b + b_1)} \sqrt{2 (b^2 + 4 bb_1 + b_1{}^2)}$$

CIRCLE

Axis of moments
through center

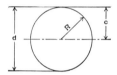

$$A = \frac{\pi d^2}{4} = \pi R^2 = .785398\ d^2 = 3.141593\ R^2$$

$$c = \frac{d}{2} = R$$

$$I = \frac{\pi d^4}{64} = \frac{\pi R^4}{4} = .049087\ d^4 = .785398\ R^4$$

$$S = \frac{\pi d^3}{32} = \frac{\pi R^3}{4} = .098175\ d^3 = .785398\ R^3$$

$$r = \frac{d}{4} = \frac{R}{2}$$

$$Z = \frac{d^3}{6}$$

HOLLOW CIRCLE

Axis of moments
through center

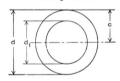

$$A = \frac{\pi(d^2 - d_1{}^2)}{4} = .785398\ (d^2 - d_1{}^2)$$

$$c = \frac{d}{2}$$

$$I = \frac{\pi(d^4 - d_1{}^4)}{64} = .049087\ (d^4 - d_1{}^4)$$

$$S = \frac{\pi(d^4 - d_1{}^4)}{32d} = .098175\ \frac{d^4 - d_1{}^4}{d}$$

$$r = \frac{\sqrt{d^2 + d_1{}^2}}{4}$$

$$Z = \frac{d^3}{6} - \frac{d_1{}^3}{6}$$

HALF CIRCLE

Axis of moments through
center of gravity

$$A = \frac{\pi R^2}{2} = 1.570796\ R^2$$

$$c = R \left(1 - \frac{4}{3\pi}\right) = .575587\ R$$

$$I = R^4 \left(\frac{\pi}{8} - \frac{8}{9\pi}\right) = .109757\ R^4$$

$$S = \frac{R^3}{24} \frac{(9\pi^2 - 64)}{(3\pi - 4)} = .190687\ R^3$$

$$r = R \frac{\sqrt{9\pi^2 - 64}}{6\pi} = .264336\ R$$

PROPERTIES OF GEOMETRIC SECTIONS

PARABOLA OF nth DEGREE

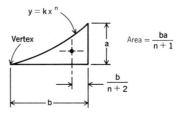

$$\text{Area} = \frac{ba}{n+1}$$

PARABOLA OF nth DEGREE

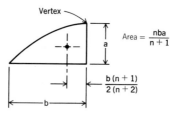

$$\text{Area} = \frac{nba}{n+1}$$

PROPERTIES OF GEOMETRIC SECTIONS

* HALF ELLIPSE

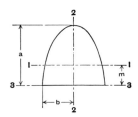

$$A = \frac{1}{2}\pi ab$$

$$m = \frac{4a}{3\pi}$$

$$I_1 = a^3b\left(\frac{\pi}{8} - \frac{8}{9\pi}\right)$$

$$I_2 = \frac{1}{8}\pi ab^3$$

$$I_3 = \frac{1}{8}\pi a^3b$$

* QUARTER ELLIPSE

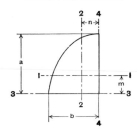

$$A = \frac{1}{4}\pi ab$$

$$m = \frac{4a}{3\pi}$$

$$n = \frac{4b}{3\pi}$$

$$I_1 = a^3b\left(\frac{\pi}{16} - \frac{4}{9\pi}\right)$$

$$I_2 = ab^3\left(\frac{\pi}{16} - \frac{4}{9\pi}\right)$$

$$I_3 = \frac{1}{16}\pi a^3b$$

$$I_4 = \frac{1}{16}\pi ab^3$$

* ELLIPTIC COMPLEMENT

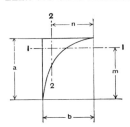

$$A = ab\left(1 - \frac{\pi}{4}\right)$$

$$m = \frac{a}{6\left(1 - \frac{\pi}{4}\right)}$$

$$n = \frac{b}{6\left(1 - \frac{\pi}{4}\right)}$$

$$I_1 = a^3b\left(\frac{1}{3} - \frac{\pi}{16} - \frac{1}{36\left(1 - \frac{\pi}{4}\right)}\right)$$

$$I_2 = ab^3\left(\frac{1}{3} - \frac{\pi}{16} - \frac{1}{36\left(1 - \frac{\pi}{4}\right)}\right)$$

* To obtain properties of half circle, quarter circle and circular complement substitute a = b = R.

PROPERTIES OF GEOMETRIC SECTIONS
AND STRUCTURAL SHAPES

REGULAR POLYGON

Axis of moments
through center

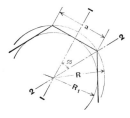

n = Number of sides

$\phi = \dfrac{180°}{n}$

$a = 2\sqrt{R^2 - R_1^2}$

$R = \dfrac{a}{2 \sin \phi}$

$R_1 = \dfrac{a}{2 \tan \phi}$

$A = \dfrac{1}{4} n a^2 \cot \phi = \dfrac{1}{2} n R^2 \sin 2\phi = n R_1^2 \tan \phi$

$I_1 = I_2 = \dfrac{A(6R^2 - a^2)}{24} = \dfrac{A(12R_1^2 + a^2)}{48}$

$r_1 = r_2 = \sqrt{\dfrac{6R^2 - a^2}{24}} = \sqrt{\dfrac{12R_1^2 + a^2}{48}}$

ANGLE

Axis of moments through
center of gravity

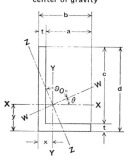

Z-Z is axis of minimum I

$\tan 2\theta = \dfrac{2K}{I_y - I_x}$

$A = t(b + c)$ $x = \dfrac{b^2 + ct}{2(b + c)}$ $y = \dfrac{d^2 + at}{2(b + c)}$

K = Product of Inertia about X-X & Y-Y

$= \dfrac{abcdt}{\pm 4(b + c)}$

$I_x = \dfrac{1}{3} \left(t(d - y)^3 + by^3 - a(y - t)^3 \right)$

$I_y = \dfrac{1}{3} \left(t(b - x)^3 + dx^3 - c(x - t)^3 \right)$

$I_z = I_x \sin^2\theta + I_y \cos^2\theta + K \sin 2\theta$

$I_w = I_x \cos^2\theta + I_y \sin^2\theta - K \sin 2\theta$

K is negative when heel of angle, with respect
to c. g., is in 1st or 3rd quadrant, positive
when in 2nd or 4th quadrant.

BEAMS AND CHANNELS

Transverse force oblique
through center of gravity

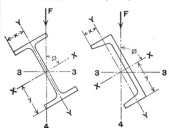

$I_3 = I_x \sin^2\phi + I_y \cos^2\phi$

$I_4 = I_x \cos^2\phi + I_y \sin^2\phi$

$= M \left(\dfrac{y}{I_x} \sin\phi + \dfrac{x}{I_y} \cos\phi \right)$

where M is bending moment due to force F.

TRIGONOMETRIC FORMULAS

TRIGONOMETRIC FUNCTIONS

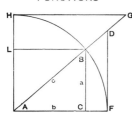

Radius AF $= 1$
$= \sin^2 A + \cos^2 A = \sin A \operatorname{cosec} A$
$= \cos A \sec A = \tan A \cot A$

Sine A $= \dfrac{\cos A}{\cot A} = \dfrac{1}{\operatorname{cosec} A} = \cos A \tan A = \sqrt{1-\cos^2 A} = BC$

Cosine A $= \dfrac{\sin A}{\tan A} = \dfrac{1}{\sec A} = \sin A \cot A = \sqrt{1-\sin^2 A} = AC$

Tangent A $= \dfrac{\sin A}{\cos A} = \dfrac{1}{\cot A} = \sin A \sec A \qquad = FD$

Cotangent A $= \dfrac{\cos A}{\sin A} = \dfrac{1}{\tan A} = \cos A \operatorname{cosec} A \qquad = HG$

Secant A $= \dfrac{\tan A}{\sin A} = \dfrac{1}{\cos A} \qquad = AD$

Cosecant A $= \dfrac{\cot A}{\cos A} = \dfrac{1}{\sin A} \qquad = AG$

RIGHT ANGLED TRIANGLES

$a^2 = c^2 - b^2$
$b^2 = c^2 - a^2$
$c^2 = a^2 + b^2$

Known	Required					
	A	B	a	b	c	Area
a, b	$\tan A = \dfrac{a}{b}$	$\tan B = \dfrac{b}{a}$			$\sqrt{a^2+b^2}$	$\dfrac{ab}{2}$
a, c	$\sin A = \dfrac{a}{c}$	$\cos B = \dfrac{a}{c}$		$\sqrt{c^2-a^2}$		$\dfrac{a\sqrt{c^2-a^2}}{2}$
A, a		$90° - A$		$a \cot A$	$\dfrac{a}{\sin A}$	$\dfrac{a^2 \cot A}{2}$
A, b		$90° - A$	$b \tan A$		$\dfrac{b}{\cos A}$	$\dfrac{b^2 \tan A}{2}$
A, c		$90° - A$	$c \sin A$	$c \cos A$		$\dfrac{c^2 \sin 2A}{4}$

OBLIQUE ANGLED TRIANGLES

$s = \dfrac{a+b+c}{2}$

$K = \sqrt{\dfrac{(s-a)(s-b)(s-c)}{s}}$

$a^2 = b^2 + c^2 - 2bc \cos A$
$b^2 = a^2 + c^2 - 2ac \cos B$
$c^2 = a^2 + b^2 - 2ab \cos C$

Known	Required					
	A	B	C	b	c	Area
a, b, c	$\tan\frac{1}{2}A = \dfrac{K}{s-a}$	$\tan\frac{1}{2}B = \dfrac{K}{s-b}$	$\tan\frac{1}{2}C = \dfrac{K}{s-c}$			$\sqrt{s(s-a)(s-b)(s-c)}$
a, A, B			$180°-(A+B)$	$\dfrac{a \sin B}{\sin A}$	$\dfrac{a \sin C}{\sin A}$	
a, b, A		$\sin B = \dfrac{b \sin A}{a}$			$\dfrac{b \sin C}{\sin B}$	
a, b, C	$\tan A = \dfrac{a \sin C}{b - a \cos C}$				$\sqrt{a^2+b^2-2ab\cos C}$	$\dfrac{ab \sin C}{2}$

I

W SHAPES
Properties for designing

Designation	Area A	Depth d	Flange Width b_f	Flange Thickness t_f	Web Thickness t_w	Axis X-X I	Axis X-X S	Axis X-X r	Axis Y-Y I	Axis Y-Y S	Axis Y-Y r
	In.²	In.	In.	In.	In.	In.⁴	In.³	In.	In.⁴	In.³	In.
W 36×300	88.3	36.72	16.655	1.680	0.945	20300	1110	15.2	1300	156	3.83
×280	82.4	36.50	16.595	1.570	0.885	18900	1030	15.1	1200	144	3.81
×260	76.5	36.24	16.551	1.440	0.841	17300	952	15.0	1090	132	3.77
×245	72.1	36.06	16.512	1.350	0.802	16100	894	15.0	1010	123	3.75
×230	67.7	35.88	16.471	1.260	0.761	15000	837	14.9	940	114	3.73
W 36×194	57.2	36.48	12.117	1.260	0.770	12100	665	14.6	375	61.9	2.56
×182	53.6	36.32	12.072	1.180	0.725	11300	622	14.5	347	57.5	2.55
×170	50.0	36.16	12.027	1.100	0.680	10500	580	14.5	320	53.2	2.53
×160	47.1	36.00	12.000	1.020	0.653	9760	542	14.4	295	49.1	2.50
×150	44.2	35.84	11.972	0.940	0.625	9030	504	14.3	270	45.0	2.47
×135	39.8	35.55	11.945	0.794	0.598	7820	440	14.0	226	37.9	2.39
W 33×240	70.6	33.50	15.865	1.400	0.830	13600	813	13.9	933	118	3.64
×220	64.8	33.25	15.810	1.275	0.775	12300	742	13.8	841	106	3.60
×200	58.9	33.00	15.750	1.150	0.715	11100	671	13.7	750	95.2	3.57
W 33×152	44.8	33.50	11.565	1.055	0.635	8160	487	13.5	273	47.2	2.47
×141	41.6	33.31	11.535	0.960	0.605	7460	448	13.4	246	42.7	2.43
×130	38.3	33.10	11.510	0.855	0.580	6710	406	13.2	218	37.9	2.38
×118	34.8	32.86	11.484	0.738	0.554	5900	359	13.0	187	32.5	2.32
W 30×210	61.9	30.38	15.105	1.315	0.775	9890	651	12.6	757	100	3.50
×190	56.0	30.12	15.040	1.185	0.710	8850	587	12.6	673	89.5	3.47
×172	50.7	29.88	14.985	1.065	0.655	7910	530	12.5	598	79.8	3.43
W 30×132	38.9	30.30	10.551	1.000	0.615	5760	380	12.2	196	37.2	2.25
×124	36.5	30.16	10.521	0.930	0.585	5360	355	12.1	181	34.4	2.23
×116	34.2	30.00	10.500	0.850	0.564	4930	329	12.0	164	31.3	2.19
×108	31.8	29.82	10.484	0.760	0.548	4470	300	11.9	146	27.9	2.15
× 99	29.1	29.64	10.458	0.670	0.522	4000	270	11.7	128	24.5	2.10

I

W SHAPES
Properties for designing

Designation	Area A	Depth d	Flange Width b_f	Flange Thickness t_f	Web Thickness t_w	Axis X-X I	Axis X-X S	Axis X-X r	Axis Y-Y I	Axis Y-Y S	Axis Y-Y r
	In.²	In.	In.	In.	In.	in.⁴	In.³	In.	In.⁴	In.³	In.
W 27×177	52.2	27.31	14.090	1.190	0.725	6740	494	11.4	556	78.9	3.26
×160	47.1	27.08	14.023	1.075	0.658	6030	446	11.3	495	70.6	3.24
×145	42.7	26.88	13.965	0.975	0.600	5430	404	11.3	443	63.5	3.22
W 27×114	33.6	27.28	10.070	0.932	0.570	4090	300	11.0	159	31.6	2.18
×102	30.0	27.07	10.018	0.827	0.518	3610	267	11.0	139	27.7	2.15
× 94	27.7	26.91	9.990	0.747	0.490	3270	243	10.9	124	24.9	2.12
× 84	24.8	26.69	9.963	0.636	0.463	2830	212	10.7	105	21.1	2.06
W 24×160	47.1	24.72	14.091	1.135	0.656	5120	414	10.4	530	75.2	3.35
×145	42.7	24.49	14.043	1.020	0.608	4570	373	10.3	471	67.1	3.32
×130	38.3	24.25	14.000	0.900	0.565	4020	332	10.2	412	58.9	3.28
W 24×120	35.4	24.31	12.088	0.930	0.556	3650	300	10.2	274	45.4	2.78
×110	32.5	24.16	12.042	0.855	0.510	3330	276	10.1	249	41.4	2.77
×100	29.5	24.00	12.000	0.775	0.468	3000	250	10.1	223	37.2	2.75
W 24× 94	27.7	24.29	9.061	0.872	0.516	2690	221	9.86	108	23.9	1.98
× 84	24.7	24.09	9.015	0.772	0.470	2370	197	9.79	94.5	21.0	1.95
× 76	22.4	23.91	8.985	0.682	0.440	2100	176	9.69	82.6	18.4	1.92
× 68	20.0	23.71	8.961	0.582	0.416	1820	153	9.53	70.0	15.6	1.87
W 24× 61	18.0	23.72	7.023	0.591	0.419	1540	130	9.25	34.3	9.76	1.38
× 55	16.2	23.55	7.000	0.503	0.396	1340	114	9.10	28.9	8.25	1.34
W 21×142	41.8	21.46	13.132	1.095	0.659	3410	317	9.03	414	63.0	3.15
×127	37.4	21.24	13.061	0.985	0.588	3020	284	8.99	366	56.1	3.13
×112	33.0	21.00	13.000	0.865	0.527	2620	250	8.92	317	48.8	3.10
W 21× 96	28.3	21.14	9.038	0.935	0.575	2100	198	8.61	115	25.5	2.02
× 82	24.2	20.86	8.962	0.795	0.499	1760	169	8.53	95.6	21.3	1.99
W 21× 73	21.5	21.24	8.295	0.740	0.455	1600	151	8.64	70.6	17.0	1.81
× 68	20.0	21.13	8.270	0.685	0.430	1480	140	8.60	64.7	15.7	1.80
× 62	18.3	20.99	8.240	0.615	0.400	1330	127	8.54	57.5	13.9	1.77
× 55	16.2	20.80	8.215	0.522	0.375	1140	110	8.40	48.3	11.8	1.73
W 21× 49	14.4	20.82	6.520	0.532	0.368	971	93.3	8.21	24.7	7.57	1.31
× 44	13.0	20.66	6.500	0.451	0.348	843	81.6	8.07	20.7	6.38	1.27

W SHAPES
Properties for designing

Designation	Area A	Depth d	Flange Width b_f	Flange Thick-ness t_f	Web Thick-ness t_w	Axis X-X I	Axis X-X S	Axis X-X r	Axis Y-Y I	Axis Y-Y S	Axis Y-Y r
	In.²	In.	In.	In.	In.	In.⁴	In.³	In.	In.⁴	In.³	In.
W 18×114	33.5	18.48	11.833	0.991	0.595	2040	220	7.79	274	46.3	2.86
×105	30.9	18.32	11.792	0.911	0.554	1850	202	7.75	249	42.3	2.84
× 96	28.2	18.16	11.750	0.831	0.512	1680	185	7.70	225	38.3	2.82
W 18× 85	25.0	18.32	8.838	0.911	0.526	1440	157	7.57	105	23.8	2.05
× 77	22.7	18.16	8.787	0.831	0.475	1290	142	7.54	94.1	21.4	2.04
× 70	20.6	18.00	8.750	0.751	0.438	1160	129	7.50	84.0	19.2	2.02
× 64	18.9	17.87	8.715	0.686	0.403	1050	118	7.46	75.8	17.4	2.00
W 18× 60	17.7	18.25	7.558	0.695	0.416	986	108	7.47	50.1	13.3	1.68
× 55	16.2	18.12	7.532	0.630	0.390	891	98.4	7.42	45.0	11.9	1.67
× 50	14.7	18.00	7.500	0.570	0.358	802	89.1	7.38	40.2	10.7	1.65
× 45	13.2	17.86	7.477	0.499	0.335	706	79.0	7.30	34.8	9.32	1.62
W 18× 40	11.8	17.90	6.018	0.524	0.316	612	68.4	7.21	19.1	6.34	1.27
× 35	10.3	17.71	6.000	0.429	0.298	513	57.9	7.05	15.5	5.16	1.23
W 16× 96	28.2	16.32	11.533	0.875	0.535	1360	166	6.93	224	38.8	2.82
× 88	25.9	16.16	11.502	0.795	0.504	1220	151	6.87	202	35.1	2.79
W 16× 78	23.0	16.32	8.586	0.875	0.529	1050	128	6.75	92.5	21.6	2.01
× 71	20.9	16.16	8.543	0.795	0.486	941	116	6.71	82.8	19.4	1.99
× 64	18.8	16.00	8.500	0.715	0.443	836	104	6.66	73.3	17.3	1.97
× 58	17.1	15.86	8.464	0.645	0.407	748	94.4	6.62	65.3	15.4	1.96
W 16× 50	14.7	16.25	7.073	0.628	0.380	657	80.8	6.68	37.1	10.5	1.59
× 45	13.3	16.12	7.039	0.563	0.346	584	72.5	6.64	32.8	9.32	1.57
× 40	11.8	16.00	7.000	0.503	0.307	517	64.6	6.62	28.8	8.23	1.56
× 36	10.6	15.85	6.992	0.428	0.299	447	56.5	6.50	24.4	6.99	1.52
W 16× 31	9.13	15.84	5.525	0.442	0.275	374	47.2	6.40	12.5	4.51	1.17
× 26	7.67	15.65	5.500	0.345	0.250	300	38.3	6.25	9.59	3.49	1.12

I

W SHAPES
Properties for designing

Designation	Area A	Depth d	Flange Width b_f	Flange Thickness t_f	Web Thickness t_w	Axis X-X I	Axis X-X S	Axis X-X r	Axis Y-Y I	Axis Y-Y S	Axis Y-Y r
	In.²	In.	In.	In.	In.	In.⁴	In.³	In.	In.⁴	In.³	In.
W 14×730	215	22.44	17.889	4.910	3.069	14400	1280	8.18	4720	527	4.69
×665	196	21.67	17.646	4.522	2.826	12500	1150	7.99	4170	472	4.62
×605	178	20.94	17.418	4.157	2.598	10900	1040	7.81	3680	423	4.55
×550	162	20.26	17.206	3.818	2.386	9450	933	7.64	3260	378	4.49
×500	147	19.63	17.008	3.501	2.188	8250	840	7.49	2880	339	4.43
×455	134	19.05	16.828	3.213	2.008	7220	758	7.35	2560	304	4.37
W 14×426	125	18.69	16.695	3.033	1.875	6610	707	7.26	2360	283	4.34
×398	117	18.31	16.590	2.843	1.770	6010	657	7.17	2170	262	4.31
×370	109	17.94	16.475	2.658	1.655	5450	608	7.08	1990	241	4.27
×342	101	17.56	16.365	2.468	1.545	4910	559	6.99	1810	221	4.24
×314	92.3	17.19	16.235	2.283	1.415	4400	512	6.90	1630	201	4.20
×287	84.4	16.81	16.130	2.093	1.310	3910	465	6.81	1470	182	4.17
×264	77.6	16.50	16.025	1.938	1.205	3530	427	6.74	1330	166	4.14
×246	72.3	16.25	15.945	1.813	1.125	3230	397	6.68	1230	154	4.12
W 14×237	69.7	16.12	15.910	1.748	1.090	3080	382	6.65	1170	148	4.11
×228	67.1	16.00	15.865	1.688	1.045	2940	368	6.62	1120	142	4.10
×219	64.4	15.87	15.825	1.623	1.005	2800	353	6.59	1070	136	4.08
×211	62.1	15.75	15.800	1.563	0.980	2670	339	6.56	1030	130	4.07
×202	59.4	15.63	15.750	1.503	0.930	2540	325	6.54	980	124	4.06
×193	56.7	15.50	15.710	1.438	0.890	2400	310	6.51	930	118	4.05
×184	54.1	15.38	15.660	1.378	0.840	2270	296	6.49	883	113	4.04
×176	51.7	15.25	15.640	1.313	0.820	2150	282	6.45	838	107	4.02
×167	49.1	15.12	15.600	1.248	0.780	2020	267	6.42	790	101	4.01
×158	46.5	15.00	15.550	1.188	0.730	1900	253	6.40	745	95.8	4.00
×150	44.1	14.88	15.515	1.128	0.695	1790	240	6.37	703	90.6	3.99
×142	41.8	14.75	15.500	1.063	0.680	1670	227	6.32	660	85.2	3.97
W 14×320	94.1	16.81	16.710	2.093	1.890	4140	493	6.63	1640	196	4.17

I

W SHAPES
Properties for designing

Designation	Area A	Depth d	Flange Width b_f	Flange Thickness t_f	Web Thickness t_w	Axis X-X I	Axis X-X S	Axis X-X r	Axis Y-Y I	Axis Y-Y S	Axis Y-Y r
	In.²	In.	In.	In.	In.	In.⁴	In.³	In.	In.⁴	In.³	In.
W 14×136	40.0	14.75	14.740	1.063	0.660	1590	216	6.31	568	77.0	3.77
×127	37.3	14.62	14.690	0.998	0.610	1480	202	6.29	528	71.8	3.76
×119	35.0	14.50	14.650	0.938	0.570	1370	189	6.26	492	67.1	3.75
×111	32.7	14.37	14.620	0.873	0.540	1270	176	6.23	455	62.2	3.73
×103	30.3	14.25	14.575	0.813	0.495	1170	164	6.21	420	57.6	3.72
× 95	27.9	14.12	14.545	0.748	0.465	1060	151	6.17	384	52.8	3.71
× 87	25.6	14.00	14.500	0.688	0.420	967	138	6.15	350	48.2	3.70
W 14× 84	24.7	14.18	12.023	0.778	0.451	928	131	6.13	225	37.5	3.02
× 78	22.9	14.06	12.000	0.718	0.428	851	121	6.09	207	34.5	3.00
W 14× 74	21.8	14.19	10.072	0.783	0.450	797	112	6.05	133	26.5	2.48
× 68	20.0	14.06	10.040	0.718	0.418	724	103	6.02	121	24.1	2.46
× 61	17.9	13.91	10.000	0.643	0.378	641	92.2	5.98	107	21.5	2.45
W 14× 53	15.6	13.94	8.062	0.658	0.370	542	77.8	5.90	57.5	14.3	1.92
× 48	14.1	13.81	8.031	0.593	0.339	485	70.2	5.86	51.3	12.8	1.91
× 43	12.6	13.68	8.000	0.528	0.308	429	62.7	5.82	45.1	11.3	1.89
W 14× 38	11.2	14.12	6.776	0.513	0.313	386	54.7	5.88	26.6	7.86	1.54
× 34	10.0	14.00	6.750	0.453	0.287	340	48.6	5.83	23.3	6.89	1.52
× 30	8.83	13.86	6.733	0.383	0.270	290	41.9	5.74	19.5	5.80	1.49
W 14× 26	7.67	13.89	5.025	0.418	0.255	244	35.1	5.64	8.86	3.53	1.08
× 22	6.49	13.72	5.000	0.335	0.230	198	28.9	5.53	7.00	2.80	1.04

I

W SHAPES
Properties for designing

Designation	Area A	Depth d	Flange Width b_f	Flange Thickness t_f	Web Thickness t_w	Axis X-X I	Axis X-X S	Axis X-X r	Axis Y-Y I	Axis Y-Y S	Axis Y-Y r
	In.²	In.	In.	In.	In.	In.⁴	In.³	In.	In.⁴	In.³	In.
W 12×190	55.9	14.38	12.670	1.736	1.060	1890	263	5.82	590	93.1	3.25
×161	47.4	13.88	12.515	1.486	0.905	1540	222	5.70	486	77.7	3.20
×133	39.1	13.38	12.365	1.236	0.755	1220	183	5.59	390	63.1	3.16
×120	35.3	13.12	12.320	1.106	0.710	1070	163	5.51	345	56.0	3.13
×106	31.2	12.88	12.230	0.986	0.620	931	145	5.46	301	49.2	3.11
× 99	29.1	12.75	12.192	0.921	0.582	859	135	5.43	278	45.7	3.09
× 92	27.1	12.62	12.155	0.856	0.545	789	125	5.40	256	42.2	3.08
× 85	25.0	12.50	12.105	0.796	0.495	723	116	5.38	235	38.9	3.07
× 79	23.2	12.38	12.080	0.736	0.470	663	107	5.34	216	35.8	3.05
× 72	21.2	12.25	12.040	0.671	0.430	597	97.5	5.31	195	32.4	3.04
× 65	19.1	12.12	12.000	0.606	0.390	533	88.0	5.28	175	29.1	3.02
W 12× 58	17.1	12.19	10.014	0.641	0.359	476	78.1	5.28	107	21.4	2.51
× 53	15.6	12.06	10.000	0.576	0.345	426	70.7	5.23	96.1	19.2	2.48
W 12× 50	14.7	12.19	8.077	0.641	0.371	395	64.7	5.18	56.4	14.0	1.96
× 45	13.2	12.06	8.042	0.576	0.336	351	58.2	5.15	50.0	12.4	1.94
× 40	11.8	11.94	8.000	0.516	0.294	310	51.9	5.13	44.1	11.0	1.94
W 12× 36	10.6	12.24	6.565	0.540	0.305	281	46.0	5.15	25.5	7.77	1.55
× 31	9.13	12.09	6.525	0.465	0.265	239	39.5	5.12	21.6	6.61	1.54
× 27	7.95	11.96	6.497	0.400	0.237	204	34.2	5.07	18.3	5.63	1.52
W 12× 22	6.47	12.31	4.030	0.424	0.260	156	25.3	4.91	4.64	2.31	0.847
× 19	5.59	12.16	4.007	0.349	0.237	130	21.3	4.82	3.76	1.88	0.820
× 16.5	4.87	12.00	4.000	0.269	0.230	105	17.6	4.65	2.88	1.44	0.770
× 14	4.12	11.91	3.968	0.224	0.198	88.0	14.8	4.62	2.34	1.18	0.754

W SHAPES
Properties for designing

Designation	Area A	Depth d	Flange Width b_f	Flange Thickness t_f	Web Thickness t_w	Axis X-X I	Axis X-X S	Axis X-X r	Axis Y-Y I	Axis Y-Y S	Axis Y-Y r
	In.²	In.	In.	In.	In.	In.⁴	In.³	In.	In.⁴	In.³	In.
W 10×112	32.9	11.38	10.415	1.248	0.755	719	126	4.67	235	45.2	2.67
×100	29.4	11.12	10.345	1.118	0.685	625	112	4.61	207	39.9	2.65
× 89	26.2	10.88	10.275	0.998	0.615	542	99.7	4.55	181	35.2	2.63
× 77	22.7	10.62	10.195	0.868	0.535	457	86.1	4.49	153	30.1	2.60
× 72	21.2	10.50	10.170	0.808	0.510	421	80.1	4.46	142	27.9	2.59
× 66	19.4	10.38	10.117	0.748	0.457	382	73.7	4.44	129	25.5	2.58
× 60	17.7	10.25	10.075	0.683	0.415	344	67.1	4.41	116	23.1	2.57
× 54	15.9	10.12	10.028	0.618	0.368	306	60.4	4.39	104	20.7	2.56
× 49	14.4	10.00	10.000	0.558	0.340	273	54.6	4.35	93.0	18.6	2.54
W 10× 45	13.2	10.12	8.022	0.618	0.350	249	49.1	4.33	53.2	13.3	2.00
× 39	11.5	9.94	7.990	0.528	0.318	210	42.2	4.27	44.9	11.2	1.98
× 33	9.71	9.75	7.964	0.433	0.292	171	35.0	4.20	36.5	9.16	1.94
W 10× 29	8.54	10.22	5.799	0.500	0.289	158	30.8	4.30	16.3	5.61	1.38
× 25	7.36	10.08	5.762	0.430	0.252	133	26.5	4.26	13.7	4.76	1.37
× 21	6.20	9.90	5.750	0.340	0.240	107	21.5	4.15	10.8	3.75	1.32
W 10× 19	5.61	10.25	4.020	0.394	0.250	96.3	18.8	4.14	4.28	2.13	0.874
× 17	4.99	10.12	4.010	0.329	0.240	81.9	16.2	4.05	3.55	1.77	0.844
× 15	4.41	10.00	4.000	0.269	0.230	68.9	13.8	3.95	2.88	1.44	0.809
× 11.5	3.39	9.87	3.950	0.204	0.180	52.0	10.5	3.92	2.10	1.06	0.787

W SHAPES
Properties for designing

Designation	Area A	Depth d	Flange Width b_f	Flange Thickness t_f	Web Thickness t_w	Axis X-X I	Axis X-X S	Axis X-X r	Axis Y-Y I	Axis Y-Y S	Axis Y-Y r
	In.²	In.	In.	In.	In.	In.⁴	In.³	In.	In.⁴	In.³	In.
W 8×67	19.7	9.00	8.287	0.933	0.575	272	60.4	3.71	88.6	21.4	2.12
×58	17.1	8.75	8.222	0.808	0.510	227	52.0	3.65	74.9	18.2	2.10
×48	14.1	8.50	8.117	0.683	0.405	184	43.2	3.61	60.9	15.0	2.08
×40	11.8	8.25	8.077	0.558	0.365	146	35.5	3.53	49.0	12.1	2.04
×35	10.3	8.12	8.027	0.493	0.315	126	31.1	3.50	42.5	10.6	2.03
×31	9.12	8.00	8.000	0.433	0.288	110	27.4	3.47	37.0	9.24	2.01
W 8×28	8.23	8.06	6.540	0.463	0.285	97.8	24.3	3.45	21.6	6.61	1.62
×24	7.06	7.93	6.500	0.398	0.245	82.5	20.8	3.42	18.2	5.61	1.61
W 8×20	5.89	8.14	5.268	0.378	0.248	69.4	17.0	3.43	9.22	3.50	1.25
×17	5.01	8.00	5.250	0.308	0.230	56.6	14.1	3.36	7.44	2.83	1.22
W 8×15	4.43	8.12	4.015	0.314	0.245	48.1	11.8	3.29	3.40	1.69	0.876
×13	3.83	8.00	4.000	0.254	0.230	39.6	9.90	3.21	2.72	1.36	0.842
×10	2.96	7.90	3.940	0.204	0.170	30.8	7.80	3.23	2.08	1.06	0.839
W 6×25	7.35	6.37	6.080	0.456	0.320	53.3	16.7	2.69	17.1	5.62	1.53
×20	5.88	6.20	6.018	0.367	0.258	41.5	13.4	2.66	13.3	4.43	1.51
×15.5	4.56	6.00	5.995	0.269	0.235	30.1	10.0	2.57	9.67	3.23	1.46
W 6×16	4.72	6.25	4.030	0.404	0.260	31.7	10.2	2.59	4.42	2.19	0.967
×12	3.54	6.00	4.000	0.279	0.230	21.7	7.25	2.48	2.98	1.49	0.918
× 8.5	2.51	5.83	3.940	0.194	0.170	14.8	5.08	2.43	1.98	1.01	0.889
W 5×18.5	5.43	5.12	5.025	0.420	0.265	25.4	9.94	2.16	8.89	3.54	1.28
×16	4.70	5.00	5.000	0.360	0.240	21.3	8.53	2.13	7.51	3.00	1.26
W 4×13	3.82	4.16	4.060	0.345	0.280	11.3	5.45	1.72	3.76	1.85	0.991

M SHAPES
Properties for designing

Designation	Area A	Depth d	Flange Width b_f	Flange Thickness t_f	Web Thickness t_w	Axis X-X I	Axis X-X S	Axis X-X r	Axis Y-Y I	Axis Y-Y S	Axis Y-Y r
	In.²	In.	In.	In.	In.	In.⁴	In.³	In.	In.⁴	In.³	In.
M 14× 17.2	5.05	14.00	4.000	0.272	0.210	147	21.1	5.40	2.65	1.33	0.725
M 12× 11.8	3.47	12.00	3.065	0.225	0.177	71.9	12.0	4.55	0.980	0.639	0.532
M 10× 29.1	8.56	9.88	5.937	0.389	0.427	131	26.6	3.92	11.2	3.76	1.14
× 22.9	6.73	9.88	5.752	0.389	0.242	117	23.6	4.16	10.0	3.48	1.22
M 10× 9	2.65	10.00	2.690	0.206	0.157	38.8	7.76	3.83	0.609	0.453	0.480
M 8× 34.3	10.1	8.00	8.003	0.459	0.378	116	29.1	3.40	34.9	8.73	1.86
× 32.6	9.58	8.00	7.940	0.459	0.315	114	28.4	3.44	34.1	8.58	1.89
M 8× 22.5	6.60	8.00	5.395	0.353	0.375	68.2	17.1	3.22	7.48	2.77	1.06
× 18.5	5.44	8.00	5.250	0.353	0.230	62.0	15.5	3.38	6.82	2.60	1.12
M 8× 6.5	1.92	8.00	2.281	0.189	0.135	18.5	4.62	3.10	0.343	0.301	0.423
M 7× 5.5	1.62	7.00	2.080	0.180	0.128	12.0	3.44	2.73	0.249	0.239	0.392
M 6× 22.5	6.62	6.00	6.060	0.379	0.372	41.2	13.7	2.49	12.4	4.08	1.37
× 20	5.89	6.00	5.938	0.379	0.250	39.0	13.0	2.57	11.6	3.90	1.40
M 6× 4.4	1.29	6.00	1.844	0.171	0.114	7.20	2.40	2.36	0.165	0.179	0.358
M 5× 18.9	5.55	5.00	5.003	0.416	0.316	24.1	9.63	2.08	7.86	3.14	1.19
M 4× 13.8	4.06	4.00	4.000	0.371	0.313	10.8	5.42	1.63	3.58	1.79	0.939
× 13	3.81	4.00	3.940	0.371	0.254	10.5	5.24	1.66	3.36	1.71	0.939

I

S SHAPES
Properties for designing

Designation	Area A	Depth d	Flange Width b_f	Flange Thickness t_f	Web Thickness t_w	Axis X-X I	Axis X-X S	Axis X-X r	Axis Y-Y I	Axis Y-Y S	Axis Y-Y r
	In.²	In.	In.	In.	In.	In.⁴	In.³	In.	In.⁴	In.³	In.
S 24×120	35.3	24.00	8.048	1.102	0.798	3030	252	9.26	84.2	20.9	1.54
×105.9	31.1	24.00	7.875	1.102	0.625	2830	236	9.53	78.2	19.8	1.58
S 24×100	29.4	24.00	7.247	0.871	0.747	2390	199	9.01	47.8	13.2	1.27
× 90	26.5	24.00	7.124	0.871	0.624	2250	187	9.22	44.9	12.6	1.30
× 79.9	23.5	24.00	7.001	0.871	0.501	2110	175	9.47	42.3	12.1	1.34
S 20× 95	27.9	20.00	7.200	0.916	0.800	1610	161	7.60	49.7	13.8	1.33
× 85	25.0	20.00	7.053	0.916	0.653	1520	152	7.79	46.2	13.1	1.36
S 20× 75	22.1	20.00	6.391	0.789	0.641	1280	128	7.60	29.6	9.28	1.16
× 65.4	19.2	20.00	6.250	0.789	0.500	1180	118	7.84	27.4	8.77	1.19
S 18× 70	20.6	18.00	6.251	0.691	0.711	926	103	6.71	24.1	7.72	1.08
× 54.7	16.1	18.00	6.001	0.691	0.461	804	89.4	7.07	20.8	6.94	1.14
S 15× 50	14.7	15.00	5.640	0.622	0.550	486	64.8	5.75	15.7	5.57	1.03
× 42.9	12.6	15.00	5.501	0.622	0.411	447	59.6	5.95	14.4	5.23	1.07
S 12× 50	14.7	12.00	5.477	0.659	0.687	305	50.8	4.55	15.7	5.74	1.03
× 40.8	12.0	12.00	5.252	0.659	0.462	272	45.4	4.77	13.6	5.16	1.06
S 12× 35	10.3	12.00	5.078	0.544	0.428	229	38.2	4.72	9.87	3.89	0.980
× 31.8	9.35	12.00	5.000	0.544	0.350	218	36.4	4.83	9.36	3.74	1.00
S 10× 35	10.3	10.00	4.944	0.491	0.594	147	29.4	3.78	8.36	3.38	0.901
× 25.4	7.46	10.00	4.661	0.491	0.311	124	24.7	4.07	6.79	2.91	0.954
S 8× 23	6.77	8.00	4.171	0.425	0.441	64.9	16.2	3.10	4.31	2.07	0.798
× 18.4	5.41	8.00	4.001	0.425	0.271	57.6	14.4	3.26	3.73	1.86	0.831
S 7× 20	5.88	7.00	3.860	0.392	0.450	42.4	12.1	2.69	3.17	1.64	0.734
× 15.3	4.50	7.00	3.662	0.392	0.252	36.7	10.5	2.86	2.64	1.44	0.766
S 6× 17.25	5.07	6.00	3.565	0.359	0.465	26.3	8.77	2.28	2.31	1.30	0.675
× 12.5	3.67	6.00	3.332	0.359	0.232	22.1	7.37	2.45	1.82	1.09	0.705
S 5× 14.75	4.34	5.00	3.284	0.326	0.494	15.2	6.09	1.87	1.67	1.01	0.620
× 10	2.94	5.00	3.004	0.326	0.214	12.3	4.92	2.05	1.22	0.809	0.643
S 4× 9.5	2.79	4.00	2.796	0.293	0.326	6.79	3.39	1.56	0.903	0.646	0.569
× 7.7	2.26	4.00	2.663	0.293	0.193	6.08	3.04	1.64	0.764	0.574	0.581
S 3× 7.5	2.21	3.00	2.509	0.260	0.349	2.93	1.95	1.15	0.586	0.468	0.516
× 5.7	1.67	3.00	2.330	0.260	0.170	2.52	1.68	1.23	0.455	0.390	0.522

CHANNELS
AMERICAN STANDARD
Properties for designing

Designation	Area A	Depth d	Flange Width b_f	Flange Average thickness t_f	Web thickness t_w	$\dfrac{d}{A_f}$	Axis X-X I	Axis X-X S	Axis X-X r
	In.²	In.	In.	In.	In.		In.⁴	In.³	In.
C 15×50	14.7	15.00	3.716	0.650	0.716	6.21	404	53.8	5.24
×40	11.8	15.00	3.520	0.650	0.520	6.56	349	46.5	5.44
×33.9	9.96	15.00	3.400	0.650	0.400	6.79	315	42.0	5.62
C 12×30	8.82	12.00	3.170	0.501	0.510	7.55	162	27.0	4.29
×25	7.35	12.00	3.047	0.501	0.387	7.85	144	24.1	4.43
×20.7	6.09	12.00	2.942	0.501	0.282	8.13	129	21.5	4.61
C 10×30	8.82	10.00	3.033	0.436	0.673	7.55	103	20.7	3.42
×25	7.35	10.00	2.886	0.436	0.526	7.94	91.2	18.2	3.52
×20	5.88	10.00	2.739	0.436	0.379	8.36	78.9	15.8	3.66
×15.3	4.49	10.00	2.600	0.436	0.240	8.81	67.4	13.5	3.87
C 9×20	5.88	9.00	2.648	0.413	0.448	8.22	60.9	13.5	3.22
×15	4.41	9.00	2.485	0.413	0.285	8.76	51.0	11.3	3.40
×13.4	3.94	9.00	2.433	0.413	0.233	8.95	47.9	10.6	3.48
C 8×18.75	5.51	8.00	2.527	0.390	0.487	8.12	44.0	11.0	2.82
×13.75	4.04	8.00	2.343	0.390	0.303	8.75	36.1	9.03	2.99
×11.5	3.38	8.00	2.260	0.390	0.220	9.08	32.6	8.14	3.11
C 7×14.75	4.33	7.00	2.299	0.366	0.419	8.31	27.2	7.78	2.51
×12.25	3.60	7.00	2.194	0.366	0.314	8.71	24.2	6.93	2.60
× 9.8	2.87	7.00	2.090	0.366	0.210	9.14	21.3	6.08	2.72
C 6×13	3.83	6.00	2.157	0.343	0.437	8.10	17.4	5.80	2.13
×10.5	3.09	6.00	2.034	0.343	0.314	8.59	15.2	5.06	2.22
× 8.2	2.40	6.00	1.920	0.343	0.200	9.10	13.1	4.38	2.34
C 5× 9	2.64	5.00	1.885	0.320	0.325	8.29	8.90	3.56	1.83
× 6.7	1.97	5.00	1.750	0.320	0.190	8.93	7.49	3.00	1.95
C 4× 7.25	2.13	4.00	1.721	0.296	0.321	7.84	4.59	2.29	1.47
× 5.4	1.59	4.00	1.584	0.296	0.184	8.52	3.85	1.93	1.56
C 3× 6	1.76	3.00	1.596	0.273	0.356	6.87	2.07	1.38	1.08
× 5	1.47	3.00	1.498	0.273	0.258	7.32	1.85	1.24	1.12
× 4.1	1.21	3.00	1.410	0.273	0.170	7.78	1.66	1.10	1.17

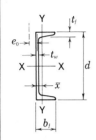

CHANNELS
AMERICAN STANDARD
Properties for designing

Nominal Weight per Ft.	Axis Y-Y			$\bar{x}$	Shear Center Location E_0	Torsional Constant J	Warping Constant C_w
	I	S	r				
	In.⁴	In.³	In.	In.	In.	In.⁴	In.⁶
50	11.0	3.78	0.867	0.799	0.941	2.66	492
40	9.23	3.36	0.886	0.778	1.03	1.46	410
33.9	8.13	3.11	0.904	0.787	1.10	1.01	358
30	5.14	2.06	0.763	0.674	0.873	0.865	151
25	4.47	1.88	0.780	0.674	0.940	0.541	131
20.7	3.88	1.73	0.799	0.698	1.01	0.371	112
30	3.94	1.65	0.669	0.649	0.705	1.22	79.5
25	3.36	1.48	0.676	0.617	0.757	0.690	68.4
20	2.81	1.32	0.691	0.606	0.826	0.370	57.0
15.3	2.28	1.16	0.713	0.634	0.916	0.211	45.5
20	2.42	1.17	0.642	0.583	0.739	0.429	39.5
15	1.93	1.01	0.661	0.586	0.824	0.209	31.0
13.4	1.76	0.962	0.668	0.601	0.859	0.169	28.2
18.75	1.98	1.01	0.599	0.565	0.674	0.436	25.1
13.75	1.53	0.853	0.615	0.553	0.756	0.187	19.3
11.5	1.32	0.781	0.625	0.571	0.807	0.131	16.5
14.75	1.38	0.779	0.564	0.532	0.651	0.268	13.1
12.25	1.17	0.702	0.571	0.525	0.695	0.161	11.2
9.8	0.968	0.625	0.581	0.541	0.752	0.100	9.16
13	1.05	0.642	0.525	0.514	0.599	0.241	7.21
10.5	0.865	0.564	0.529	0.500	0.643	0.131	5.94
8.2	0.692	0.492	0.537	0.512	0.699	0.075	4.73
9	0.632	0.449	0.489	0.478	0.590	0.109	2.93
6.7	0.478	0.378	0.493	0.484	0.647	0.055	2.22
7.25	0.432	0.343	0.450	0.459	0.546	0.082	1.24
5.4	0.319	0.283	0.449	0.458	0.594	0.040	0.923
6	0.305	0.268	0.416	0.455	0.500	0.073	0.463
5	0.247	0.233	0.410	0.438	0.521	0.043	0.380
4.1	0.197	0.202	0.404	0.437	0.546	0.027	0.307

CHANNELS
MISCELLANEOUS
Properties for designing

Designation	Area A	Depth d	Flange Width b_f	Flange Average thickness t_f	Web thickness t_w	$\dfrac{d}{A_f}$	Axis X-X I	S	r
	In.²	In.	In.	In.	In.		In.⁴	In.³	In.
MC 18×58	17.1	18.00	4.200	0.625	0.700	6.86	676	75.1	6.29
×51.9	15.3	18.00	4.100	0.625	0.600	7.02	627	69.7	6.41
×45.8	13.5	18.00	4.000	0.625	0.500	7.20	578	64.3	6.56
×42.7	12.6	18.00	3.950	0.625	0.450	7.29	554	61.6	6.64
MC 13×50	14.7	13.00	4.412	0.610	0.787	4.83	314	48.4	4.62
×40	11.8	13.00	4.185	0.610	0.560	5.09	273	42.0	4.82
×35	10.3	13.00	4.072	0.610	0.447	5.23	252	38.8	4.95
×31.8	9.35	13.00	4.000	0.610	0.375	5.33	239	36.8	5.06
MC 12×50	14.7	12.00	4.135	0.700	0.835	4.15	269	44.9	4.28
×45	13.2	12.00	4.012	0.700	0.712	4.27	252	42.0	4.36
×40	11.8	12.00	3.890	0.700	0.590	4.41	234	39.0	4.46
×35	10.3	12.00	3.767	0.700	0.467	4.55	216	36.1	4.59
MC 12×37	10.9	12.00	3.600	0.600	0.600	5.56	205	34.2	4.34
×32.9	9.67	12.00	3.500	0.600	0.500	5.71	191	31.8	4.44
×30.9	9.07	12.00	3.450	0.600	0.450	5.80	183	30.6	4.50
MC 12×10.6	3.10	12.00	1.500	0.309	0.190	25.9	55.4	9.23	4.22
MC 10×41.1	12.1	10.00	4.321	0.575	0.796	4.02	158	31.5	3.61
×33.6	9.87	10.00	4.100	0.575	0.575	4.24	139	27.8	3.75
×28.5	8.37	10.00	3.950	0.575	0.425	4.40	127	25.3	3.89
MC 10×28.3	8.32	10.00	3.502	0.575	0.477	4.97	118	23.6	3.77
×25.3	7.43	10.00	3.550	0.500	0.425	5.63	107	21.4	3.79
×24.9	7.32	10.00	3.402	0.575	0.377	5.11	110	22.0	3.87
×21.9	6.43	10.00	3.450	0.500	0.325	5.80	98.5	19.7	3.91
MC 10× 8.4	2.46	10.00	1.500	0.280	0.170	23.8	32.0	6.40	3.61
MC 10× 6.5	1.91	10.00	1.127	0.202	0.152	43.8	22.1	4.42	3.40

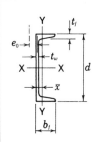

CHANNELS
MISCELLANEOUS
Properties for designing

Nominal Weight per Ft.	Axis Y-Y			$\bar{x}$	Shear Center Location E_0	Torsional Constant J	Warping Constant C_w
	I	S	r				
	In.⁴	In.³	In.	In.	In.	In.⁴	In.⁶
58	17.8	5.32	1.02	0.862	1.04	2.81	1070
51.9	16.4	5.07	1.04	0.858	1.10	2.03	986
45.8	15.1	4.82	1.06	0.866	1.16	1.45	898
42.7	14.4	4.69	1.07	0.877	1.19	1.23	852
50	16.5	4.79	1.06	0.974	1.21	2.97	558
40	13.7	4.26	1.08	0.964	1.31	1.55	462
35	12.3	3.99	1.10	0.980	1.38	1.13	413
31.8	11.4	3.81	1.11	1.00	1.43	0.938	380
50	17.4	5.65	1.09	1.05	1.16	3.23	411
45	15.8	5.33	1.09	1.04	1.20	2.34	374
40	14.3	5.00	1.10	1.04	1.25	1.70	336
35	12.7	4.67	1.11	1.05	1.30	1.25	298
37	9.81	3.59	0.950	0.866	1.05	1.43	247
32.9	8.91	3.39	0.960	0.867	1.09	1.04	224
30.9	8.46	3.28	0.966	0.873	1.12	0.895	212
10.6	0.382	0.310	0.351	0.269	0.379	0.060	11.7
41.1	15.8	4.88	1.14	1.09	1.26	2.26	269
33.6	13.2	4.38	1.16	1.08	1.35	1.20	224
28.5	11.4	4.02	1.17	1.12	1.43	0.792	193
28.3	8.21	3.20	0.993	0.933	1.17	0.842	139
25.3	7.61	2.89	1.01	0.918	1.19	0.576	132
24.9	7.32	2.99	1.00	0.954	1.22	0.633	124
21.9	6.74	2.70	1.02	0.954	1.25	0.417	116
8.4	0.328	0.270	0.365	0.284	0.417	0.041	7.01
6.5	0.112	0.118	0.242	0.180	0.243	0.018	2.46

CHANNELS
MISCELLANEOUS
Properties for designing

Designation	Area A	Depth d	Flange Width b_f	Flange Average thickness t_f	Web thickness t_w	$\dfrac{d}{A_f}$	Axis X-X I	Axis X-X S	Axis X-X r
	In.²	In.	In.	In.	In.		In.⁴	In.³	In.
MC 9×25.4	7.47	9.00	3.500	0.550	0.450	4.68	88.0	19.6	3.43
×23.9	7.02	9.00	3.450	0.550	0.400	4.74	85.0	18.9	3.48
MC 8×22.8	6.70	8.00	3.502	0.525	0.427	4.35	63.8	16.0	3.09
×21.4	6.28	8.00	3.450	0.525	0.375	4.42	61.6	15.4	3.13
MC 8×20	5.88	8.00	3.025	0.500	0.400	5.29	54.5	13.6	3.05
×18.7	5.50	8.00	2.978	0.500	0.353	5.37	52.5	13.1	3.09
MC 8×8.5	2.50	8.00	1.874	0.311	0.179	13.7	23.3	5.83	3.05
MC 7×22.7	6.67	7.00	3.603	0.500	0.503	3.89	47.5	13.6	2.67
×19.1	5.61	7.00	3.452	0.500	0.352	4.06	43.2	12.3	2.77
MC 7×17.6	5.17	7.00	3.000	0.475	0.375	4.91	37.6	10.8	2.70
MC 6×18	5.29	6.00	3.504	0.475	0.379	3.60	29.7	9.91	2.37
×15.3	4.50	6.00	3.500	0.385	0.340	4.45	25.4	8.47	2.38
MC 6×16.3	4.79	6.00	3.000	0.475	0.375	4.21	26.0	8.68	2.33
×15.1	4.44	6.00	2.941	0.475	0.316	4.29	25.0	8.32	2.37
MC 6×12	3.53	6.00	2.497	0.375	0.310	6.41	18.7	6.24	2.30
MC 3× 9	2.65	3.00	2.122	0.351	0.497	4.02	3.15	2.10	1.09
× 7.1	2.09	3.00	1.938	0.351	0.312	4.40	2.73	1.82	1.14

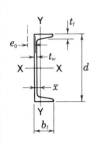

CHANNELS
MISCELLANEOUS
Properties for designing

Nominal Weight per Ft.	Axis Y-Y			$\bar{x}$	Shear Center Location E_0	Torsional Constant J	Warping Constant C_w
	I	S	r				
	In.⁴	In.³	In.	In.	In.	In.⁴	In.⁶
25.4	7.65	3.02	1.01	0.970	1.21	0.692	104
23.9	7.22	2.93	1.01	0.981	1.24	0.599	98.2
22.8	7.07	2.84	1.03	1.01	1.26	0.573	75.4
21.4	6.64	2.74	1.03	1.02	1.28	0.495	70.9
20	4.47	2.05	0.872	0.840	1.04	0.442	47.9
18.7	4.20	1.97	0.874	0.849	1.07	0.381	45.1
8.5	0.628	0.434	0.501	0.428	0.631	0.059	8.22
22.7	7.29	2.85	1.05	1.04	1.26	0.626	58.5
19.1	6.11	2.57	1.04	1.08	1.33	0.407	49.4
17.6	4.01	1.89	0.881	0.873	1.08	0.354	32.5
18	5.93	2.48	1.06	1.12	1.36	0.379	34.7
15.3	4.97	2.03	1.05	1.05	1.33	0.224	30.2
16.3	3.82	1.84	0.892	0.927	1.12	0.336	22.2
15.1	3.51	1.75	0.889	0.940	1.14	0.285	20.5
12	1.87	1.04	0.728	0.704	0.880	0.155	11.3
9	0.967	0.677	0.604	0.694	0.703	0.175	1.21
7.1	0.712	0.561	0.583	0.669	0.730	0.084	0.923

ANGLES
Equal legs

Properties for designing

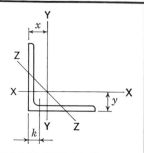

Size and Thickness	k	Weight per Foot	Area	AXIS X-X AND AXIS Y-Y				AXIS Z-Z
				I	S	r	x or y	r
In.	In.	Lb.	In.²	In.⁴	In.³	In.	In.	In.
L 8 × 8 × 1⅛	1¾	56.9	16.7	98.0	17.5	2.42	2.41	1.56
1	1⅝	51.0	15.0	89.0	15.8	2.44	2.37	1.56
⅞	1½	45.0	13.2	79.6	14.0	2.45	2.32	1.57
¾	1⅜	38.9	11.4	69.7	12.2	2.47	2.28	1.58
⅝	1¼	32.7	9.61	59.4	10.3	2.49	2.23	1.58
9⁄16	1³⁄16	29.6	8.68	54.1	9.34	2.50	2.21	1.59
½	1⅛	26.4	7.75	48.6	8.36	2.50	2.19	1.59
L 6 × 6 × 1	1½	37.4	11.0	35.5	8.57	1.80	1.86	1.17
⅞	1⅜	33.1	9.73	31.9	7.63	1.81	1.82	1.17
¾	1¼	28.7	8.44	28.2	6.66	1.83	1.78	1.17
⅝	1⅛	24.2	7.11	24.2	5.66	1.84	1.73	1.18
9⁄16	1¹⁄16	21.9	6.43	22.1	5.14	1.85	1.71	1.18
½	1	19.6	5.75	19.9	4.61	1.86	1.68	1.18
7⁄16	15⁄16	17.2	5.06	17.7	4.08	1.87	1.66	1.19
⅜	⅞	14.9	4.36	15.4	3.53	1.88	1.64	1.19
5⁄16	13⁄16	12.4	3.65	13.0	2.97	1.89	1.62	1.20
L 5 × 5 × ⅞	1⅜	27.2	7.98	17.8	5.17	1.49	1.57	.973
¾	1¼	23.6	6.94	15.7	4.53	1.51	1.52	.975
⅝	1⅛	20.0	5.86	13.6	3.86	1.52	1.48	.978
½	1	16.2	4.75	11.3	3.16	1.54	1.43	.983
7⁄16	15⁄16	14.3	4.18	10.0	2.79	1.55	1.41	.986
⅜	⅞	12.3	3.61	8.74	2.42	1.56	1.39	.990
5⁄16	13⁄16	10.3	3.03	7.42	2.04	1.57	1.37	.994
L 4 × 4 × ¾	1⅛	18.5	5.44	7.67	2.81	1.19	1.27	.778
⅝	1	15.7	4.61	6.66	2.40	1.20	1.23	.779
½	⅞	12.8	3.75	5.56	1.97	1.22	1.18	.782
7⁄16	13⁄16	11.3	3.31	4.97	1.75	1.23	1.16	.785
⅜	¾	9.8	2.86	4.36	1.52	1.23	1.14	.788
5⁄16	11⁄16	8.2	2.40	3.71	1.29	1.24	1.12	.791
¼	⅝	6.6	1.94	3.04	1.05	1.25	1.09	.795

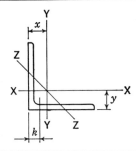

ANGLES
Equal legs

Properties for designing

Size and Thickness	k	Weight per Foot	Area	AXIS X-X AND AXIS Y-Y				AXIS Z-Z
				I	S	r	x or y	r
In.	In.	Lb.	In.²	In.⁴	In.³	In.	In.	In.
L 3½ × 3½ × ½	⅞	11.1	3.25	3.64	1.49	1.06	1.06	.683
⁷⁄₁₆	¹³⁄₁₆	9.8	2.87	3.26	1.32	1.07	1.04	.684
⅜	¾	8.5	2.48	2.87	1.15	1.07	1.01	.687
⁵⁄₁₆	¹¹⁄₁₆	7.2	2.09	2.45	.976	1.08	.990	.690
¼	⅝	5.8	1.69	2.01	.794	1.09	.968	.694
L 3 × 3 × ½	¹³⁄₁₆	9.4	2.75	2.22	1.07	.898	.932	.584
⁷⁄₁₆	¾	8.3	2.43	1.99	.954	.905	.910	.585
⅜	¹¹⁄₁₆	7.2	2.11	1.76	.833	.913	.888	.587
⁵⁄₁₆	⅝	6.1	1.78	1.51	.707	.922	.865	.589
¼	⁹⁄₁₆	4.9	1.44	1.24	.577	.930	.842	.592
³⁄₁₆	½	3.71	1.09	.962	.441	.939	.820	.596
L 2½ × 2½ × ½	¹³⁄₁₆	7.7	2.25	1.23	.724	.739	.806	.487
⅜	¹¹⁄₁₆	5.9	1.73	.984	.566	.753	.762	.487
⁵⁄₁₆	⅝	5.0	1.46	.849	.482	.761	.740	.489
¼	⁹⁄₁₆	4.1	1.19	.703	.394	.769	.717	.491
³⁄₁₆	½	3.07	.902	.547	.303	.778	.694	.495
L 2 × 2 × ⅜	¹¹⁄₁₆	4.7	1.36	.479	.351	.594	.636	.389
⁵⁄₁₆	⅝	3.92	1.15	.416	.300	.601	.614	.390
¼	⁹⁄₁₆	3.19	.938	.348	.247	.609	.592	.391
³⁄₁₆	½	2.44	.715	.272	.190	.617	.569	.394
⅛	⁷⁄₁₆	1.65	.484	.190	.131	.626	.546	.398
L 1¾ × 1¾ × ¼	½	2.77	.813	.227	.186	.529	.529	.341
³⁄₁₆	⁷⁄₁₆	2.12	.621	.179	.144	.537	.506	.343
⅛	⅜	1.44	.422	.126	.099	.546	.484	.347
L 1½ × 1½ × ¼	⁷⁄₁₆	2.34	.688	.139	.134	.449	.466	.292
³⁄₁₆	⅜	1.80	.527	.110	.104	.457	.444	.293
⁵⁄₃₂	⅜	1.52	.444	.094	.088	.461	.433	.295
⅛	⁵⁄₁₆	1.23	.359	.078	.072	.465	.421	.296
L 1¼ × 1¼ × ¼	⁷⁄₁₆	1.92	.563	.077	.091	.369	.403	.243
³⁄₁₆	⅜	1.48	.434	.061	.071	.377	.381	.244
⅛	⁵⁄₁₆	1.01	.297	.044	.049	.385	.359	.246
L 1 × 1 × ¼	⅜	1.49	.438	.037	.056	.290	.339	.196
³⁄₁₆	⁵⁄₁₆	1.16	.340	.030	.044	.297	.318	.195
⅛	¼	.80	.234	.022	.031	.304	.296	.196

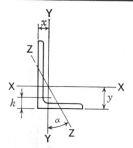

ANGLES
Unequal legs
Properties for designing

Size and Thickness	k	Weight per Foot	Area	AXIS X-X				AXIS Y-Y				AXIS Z-Z	
				I	S	r	y	I	S	r	x	r	Tan α
In.	In.	Lb.	In.²	In.⁴	In.³	In.	In.	In.⁴	In.³	In.	In.	In.	
L 9 × 4 × 1	1½	40.8	12.0	97.0	17.6	2.84	3.50	12.0	4.00	1.00	1.00	.834	.203
⅞	1⅜	36.1	10.6	86.8	15.7	2.86	3.45	10.8	3.56	1.01	.953	.836	.208
¾	1¼	31.3	9.19	76.1	13.6	2.88	3.41	9.63	3.11	1.02	.906	.841	.212
⅝	1⅛	26.3	7.73	64.9	11.5	2.90	3.36	8.32	2.65	1.04	.858	.847	.216
⁹⁄₁₆	1¹⁄₁₆	23.8	7.00	59.1	10.4	2.91	3.33	7.63	2.41	1.04	.834	.850	.218
½	1	21.3	6.25	53.2	9.34	2.92	3.31	6.92	2.17	1.05	.810	.854	.220
L 8 × 6 × 1	1½	44.2	13.0	80.8	15.1	2.49	2.65	38.8	8.92	1.73	1.65	1.28	.543
⅞	1⅜	39.1	11.5	72.3	13.4	2.51	2.61	34.9	7.94	1.74	1.61	1.28	.547
¾	1¼	33.8	9.94	63.4	11.7	2.53	2.56	30.7	6.92	1.76	1.56	1.29	.551
⅝	1⅛	28.5	8.36	54.1	9.87	2.54	2.52	26.3	5.88	1.77	1.52	1.29	.554
⁹⁄₁₆	1¹⁄₁₆	25.7	7.56	49.3	8.95	2.55	2.50	24.0	5.34	1.78	1.50	1.30	.556
½	1	23.0	6.75	44.3	8.02	2.56	2.47	21.7	4.79	1.79	1.47	1.30	.558
⁷⁄₁₆	¹⁵⁄₁₆	20.2	5.93	39.2	7.07	2.57	2.45	19.3	4.23	1.80	1.45	1.31	.560
L 8 × 4 × 1	1½	37.4	11.0	69.6	14.1	2.52	3.05	11.6	3.94	1.03	1.05	.846	.247
⅞	1⅜	33.1	9.73	62.5	12.5	2.53	3.00	10.5	3.51	1.04	.999	.848	.253
¾	1¼	28.7	8.44	54.9	10.9	2.55	2.95	9.36	3.07	1.05	.953	.852	.258
⅝	1⅛	24.2	7.11	46.9	9.21	2.57	2.91	8.10	2.62	1.07	.906	.857	.262
⁹⁄₁₆	1¹⁄₁₆	21.9	6.43	42.8	8.35	2.58	2.88	7.43	2.38	1.07	.882	.861	.265
½	1	19.6	5.75	38.5	7.49	2.59	2.86	6.74	2.15	1.08	.859	.865	.267
⁷⁄₁₆	¹⁵⁄₁₆	17.2	5.06	34.1	6.60	2.60	2.83	6.02	1.90	1.09	.835	.869	.269
L 7 × 4 × ⅞	1⅜	30.2	8.86	42.9	9.65	2.20	2.55	10.2	3.46	1.07	1.05	.856	.318
¾	1¼	26.2	7.69	37.8	8.42	2.22	2.51	9.05	3.03	1.09	1.01	.860	.324
⅝	1⅛	22.1	6.48	32.4	7.14	2.24	2.46	7.84	2.58	1.10	.963	.865	.329
⁹⁄₁₆	1¹⁄₁₆	20.0	5.87	29.6	6.48	2.24	2.44	7.19	2.35	1.11	.940	.868	.332
½	1	17.9	5.25	26.7	5.81	2.25	2.42	6.53	2.12	1.11	.917	.872	.335
⁷⁄₁₆	¹⁵⁄₁₆	15.8	4.62	23.7	5.13	2.26	2.39	5.83	1.88	1.12	.893	.876	.337
⅜	⅞	13.6	3.98	20.6	4.44	2.27	2.37	5.10	1.63	1.13	.870	.880	.340

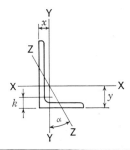

ANGLES
Unequal legs
Properties for designing

Size and Thickness	k	Weight per Foot	Area	AXIS X-X				AXIS Y-Y				AXIS Z-Z	
				I	S	r	y	I	S	r	x	r	Tan α
In.	In.	Lb.	In.²	In.⁴	In.³	In.	In.	In.⁴	In.³	In.	In.	In.	
L 6 × 4 × ⅞	1⅜	27.2	7.98	27.7	7.15	1.86	2.12	9.75	3.39	1.11	1.12	.857	.421
¾	1¼	23.6	6.94	24.5	6.25	1.88	2.08	8.68	2.97	1.12	1.08	.860	.428
⅝	1⅛	20.0	5.86	21.1	5.31	1.90	2.03	7.52	2.54	1.13	1.03	.864	.435
9/16	1 1/16	18.1	5.31	19.3	4.83	1.90	2.01	6.91	2.31	1.14	1.01	.866	.438
½	1	16.2	4.75	17.4	4.33	1.91	1.99	6.27	2.08	1.15	.987	.870	.440
7/16	15/16	14.3	4.18	15.5	3.83	1.92	1.96	5.60	1.85	1.16	.964	.873	.443
⅜	⅞	12.3	3.61	13.5	3.32	1.93	1.94	4.90	1.60	1.17	.941	.877	.446
5/16	13/16	10.3	3.03	11.4	2.79	1.94	1.92	4.18	1.35	1.17	.918	.882	.448
¼	¾	8.3	2.44	9.27	2.26	1.95	1.89	3.41	1.10	1.18	.894	.887	.451
L 6 × 3½ × ½	1	15.3	4.50	16.6	4.24	1.92	2.08	4.25	1.59	.972	.833	.759	.344
⅜	⅞	11.7	3.42	12.9	3.24	1.94	2.04	3.34	1.23	.988	.787	.767	.350
5/16	13/16	9.8	2.87	10.9	2.73	1.95	2.01	2.85	1.04	.996	.763	.772	.352
¼	¾	7.9	2.31	8.86	2.21	1.96	1.99	2.34	0.847	1.01	.740	.777	.355
L 5 × 3½ × ¾	1¼	19.8	5.81	13.9	4.28	1.55	1.75	5.55	2.22	.977	.996	.748	.464
⅝	1⅛	16.8	4.92	12.0	3.65	1.56	1.70	4.83	1.90	.991	.951	.751	.472
½	1	13.6	4.00	9.99	2.99	1.58	1.66	4.05	1.56	1.01	.906	.755	.479
7/16	15/16	12.0	3.53	8.90	2.64	1.59	1.63	3.63	1.39	1.01	.883	.758	.482
⅜	⅞	10.4	3.05	7.78	2.29	1.60	1.61	3.18	1.21	1.02	.861	.762	.486
5/16	13/16	8.7	2.56	6.60	1.94	1.61	1.59	2.72	1.02	1.03	.838	.766	.489
¼	¾	7.0	2.06	5.39	1.57	1.62	1.56	2.23	.830	1.04	.814	.770	.492
L 5 × 3 × ½	1	12.8	3.75	9.45	2.91	1.59	1.75	2.58	1.15	.829	.750	.648	.357
7/16	15/16	11.3	3.31	8.43	2.58	1.60	1.73	2.32	1.02	.837	.727	.651	.361
⅜	⅞	9.8	2.86	7.37	2.24	1.61	1.70	2.04	.888	.845	.704	.654	.364
5/16	13/16	8.2	2.40	6.26	1.89	1.61	1.68	1.75	.753	.853	.681	.658	.368
¼	¾	6.6	1.94	5.11	1.53	1.62	1.66	1.44	.614	.861	.657	.663	.371

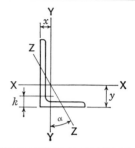

ANGLES
Unequal legs
Properties for designing

Size and Thickness	k	Weight per Foot	Area	AXIS X-X				AXIS Y-Y				AXIS Z-Z	
				I	S	r	y	I	S	r	x	r	Tan α
In.	In.	Lb.	In.²	In.⁴	In.³	In.	In.	In.⁴	In.³	In.	In.	In.	α
L 4 × 3½ × ⅝	1¹⁄₁₆	14.7	4.30	6.37	2.35	1.22	1.29	4.52	1.84	1.03	1.04	.719	.745
½	¹⁵⁄₁₆	11.9	3.50	5.32	1.94	1.23	1.25	3.79	1.52	1.04	1.00	.722	.750
⁷⁄₁₆	⅞	10.6	3.09	4.76	1.72	1.24	1.23	3.40	1.35	1.05	.978	.724	.753
⅜	¹³⁄₁₆	9.1	2.67	4.18	1.49	1.25	1.21	2.95	1.17	1.06	.955	.727	.755
⁵⁄₁₆	¾	7.7	2.25	3.56	1.26	1.26	1.18	2.55	.994	1.07	.932	.730	.757
¼	¹¹⁄₁₆	6.2	1.81	2.91	1.03	1.27	1.16	2.09	.808	1.07	.909	.734	.759
L 4 × 3 × ⅝	1¹⁄₁₆	13.6	3.98	6.03	2.30	1.23	1.37	2.87	1.35	.849	.871	.637	.534
½	¹⁵⁄₁₆	11.1	3.25	5.05	1.89	1.25	1.33	2.42	1.12	.864	.827	.639	.543
⁷⁄₁₆	⅞	9.8	2.87	4.52	1.68	1.25	1.30	2.18	.992	.871	.804	.641	.547
⅜	¹³⁄₁₆	8.5	2.48	3.96	1.46	1.26	1.28	1.92	.866	.879	.782	.644	.551
⁵⁄₁₆	¾	7.2	2.09	3.38	1.23	1.27	1.26	1.65	.734	.887	.759	.647	.554
¼	¹¹⁄₁₆	5.8	1.69	2.77	1.00	1.28	1.24	1.36	.599	.896	.736	.651	.558
L 3½ × 3 × ½	¹⁵⁄₁₆	10.2	3.00	3.45	1.45	1.07	1.13	2.33	1.10	.881	.875	.621	.714
⁷⁄₁₆	⅞	9.1	2.65	3.10	1.29	1.08	1.10	2.09	.975	.889	.853	.622	.718
⅜	¹³⁄₁₆	7.9	2.30	2.72	1.13	1.09	1.08	1.85	.851	.897	.830	.625	.721
⁵⁄₁₆	¾	6.6	1.93	2.33	.954	1.10	1.06	1.58	.722	.905	.808	.627	.724
¼	¹¹⁄₁₆	5.4	1.56	1.91	.776	1.11	1.04	1.30	.589	.914	.785	.631	.727
L 3½ × 2½ × ½	¹⁵⁄₁₆	9.4	2.75	3.24	1.41	1.09	1.20	1.36	.760	.704	.705	.534	.486
⁷⁄₁₆	⅞	8.3	2.43	2.91	1.26	1.09	1.18	1.23	.677	.711	.682	.535	.491
⅜	¹³⁄₁₆	7.2	2.11	2.56	1.09	1.10	1.16	1.09	.592	.719	.660	.537	.496
⁵⁄₁₆	¾	6.1	1.78	2.19	.927	1.11	1.14	.939	.504	.727	.637	.540	.501
¼	¹¹⁄₁₆	4.9	1.44	1.80	.755	1.12	1.11	.777	.412	.735	.614	.544	.506
L 3 × 2½ × ½	⅞	8.5	2.50	2.08	1.04	.913	1.00	1.30	.744	.722	.750	.520	.667
⁷⁄₁₆	¹³⁄₁₆	7.6	2.21	1.88	.928	.920	.978	1.18	.664	.729	.728	.521	.672
⅜	¾	6.6	1.92	1.66	.810	.928	.956	1.04	.581	.736	.706	.522	.676
⁵⁄₁₆	¹¹⁄₁₆	5.6	1.62	1.42	.688	.937	.933	.898	.494	.744	.683	.525	.680
¼	⅝	4.5	1.31	1.17	.561	.945	.911	.743	.404	.753	.661	.528	.684
³⁄₁₆	⁹⁄₁₆	3.39	.996	.907	.430	.954	.888	.577	.310	.761	.638	.533	.688

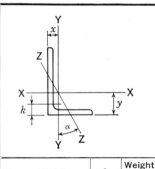

ANGLES
Unequal legs
Properties for designing

Size and Thickness	k	Weight per Foot	Area	AXIS X-X				AXIS Y-Y				AXIS Z-Z	
				I	S	r	y	I	S	r	x	r	Tan α
In.	In.	Lb.	In.²	In.⁴	In.³	In.	In.	In.⁴	In.³	In.	In.	In.	
L 3 × 2 × ½	¹³⁄₁₆	7.7	2.25	1.92	1.00	.924	1.08	.672	.474	.546	.583	.428	.414
⁷⁄₁₆	¾	6.8	2.00	1.73	.894	.932	1.06	.609	.424	.553	.561	.429	.421
⅜	¹¹⁄₁₆	5.9	1.73	1.53	.781	.940	1.04	.543	.371	.559	.539	.430	.428
⁵⁄₁₆	⅝	5.0	1.46	1.32	.664	.948	1.02	.470	.317	.567	.516	.432	.435
¼	⁹⁄₁₆	4.1	1.19	1.09	.542	.957	.993	.392	.260	.574	.493	.435	.440
³⁄₁₆	½	3.07	.902	.842	.415	.966	.970	.307	.200	.583	.470	.439	.446
L 2½ × 2 × ⅜	¹¹⁄₁₆	5.3	1.55	.912	.547	.768	.831	.514	.363	.577	.581	.420	.614
⁵⁄₁₆	⅝	4.5	1.31	.788	.466	.776	.809	.446	.310	.584	.559	.422	.620
¼	⁹⁄₁₆	3.62	1.06	.654	.381	.784	.787	.372	.254	.592	.537	.424	.626
³⁄₁₆	½	2.75	.809	.509	.293	.793	.764	.291	.196	.600	.514	.427	.631
L 2½ × 1½ × ⁵⁄₁₆	⅝	3.92	1.15	.711	.444	.785	.898	.191	.174	.408	.398	.322	.349
¼	⁹⁄₁₆	3.19	.938	.591	.364	.794	.875	.161	.143	.415	.375	.324	.357
³⁄₁₆	½	2.44	.715	.461	.279	.803	.852	.127	.111	.422	.352	.327	.364
L 2 × 1½ × ¼	½	2.77	.813	.316	.236	.623	.663	.151	.139	.432	.413	.320	.543
³⁄₁₆	⁷⁄₁₆	2.12	.621	.248	.182	.632	.641	.120	.108	.440	.391	.322	.551
⅛	⅜	1.44	.422	.173	.125	.641	.618	.085	.075	.448	.368	.326	.558
L 2 × 1¼ × ¼	½	2.55	.750	.296	.229	.628	.708	.089	.097	.344	.333	.269	.378
³⁄₁₆	⁷⁄₁₆	1.96	.574	.232	.177	.636	.686	.071	.075	.351	.311	.271	.387
⅛	⅜	1.33	.391	.163	.122	.645	.663	.050	.052	.359	.287	.274	.396
L 1¾ × 1¼ × ¼	⁷⁄₁₆	2.34	.688	.202	.176	.543	.602	.085	.095	.352	.352	.267	.486
³⁄₁₆	⅜	1.80	.527	.160	.137	.551	.580	.068	.074	.359	.330	.269	.496
⅛	⁵⁄₁₆	1.23	.359	.113	.094	.560	.557	.049	.051	.368	.307	.272	.506

STRUCTURAL TEES

Cut from W shapes

Dimensions and
properties for designing

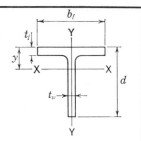

Designation	Area	Depth of Tee d	Flange Width b_f	Flange Thickness t_f	Stem Thickness t_w	$\frac{d}{t_w}$	AXIS X-X I	AXIS X-X S	AXIS X-X r	AXIS X-X y	AXIS Y-Y I	AXIS Y-Y S	AXIS Y-Y r
	In.²	In.	In.	In.	In.		In.⁴	In.³	In.	In.	In.⁴	In.³	In.
WT 18 × 150	44.1	18.36	16.655	1.680	0.945	19.4	1220	86.0	5.27	4.13	648	77.8	3.83
× 140	41.2	18.25	16.595	1.570	0.885	20.6	1130	80.0	5.25	4.06	599	72.2	3.81
× 130	38.2	18.12	16.551	1.440	0.841	21.5	1060	75.1	5.26	4.05	545	65.9	3.77
× 122.5	36.1	18.03	16.512	1.350	0.802	22.5	995	71.1	5.25	4.03	507	61.4	3.75
× 115	33.8	17.94	16.471	1.260	0.761	23.6	933	67.0	5.25	4.00	470	57.1	3.73
WT 18 × 97	28.6	18.24	12.117	1.260	0.770	23.7	905	67.4	5.63	4.81	188	31.0	2.56
× 91	26.8	18.16	12.072	1.180	0.725	25.0	845	63.1	5.61	4.77	174	28.8	2.55
× 85	25.0	18.08	12.027	1.100	0.680	26.6	786	58.8	5.60	4.73	160	26.6	2.53
× 80	23.6	18.00	12.000	1.020	0.653	27.6	742	56.0	5.61	4.75	147	24.6	2.50
× 75	22.1	17.92	11.972	0.940	0.625	28.7	698	53.1	5.62	4.78	135	22.5	2.47
× 67.5	19.9	17.78	11.945	0.794	0.598	29.7	636	49.5	5.65	4.94	113	18.9	2.39
WT 16.5 × 120	35.3	16.75	15.865	1.400	0.830	20.2	823	63.2	4.83	3.73	467	58.8	3.64
× 110	32.4	16.63	15.810	1.275	0.775	21.5	755	58.4	4.83	3.70	421	53.2	3.60
× 100	29.4	16.50	15.750	1.150	0.715	23.1	685	53.3	4.82	3.66	375	47.6	3.57
WT 16.5 × 76	22.4	16.75	11.565	1.055	0.635	26.4	592	47.4	5.15	4.26	136	23.6	2.47
× 70.5	20.8	16.66	11.535	0.960	0.605	27.5	552	44.7	5.16	4.29	123	21.3	2.43
× 65	19.2	16.55	11.510	0.855	0.580	28.5	514	42.2	5.18	4.37	109	18.9	2.38
× 59	17.4	16.43	11.484	0.738	0.554	29.7	471	39.4	5.21	4.48	93.4	16.3	2.32
WT 15 × 105	30.9	15.19	15.105	1.315	0.775	19.6	579	48.7	4.33	3.31	378	50.1	3.50
× 95	28.0	15.06	15.040	1.185	0.710	21.2	521	44.1	4.31	3.25	336	44.7	3.47
× 86	25.4	14.94	14.985	1.065	0.655	22.8	472	40.2	4.31	3.22	299	39.9	3.43
WT 15 × 66	19.4	15.15	10.551	1.000	0.615	24.6	421	37.4	4.65	3.90	98.2	18.6	2.25
× 62	18.2	15.08	10.521	0.930	0.585	25.8	395	35.3	4.65	3.89	90.5	17.2	2.23
× 58	17.1	15.00	10.500	0.850	0.564	26.6	372	33.6	4.67	3.93	82.2	15.7	2.19
× 54	15.9	14.91	10.484	0.760	0.548	27.2	350	32.1	4.69	4.02	73.2	14.0	2.15
× 49.5	14.6	14.82	10.458	0.670	0.522	28.4	323	30.1	4.71	4.10	64.1	12.3	2.10
WT 13.5 × 88.5	26.1	13.66	14.090	1.190	0.725	18.8	393	36.8	3.88	2.97	278	39.4	3.26
× 80	23.6	13.54	14.023	1.075	0.658	20.6	352	33.1	3.87	2.90	247	35.3	3.24
× 72.5	21.4	13.44	13.965	0.975	0.600	22.4	317	29.9	3.85	2.85	222	31.7	3.22

STRUCTURAL TEES
Cut from W shapes
Dimensions and properties for designing

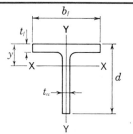

Designation	Area	Depth of Tee d	Flange Width b_f	Flange Thick-ness t_f	Stem Thick-ness t_w	$\dfrac{d}{t_w}$	AXIS X-X				AXIS Y-Y		
							I	S	r	y	I	S	r
	In.²	In.	In.	In.	In.		In.⁴	In.³	In.	In.	In.⁴	In.³	In.
WT 13.5 × 57	16.8	13.64	10.070	0.932	0.570	23.9	289	28.3	4.15	3.41	79.5	15.8	2.18
× 51	15.0	13.54	10.018	0.827	0.518	26.1	258	25.4	4.14	3.38	69.5	13.9	2.15
× 47	13.8	13.46	9.990	0.747	0.490	27.5	239	23.8	4.15	3.41	62.2	12.5	2.12
× 42	12.4	13.35	9.963	0.636	0.463	28.8	216	22.0	4.18	3.50	52.5	10.5	2.06
WT 12 × 80	23.6	12.36	14.091	1.135	0.656	18.8	272	27.6	3.40	2.50	265	37.6	3.35
× 72.5	21.4	12.25	14.043	1.020	0.608	20.1	247	25.2	3.40	2.47	236	33.6	3.32
× 65	19.2	12.13	14.000	0.900	0.565	21.5	223	23.1	3.41	2.46	206	29.4	3.28
WT 12 × 60	17.7	12.16	12.088	0.930	0.556	21.9	215	22.5	3.49	2.62	137	22.7	2.78
× 55	16.2	12.08	12.042	0.855	0.510	23.7	195	20.5	3.47	2.57	125	20.7	2.77
× 50	14.8	12.00	12.000	0.775	0.468	25.6	177	18.7	3.46	2.53	112	18.6	2.75
WT 12 × 47	13.8	12.15	9.061	0.872	0.516	23.5	186	20.3	3.67	3.00	54.2	12.0	1.98
× 42	12.4	12.05	9.015	0.772	0.470	25.6	166	18.3	3.66	2.97	47.2	10.5	1.95
× 38	11.2	11.96	8.985	0.682	0.440	27.2	151	16.9	3.68	2.99	41.3	9.20	1.92
× 34	10.0	11.86	8.961	0.582	0.416	28.5	137	15.6	3.70	3.07	35.0	7.81	1.87
WT 12 × 30.5	8.98	11.86	7.023	0.591	0.419	28.3	129	15.2	3.78	3.42	17.1	4.88	1.38
× 27.5	8.09	11.78	7.000	0.503	0.396	29.7	116	14.1	3.79	3.50	14.4	4.13	1.34
WT 10.5 × 71	20.9	10.73	13.132	1.095	0.659	16.3	177	20.8	2.92	2.18	207	31.5	3.15
× 63.5	18.7	10.62	13.061	0.985	0.588	18.1	156	18.3	2.89	2.11	183	28.0	3.13
× 56	16.5	10.50	13.000	0.865	0.527	19.9	137	16.2	2.88	2.06	159	24.4	3.10
WT 10.5 × 48	14.1	10.57	9.038	0.935	0.575	18.4	137	17.1	3.12	2.54	57.7	12.8	2.02
× 41	12.1	10.43	8.962	0.795	0.499	20.9	116	14.6	3.10	2.48	47.8	10.7	1.99
WT 10.5 × 36.5	10.7	10.62	8.295	0.740	0.455	23.3	110	13.8	3.21	2.60	35.3	8.51	1.81
× 34	10.0	10.57	8.270	0.685	0.430	24.6	103	12.9	3.20	2.59	32.4	7.83	1.80
× 31	9.13	10.50	8.240	0.615	0.400	26.2	93.8	11.9	3.21	2.58	28.7	6.97	1.77
× 27.5	8.10	10.40	8.215	0.522	0.375	27.7	84.4	10.9	3.23	2.64	24.2	5.88	1.73
WT 10.5 × 24.5	7.21	10.41	6.520	0.532	0.368	28.3	78.3	10.4	3.29	2.90	12.3	3.78	1.31
× 22	6.48	10.33	6.500	0.451	0.348	29.7	70.9	9.63	3.31	2.97	10.4	3.19	1.27

STRUCTURAL TEES
Cut from W shapes
Dimensions and properties for designing

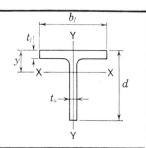

Designation	Area	Depth of Tee d	Flange Width b_f	Flange Thickness t_f	Stem Thickness t_w	$\dfrac{d}{t_w}$	AXIS X-X I	AXIS X-X S	AXIS X-X r	AXIS X-X y	AXIS Y-Y I	AXIS Y-Y S	AXIS Y-Y r
	In.²	In.	In.	In.	In.		In.⁴	In.³	In.	In.	In.⁴	In.³	In.
WT 9 × 57	16.8	9.24	11.833	0.991	0.595	15.5	103	13.9	2.48	1.85	137	23.2	2.86
× 52.5	15.4	9.16	11.792	0.911	0.554	16.5	94.0	12.8	2.47	1.82	125	21.1	2.84
× 48	14.1	9.08	11.750	0.831	0.512	17.7	85.4	11.7	2.46	1.78	112	19.1	2.82
WT 9 × 42.5	12.5	9.16	8.838	0.911	0.526	17.4	84.4	11.9	2.60	2.05	52.5	11.9	2.05
× 38.5	11.4	9.08	8.787	0.831	0.475	19.1	75.3	10.6	2.58	1.99	47.1	10.7	2.04
× 35	10.3	9.00	8.750	0.751	0.438	20.5	68.2	9.68	2.57	1.96	42.0	9.60	2.02
× 32	9.43	8.94	8.715	0.686	0.403	22.2	61.9	8.83	2.56	1.92	37.9	8.70	2.00
WT 9 × 30	8.83	9.13	7.558	0.695	0.416	21.9	64.9	9.32	2.71	2.16	25.1	6.63	1.68
× 27.5	8.10	9.06	7.532	0.630	0.390	23.2	59.6	8.64	2.71	2.16	22.5	5.97	1.67
× 25	7.36	9.00	7.500	0.570	0.358	25.1	54.0	7.86	2.71	2.13	20.1	5.35	1.65
× 22.5	6.62	8.93	7.477	0.499	0.335	26.7	49.0	7.24	2.72	2.16	17.4	4.66	1.62
WT 9 × 20	5.88	8.95	6.018	0.524	0.316	28.3	44.9	6.75	2.76	2.29	9.54	3.17	1.27
× 17.5	5.15	8.86	6.000	0.429	0.298	29.7	40.1	6.18	2.79	2.38	7.74	2.58	1.23
WT 8 × 48	14.1	8.16	11.533	0.875	0.535	15.3	64.7	9.82	2.14	1.57	112	19.4	2.82
× 44	12.9	8.08	11.502	0.795	0.504	16.0	59.5	9.11	2.14	1.55	101	17.5	2.79
WT 8 × 39	11.5	8.16	8.586	0.875	0.529	15.4	60.0	9.45	2.28	1.81	46.3	10.8	2.01
× 35.5	10.5	8.08	8.543	0.795	0.486	16.6	54.1	8.57	2.27	1.77	41.4	9.69	1.99
× 32	9.41	8.08	8.500	0.715	0.443	18.1	48.3	7.72	2.27	1.73	36.7	8.63	1.97
× 29	8.53	7.93	8.464	0.645	0.407	19.5	43.6	7.01	2.26	1.71	32.6	7.71	1.96
WT 8 × 25	7.36	8.13	7.073	0.628	0.380	21.4	42.2	6.77	2.40	1.89	18.6	5.25	1.59
× 22.5	6.63	8.06	7.039	0.563	0.346	23.3	37.8	6.10	2.39	1.86	16.4	4.66	1.57
× 20	5.89	8.00	7.000	0.503	0.307	26.1	33.2	5.38	2.37	1.82	14.4	4.11	1.56
× 18	5.30	7.93	6.992	0.428	0.299	26.5	30.8	5.11	2.41	1.89	12.2	3.49	1.52
WT 8 × 15.5	4.57	7.92	5.53	0.442	0.275	28.8	27.3	4.62	2.44	2.01	6.23	2.25	1.17
× 13	3.84	7.82	5.50	0.345	0.250	31.3	23.3	4.07	2.47	2.08	4.80	1.74	1.12

STRUCTURAL TEES

Cut from W shapes

Dimensions and properties for designing

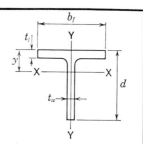

Designation	Area	Depth of Tee d	Flange		Stem Thick-ness t_w	$\dfrac{d}{t_w}$	AXIS X-X				AXIS Y-Y		
			Width b_f	Thick-ness t_f			I	S	r	y	I	S	r
	In.²	In.	In.	In.	In.		In.⁴	In.³	In.	In.	In.⁴	In.³	In.
WT 7 × 365	107.0	11.22	17.889	4.910	3.069	3.66	740	95.6	2.63	3.47	2360	264	4.69
× 332.5	97.8	10.84	17.646	4.522	2.826	3.83	623	82.2	2.52	3.25	2080	236	4.62
× 302.5	89.0	10.47	17.418	4.157	2.598	4.03	525	70.8	2.43	3.05	1840	211	4.55
× 275	80.9	10.13	17.206	3.818	2.386	4.25	444	61.1	2.34	2.86	1630	189	4.49
× 250	73.5	9.82	17.008	3.501	2.188	4.49	377	52.8	2.26	2.68	1440	169	4.43
× 227.5	66.9	9.53	16.828	3.213	2.008	4.74	322	45.9	2.19	2.51	1280	152	4.37
WT 7 × 213	62.6	9.35	16.695	3.033	1.875	4.98	288	41.4	2.14	2.40	1180	141	4.34
× 199	58.5	9.16	16.590	2.843	1.770	5.17	258	37.7	2.10	2.30	1080	131	4.31
× 185	54.4	8.97	16.475	2.658	1.655	5.42	230	34.0	2.06	2.19	993	121	4.27
× 171	50.3	8.78	16.365	2.468	1.545	5.68	204	30.5	2.02	2.09	903	110	4.24
× 157	46.2	8.60	16.235	2.283	1.415	6.07	179	27.0	1.97	1.98	816	100	4.20
× 143.5	42.2	8.41	16.130	2.093	1.310	6.42	157	24.1	1.93	1.87	733	90.9	4.17
× 132	38.8	8.25	16.025	1.938	1.205	6.85	139	21.5	1.89	1.78	666	83.1	4.14
× 123	36.2	8.13	15.945	1.813	1.125	7.22	126	19.6	1.86	1.71	613	76.9	4.12
WT 7 × 118.5	34.8	8.06	15.910	1.748	1.090	7.39	120	18.7	1.85	1.67	587	73.8	4.11
× 114	33.5	8.00	15.865	1.688	1.045	7.66	113	17.7	1.84	1.64	562	70.9	4.10
× 109.5	32.2	7.94	15.825	1.623	1.005	7.90	107	16.9	1.82	1.60	537	67.8	4.08
× 105.5	31.0	7.88	15.800	1.563	0.980	8.04	102	16.2	1.82	1.57	514	65.1	4.07
× 101	29.7	7.82	15.750	1.503	0.930	8.40	95.8	15.2	1.80	1.53	490	62.2	4.06
× 96.5	28.4	7.75	15.710	1.438	0.890	8.71	90.1	14.4	1.78	1.49	465	59.2	4.05
× 92	27.0	7.69	15.660	1.378	0.840	9.15	83.9	13.4	1.76	1.45	441	56.4	4.04
× 88	25.9	7.63	15.640	1.313	0.820	9.30	80.2	12.9	1.76	1.42	419	53.6	4.02
× 83.5	24.5	7.56	15.600	1.248	0.780	9.69	75.0	12.2	1.75	1.39	395	50.7	4.01
× 79	23.2	7.50	15.550	1.188	0.730	10.3	69.3	11.3	1.73	1.34	372	47.9	4.00
× 75	22.0	7.44	15.515	1.128	0.695	10.7	65.0	10.6	1.72	1.31	351	45.3	3.99
× 71	20.9	7.38	15.500	1.063	0.680	10.8	62.1	10.2	1.72	1.29	330	42.6	3.97
WT 7 × 160	47.1	8.41	16.710	2.093	1.890	4.45	209	33.3	2.11	2.12	818	97.8	4.17

STRUCTURAL TEES

Cut from W shapes

Dimensions and properties for designing

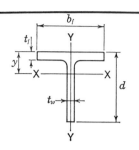

Designation	Area	Depth of Tee d	Flange		Stem Thickness t_w	$\dfrac{d}{t_w}$	AXIS X-X				AXIS Y-Y		
			Width b_f	Thickness t_f			I	S	r	y	I	S	r
	In.²	In.	In.	In.	In.		In.⁴	In.³	In.	In.	In.⁴	In.³	In.
WT 7 × 68	20.0	7.38	14.740	1.063	0.660	11.2	60.1	9.89	1.73	1.31	284	38.5	3.77
× 63.5	18.7	7.31	14.690	0.998	0.610	12.0	54.7	9.05	1.71	1.26	264	35.9	3.76
× 59.5	17.5	7.25	14.650	0.938	0.570	12.7	50.4	8.36	1.70	1.22	246	33.6	3.75
× 55.5	16.3	7.19	14.620	0.873	0.540	13.3	46.9	7.82	1.69	1.19	227	31.1	3.73
× 51.5	15.1	7.13	14.575	0.813	0.495	14.4	42.4	7.10	1.67	1.15	210	28.8	3.72
× 47.5	14.0	7.06	14.545	0.748	0.465	15.2	39.1	6.58	1.67	1.12	192	26.4	3.71
× 43.5	12.8	7.00	14.500	0.688	0.420	16.7	34.9	5.88	1.65	1.08	175	24.1	3.70
WT 7 × 42	12.4	7.09	12.023	0.778	0.451	15.7	37.4	6.36	1.74	1.21	113	18.8	3.02
× 39	11.5	7.03	12.000	0.718	0.428	16.4	34.8	5.96	1.74	1.19	103	17.2	3.00
WT 7 × 37	10.9	7.10	10.072	0.783	0.450	15.8	36.1	6.26	1.82	1.32	66.7	13.3	2.48
× 34	10.0	7.03	10.040	0.718	0.418	16.8	33.0	5.75	1.82	1.29	60.6	12.1	2.46
× 30.5	8.97	6.96	10.000	0.643	0.378	18.4	29.2	5.13	1.80	1.25	53.6	10.7	2.45
WT 7 × 26.5	7.79	6.97	8.062	0.658	0.370	18.8	27.7	4.96	1.88	1.38	28.8	7.14	1.92
× 24	7.06	6.91	8.031	0.593	0.339	20.4	24.9	4.49	1.88	1.35	25.6	6.38	1.91
× 21.5	6.32	6.84	8.000	0.528	0.308	22.2	22.2	4.02	1.87	1.33	22.6	5.64	1.89
WT 7 × 19	5.59	7.06	6.776	0.513	0.313	22.6	23.5	4.27	2.05	1.55	13.3	3.93	1.54
× 17	5.01	7.00	6.750	0.453	0.287	24.4	21.1	3.87	2.05	1.54	11.6	3.44	1.52
× 15	4.42	6.93	6.733	0.383	0.270	25.7	19.0	3.56	2.08	1.58	9.76	2.90	1.49
WT 7 × 13	3.83	6.95	5.025	0.418	0.255	27.2	17.2	3.30	2.12	1.72	4.43	1.76	1.08
× 11	3.24	6.86	5.000	0.335	0.230	29.8	14.8	2.90	2.13	1.76	3.50	1.40	1.04
WT 6 × 95	27.9	7.19	12.670	1.736	1.060	6.78	79.0	14.2	1.68	1.62	295	46.5	3.25
× 80.5	23.7	6.94	12.515	1.486	0.905	7.67	62.6	11.5	1.63	1.47	243	38.9	3.20
× 66.5	19.6	6.69	12.365	1.236	0.755	8.86	48.4	9.04	1.57	1.33	195	31.5	3.16
× 60	17.7	6.56	12.320	1.106	0.710	9.24	43.4	8.22	1.57	1.28	173	28.0	3.13
× 53	15.6	6.44	12.230	0.986	0.620	10.4	36.7	7.01	1.53	1.20	150	24.6	3.11
× 49.5	14.6	6.38	12.192	0.921	0.582	11.0	33.8	6.48	1.52	1.16	139	22.8	3.09
× 46	13.5	6.31	12.155	0.856	0.545	11.6	31.0	5.99	1.51	1.13	128	21.1	3.08
× 42.5	12.5	6.25	12.105	0.796	0.495	12.6	27.8	5.38	1.49	1.08	118	19.5	3.07
× 39.5	11.6	6.19	12.080	0.736	0.470	13.2	25.8	5.03	1.49	1.06	108	17.9	3.05
× 36	10.6	6.13	12.040	0.671	0.430	14.2	23.2	4.54	1.48	1.02	97.6	16.2	3.04
× 32.5	9.55	6.06	12.000	0.606	0.390	15.5	20.6	4.06	1.47	0.985	87.3	14.6	3.02

STRUCTURAL TEES

Cut from W shapes

Dimensions and
properties for designing

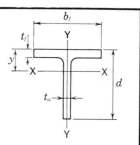

Designation	Area	Depth of Tee d	Flange		Stem Thickness t_w	$\frac{d}{t_w}$	AXIS X - X				AXIS Y - Y		
			Width b_f	Thickness t_f			I	S	r	y	I	S	r
	In.²	In.	In.	In.	In.		In.⁴	In.³	In.	In.	In.⁴	In.³	In.
WT 6 × 29	8.53	6.10	10.014	0.641	0.359	17.0	19.0	3.75	1.49	1.03	53.7	10.7	2.51
× 26.5	7.80	6.03	10.000	0.576	0.345	17.5	17.7	3.54	1.51	1.02	48.0	9.61	2.48
WT 6 × 25	7.36	6.10	8.077	0.641	0.371	16.4	18.7	3.80	1.60	1.17	28.2	6.98	1.96
× 22.5	6.62	6.03	8.042	0.576	0.336	17.9	16.6	3.40	1.59	1.13	25.0	6.22	1.94
× 20	5.89	5.97	8.000	0.516	0.294	20.3	14.4	2.94	1.56	1.08	22.0	5.51	1.94
WT 6 × 18	5.30	6.12	6.565	0.540	0.305	20.1	15.3	3.14	1.70	1.26	12.7	3.88	1.55
× 15.5	4.57	6.05	6.525	0.465	0.265	22.8	13.0	2.69	1.69	1.22	10.8	3.30	1.54
× 13.5	3.97	5.98	6.497	0.400	0.237	25.2	11.3	2.37	1.69	1.20	9.15	2.82	1.52
WT 6 × 11	3.24	6.16	4.030	0.424	0.260	23.7	11.7	2.59	1.90	1.63	2.32	1.15	0.847
× 9.5	2.80	6.08	4.007	0.349	0.237	25.7	10.2	2.30	1.91	1.65	1.88	0.938	0.820
× 8.25	2.43	6.00	4.000	0.269	0.230	26.1	9.03	2.13	1.93	1.76	1.44	0.721	0.770
× 7	2.06	5.96	3.968	0.224	0.198	30.1	7.61	1.81	1.92	1.75	1.17	0.590	0.754
WT 5 × 56	16.5	5.69	10.415	1.248	0.755	7.54	28.8	6.42	1.32	1.21	118	22.6	2.67
× 50	14.7	5.56	10.345	1.118	0.685	8.12	24.8	5.62	1.30	1.14	103	20.0	2.65
× 44.5	13.1	5.44	10.275	0.998	0.615	8.85	21.3	4.88	1.28	1.07	90.3	17.6	2.63
× 38.5	11.3	5.31	10.195	0.868	0.535	9.93	17.7	4.10	1.25	0.996	76.7	15.1	2.60
× 36	10.6	5.25	10.170	0.808	0.510	10.3	16.4	3.83	1.24	0.971	70.9	13.9	2.59
× 33	9.70	5.19	10.117	0.748	0.457	11.4	14.5	3.39	1.22	0.922	64.6	12.8	2.58
× 30	8.83	5.13	10.075	0.683	0.415	12.3	12.8	3.03	1.21	0.882	58.2	11.6	2.57
× 27	7.94	5.06	10.028	0.618	0.368	13.8	11.2	2.64	1.19	0.836	52.0	10.4	2.56
× 24.5	7.20	5.00	10.000	0.558	0.340	14.7	10.1	2.40	1.18	0.809	46.5	9.30	2.54
WT 5 × 22.5	6.62	5.06	8.022	0.618	0.350	14.5	10.3	2.48	1.25	0.910	26.6	6.63	2.00
× 19.5	5.74	4.97	7.990	0.528	0.318	15.6	8.96	2.19	1.25	0.883	22.5	5.62	1.98
× 16.5	4.85	4.88	7.964	0.433	0.292	16.7	7.80	1.95	1.27	0.875	18.2	4.58	1.94
WT 5 × 14.5	4.27	5.11	5.799	0.500	0.289	17.7	8.39	2.07	1.40	1.05	8.14	2.81	1.38
× 12.5	3.68	5.04	5.762	0.430	0.252	20.0	7.13	1.77	1.39	1.01	6.86	2.38	1.37
× 10.5	3.10	4.95	5.750	0.340	0.240	20.6	6.32	1.62	1.43	1.06	5.39	1.88	1.32

STRUCTURAL TEES

Cut from W shapes

Dimensions and properties for designing

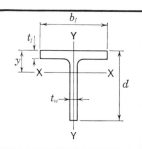

Designation	Area	Depth of Tee d	Flange Width b_f	Flange Thickness t_f	Stem Thickness t_w	$\dfrac{d}{t_w}$	AXIS X-X I	S	r	y	AXIS Y-Y I	S	r
	In.²	In.	In.	In.	In.		In.⁴	In.³	In.	In.	In.⁴	In.³	In.
WT 5 × 9.5	2.81	5.13	4.020	0.394	0.250	20.5	6.70	1.74	1.55	1.28	2.14	1.06	0.874
× 8.5	2.49	5.06	4.010	0.329	0.240	21.1	6.07	1.62	1.56	1.32	1.77	0.885	0.844
× 7.5	2.20	5.00	4.000	0.269	0.230	21.7	5.46	1.51	1.57	1.37	1.44	0.720	0.809
× 5.75	1.70	4.94	3.950	0.204	0.180	27.4	4.16	1.16	1.57	1.34	1.05	0.532	0.787
WT 4 × 33.5	9.85	4.50	8.287	0.933	0.575	7.83	10.9	3.07	1.05	0.939	44.3	10.7	2.12
× 29	8.53	4.38	8.222	0.808	0.510	8.58	9.12	2.61	1.03	0.874	37.5	9.12	2.10
× 24	7.06	4.25	8.117	0.683	0.405	10.5	6.92	2.00	0.990	0.781	30.5	7.51	2.08
× 20	5.88	4.13	8.077	0.558	0.365	11.3	5.80	1.71	0.993	0.740	24.5	6.07	2.04
× 17.5	5.15	4.06	8.027	0.493	0.315	12.9	4.88	1.45	0.973	0.694	21.3	5.30	2.03
× 15.5	4.56	4.00	8.000	0.433	0.288	13.9	4.31	1.30	0.973	0.672	18.5	4.62	2.01
WT 4 × 14	4.11	4.03	6.540	0.463	0.285	14.1	4.22	1.28	1.01	0.735	10.8	3.30	1.62
× 12	3.53	3.97	6.500	0.398	0.245	16.2	3.53	1.08	1.00	0.695	9.12	2.80	1.61
WT 4 × 10	2.95	4.07	5.268	0.378	0.248	16.4	3.67	1.13	1.12	0.825	4.61	1.75	1.25
× 8.5	2.50	4.00	5.250	0.308	0.230	17.4	3.21	1.02	1.13	0.835	3.72	1.42	1.22
WT 4 × 7.5	2.22	4.06	4.015	0.314	0.245	16.6	3.29	1.07	1.22	1.00	1.70	0.847	0.876
× 6.5	1.92	4.00	4.000	0.254	0.230	17.4	2.90	0.976	1.23	1.03	1.36	0.680	0.842
× 5	1.48	3.95	3.940	0.204	0.170	23.2	2.15	0.719	1.21	0.957	1.04	0.529	0.839
WT 3 × 12.5	3.67	3.19	6.080	0.456	0.320	9.95	2.27	0.883	0.787	0.609	8.55	2.81	1.53
× 10	2.94	3.10	6.018	0.367	0.258	12.0	1.75	0.688	0.771	0.557	6.67	2.22	1.51
× 7.75	2.28	3.00	5.995	0.269	0.235	12.8	1.44	0.591	0.795	0.559	4.83	1.61	1.46
WT 3 × 8	2.36	3.13	4.030	0.404	0.260	12.0	1.66	0.679	0.839	0.673	2.21	1.10	0.967
× 6	1.77	3.00	4.000	0.279	0.230	13.0	1.30	0.558	0.857	0.673	1.49	0.746	0.918
× 4.25	1.25	2.92	3.940	0.194	0.170	17.1	0.904	0.397	0.849	0.638	0.990	0.503	0.889
WT 2.5 × 9.25	2.72	2.56	5.025	0.420	0.265	9.66	0.980	0.471	0.601	0.481	4.45	1.77	1.28
× 8	2.35	2.50	5.000	0.360	0.240	10.4	0.840	0.411	0.598	0.457	3.75	1.50	1.26
WT 2 × 6.5	1.91	2.08	4.060	0.345	0.280	7.43	0.526	0.321	0.524	0.440	1.88	0.926	0.991

STRUCTURAL TEES
Cut from M shapes
Dimensions and properties for designing

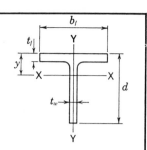

Designation	Area	Depth of Tee d	Flange Width b_f	Flange Thickness t_f	Stem Thickness t_w	$\dfrac{d}{t_w}$	AXIS X-X				AXIS Y-Y		
							I	S	r	y	I	S	r
	In.²	In.	In.	In.	In.		In.⁴	In.³	In.	In.	In.⁴	In.³	In.
MT 7 × 8.6	2.53	7.00	4.000	0.272	0.210	33.3	12.9	2.64	2.26	2.10	1.33	0.663	0.725
MT 6 × 5.9	1.73	6.00	3.065	0.225	0.177	33.9	6.60	1.60	1.95	1.89	0.490	0.320	0.532
MT 5 × 14.55	4.28	4.94	5.937	0.389	0.427	11.6	9.79	2.71	1.51	1.32	5.58	1.88	1.14
× 11.45	3.36	4.94	5.752	0.389	0.242	20.4	6.41	1.63	1.38	1.01	5.01	1.74	1.22
MT 5 × 4.5	1.32	5.00	2.690	0.206	0.157	31.8	3.46	0.997	1.62	1.53	0.305	0.227	0.480
MT 4 × 18.85	5.54	4.06	8.002	0.521	0.377	10.8	5.65	1.71	1.01	0.755	20.2	5.05	1.91
× 17.15	5.04	4.00	8.003	0.459	0.378	10.6	5.38	1.66	1.03	0.766	17.5	4.37	1.86
× 16.3	4.79	4.00	7.940	0.459	0.315	12.7	4.64	1.41	0.984	0.701	17.0	4.29	1.89
MT 4 × 11.25	3.30	4.00	5.395	0.353	0.375	10.7	4.70	1.57	1.19	1.01	3.74	1.39	1.06
× 9.25	2.72	4.00	5.250	0.353	0.230	17.4	3.24	1.01	1.09	0.804	3.41	1.30	1.12
MT 4 × 3.25	0.958	4.00	2.281	0.189	0.135	29.6	1.57	0.556	1.28	1.17	0.172	0.150	0.423
MT 3.5 × 2.75	0.809	3.50	2.080	0.180	0.128	27.3	1.01	0.404	1.12	1.01	0.124	0.120	0.392
MT 3 × 16.875	4.96	3.13	6.114	0.605	0.488	6.40	3.02	1.24	0.779	0.695	10.7	3.49	1.47
× 11.25	3.31	3.00	6.060	0.379	0.372	8.06	2.12	0.898	0.801	0.638	6.19	2.04	1.37
× 10	2.94	3.00	5.938	0.379	0.250	12.0	1.54	0.624	0.724	0.531	5.80	1.95	1.40
MT 3 × 2.2	0.646	3.00	1.844	0.171	0.114	26.3	0.577	0.267	0.945	0.836	0.083	0.090	0.358
MT 2.5 × 9.45	2.78	2.50	5.003	0.416	0.316	7.91	1.05	0.527	0.615	0.511	3.93	1.57	1.19
MT 2 × 8.15	2.40	2.10	3.938	0.472	0.312	6.73	0.587	0.359	0.495	0.465	2.22	1.13	0.962
× 6.9	2.03	2.00	4.000	0.371	0.313	6.39	0.509	0.327	0.501	0.444	1.79	0.894	0.939
× 6.5	1.90	2.00	3.940	0.371	0.254	7.87	0.431	0.271	0.475	0.410	1.68	0.853	0.939

STRUCTURAL TEES

Cut from S shapes

Dimensions and properties for designing

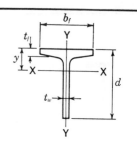

Designation	Area	Depth of Tee d	Flange Width b_f	Flange Thickness t_f	Stem Thickness t_w	$\dfrac{d}{t_w}$	AXIS X-X I	AXIS X-X S	AXIS X-X r	AXIS X-X y	AXIS Y-Y I	AXIS Y-Y S	AXIS Y-Y r
	In.²	In.	In.	In.	In.		In.⁴	In.³	In.	In.	In.⁴	In.³	In.
ST 12 × 60	17.6	12.00	8.048	1.102	0.798	15.0	245	28.9	3.72	3.52	42.1	10.5	1.54
× 52.95	15.6	12.00	7.875	1.102	0.625	19.2	205	23.3	3.63	3.19	39.1	9.92	1.58
ST 12 × 50	14.7	12.00	7.247	0.871	0.747	16.1	215	26.4	3.83	3.84	23.9	6.59	1.27
× 45	13.2	12.00	7.124	0.871	0.624	19.2	190	22.6	3.79	3.60	22.5	6.31	1.30
× 39.95	11.8	12.00	7.001	0.871	0.501	24.0	163	18.7	3.72	3.30	21.1	6.04	1.34
ST 10 × 47.5	14.0	10.00	7.200	0.916	0.800	12.5	137	19.7	3.13	3.07	24.8	6.90	1.33
× 42.5	12.5	10.00	7.053	0.916	0.653	15.3	118	16.6	3.08	2.85	23.1	6.55	1.36
ST 10 × 37.5	11.0	10.00	6.391	0.789	0.641	15.6	110	15.9	3.16	3.08	14.8	4.64	1.16
× 32.7	9.62	10.00	6.250	0.789	0.500	20.0	92.3	12.8	3.10	2.80	13.7	4.38	1.19
ST 9 × 35	10.3	9.00	6.251	0.691	0.711	12.7	84.7	14.0	2.87	2.94	12.1	3.86	1.08
× 27.35	8.04	9.00	6.001	0.691	0.461	19.5	62.4	9.61	2.79	2.50	10.4	3.47	1.14
ST 7.5 × 25	7.35	7.50	5.640	0.622	0.550	13.6	40.6	7.73	2.35	2.25	7.85	2.78	1.03
× 21.45	6.31	7.50	5.501	0.622	0.411	18.2	33.0	6.00	2.29	2.01	7.19	2.61	1.07
ST 6 × 25	7.35	6.00	5.477	0.659	0.687	8.73	25.2	6.05	1.85	1.84	7.85	2.87	1.03
× 20.4	6.00	6.00	5.252	0.659	0.462	13.0	18.9	4.28	1.78	1.58	6.78	2.58	1.06
ST 6 × 17.5	5.14	6.00	5.078	0.544	0.428	14.0	17.2	3.95	1.83	1.65	4.93	1.94	0.980
× 15.9	4.68	6.00	5.000	0.544	0.350	17.1	14.9	3.31	1.78	1.51	4.68	1.87	1.00
ST 5 × 17.5	5.15	5.00	4.944	0.491	0.594	8.42	12.5	3.63	1.56	1.56	4.18	1.69	0.901
× 12.7	3.73	5.00	4.661	0.491	0.311	16.1	7.83	2.06	1.45	1.20	3.39	1.46	0.954
ST 4 × 11.5	3.38	4.00	4.171	0.425	0.441	9.07	5.03	1.77	1.22	1.15	2.15	1.03	0.798
× 9.2	2.70	4.00	4.001	0.425	0.271	14.8	3.51	1.15	1.14	0.941	1.86	0.932	0.831
ST 3.5 × 10	2.94	3.50	3.860	0.392	0.450	7.78	3.36	1.36	1.07	1.04	1.59	0.821	0.734
× 7.65	2.25	3.50	3.662	0.392	0.252	13.9	2.19	0.816	0.987	0.817	1.32	0.720	0.766
ST 3 × 8.625	2.53	3.00	3.565	0.359	0.465	6.45	2.13	1.02	0.917	0.914	1.15	0.648	0.675
× 6.25	1.83	3.00	3.332	0.359	0.232	12.9	1.27	0.552	0.833	0.691	0.911	0.547	0.705
ST 2.5 × 7.375	2.17	2.50	3.284	0.326	0.494	5.06	1.27	0.740	0.764	0.789	0.833	0.507	0.620
× 5	1.47	2.50	3.004	0.326	0.214	11.7	0.681	0.353	0.681	0.569	0.608	0.405	0.643
ST 2 × 4.75	1.40	2.00	2.796	0.293	0.326	6.13	0.470	0.325	0.580	0.553	0.451	0.323	0.569
× 3.85	1.13	2.00	2.663	0.293	0.193	10.4	0.316	0.203	0.528	0.448	0.382	0.287	0.581
ST 1.5 × 3.75	1.10	1.50	2.509	0.260	0.349	4.30	0.204	0.191	0.430	0.432	0.293	0.234	0.516
× 2.85	0.835	1.50	2.330	0.260	0.170	8.82	0.118	0.101	0.376	0.329	0.227	0.195	0.522

PIPE
Dimensions and properties

	Dimension			Weight per Foot Lbs. Plain Ends	Properties			
Nominal Diameter In.	Outside Diameter In.	Inside Diameter In.	Wall Thickness In.		A In.²	I In.⁴	S In.³	r In.
Standard Weight								
½	.840	.622	.109	.85	.250	.017	.041	.261
¾	1.050	.824	.113	1.13	.333	.037	.071	.334
1	1.315	1.049	.133	1.68	.494	.087	.133	.421
1¼	1.660	1.380	.140	2.27	.669	.195	.235	.540
1½	1.900	1.610	.145	2.72	.799	.310	.326	.623
2	2.375	2.067	.154	3.65	1.07	.666	.561	.787
2½	2.875	2.469	.203	5.79	1.70	1.53	1.06	.947
3	3.500	3.068	.216	7.58	2.23	3.02	1.72	1.16
3½	4.000	3.548	.226	9.11	2.68	4.79	2.39	1.34
4	4.500	4.026	.237	10.79	3.17	7.23	3.21	1.51
5	5.563	5.047	.258	14.62	4.30	15.2	5.45	1.88
6	6.625	6.065	.280	18.97	5.58	28.1	8.50	2.25
8	8.625	7.981	.322	28.55	8.40	72.5	16.8	2.94
10	10.750	10.020	.365	40.48	11.9	161	29.9	3.67
12	12.750	12.000	.375	49.56	14.6	279	43.8	4.38
Extra Strong								
½	.840	.546	.147	1.09	.320	.020	.048	.250
¾	1.050	.742	.154	1.47	.433	.045	.085	.321
1	1.315	.957	.179	2.17	.639	.106	.161	.407
1¼	1.660	1.278	.191	3.00	.881	.242	.291	.524
1½	1.900	1.500	.200	3.63	1.07	.391	.412	.605
2	2.375	1.939	.218	5.02	1.48	.868	.731	.766
2½	2.875	2.323	.276	7.66	2.25	1.92	1.34	.924
3	3.500	2.900	.300	10.25	3.02	3.89	2.23	1.14
3½	4.000	3.364	.318	12.50	3.68	6.28	3.14	1.31
4	4.500	3.826	.337	14.98	4.41	9.61	4.27	1.48
5	5.563	4.813	.375	20.78	6.11	20.7	7.43	1.84
6	6.625	5.761	.432	28.57	8.40	40.5	12.2	2.19
8	8.625	7.625	.500	43.39	12.8	106	24.5	2.88
10	10.750	9.750	.500	54.74	16.1	212	39.4	3.63
12	12.750	11.750	.500	65.42	19.2	362	56.7	4.33
Double-Extra Strong								
2	2.375	1.503	.436	9.03	2.66	1.31	1.10	.703
2½	2.875	1.771	.552	13.69	4.03	2.87	2.00	.844
3	3.500	2.300	.600	18.58	5.47	5.99	3.42	1.05
4	4.500	3.152	.674	27.54	8.10	15.3	6.79	1.37
5	5.563	4.063	.750	38.55	11.3	33.6	12.1	1.72
6	6.625	4.897	.864	53.16	15.6	66.3	20.0	2.06
8	8.625	6.875	.875	72.42	21.3	162	37.6	2.76

The listed sections are available in conformance with ASTM Specification A53 Grade B or A501. Other sections are made to these specifications. Consult with pipe manufacturers or distributors for availability.

AMERICAN INSTITUTE OF STEEL CONSTRUCTION

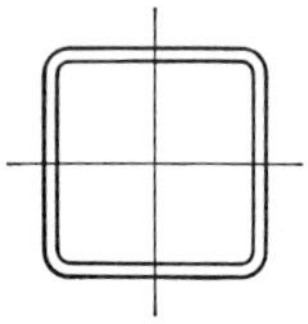

STRUCTURAL TUBING
Square
Dimensions and properties

	DIMENSIONS			PROPERTIES			
Nominal* Size	Wall Thickness		Weight per Foot	Area	I	S	r
In.	In.		Lb.	In.²	In.⁴	In.³	In.
10 × 10	.6250	⅝	73.98	21.8	304.	60.7	3.74
	.5000	½	60.95	17.9	260.	52.0	3.81
	.3750	⅜	47.03	13.8	208.	41.7	3.88
	.3125	5⁄16	‡39.74	11.7	179.	35.8	3.92
	.2500	¼	†32.23	9.48	148.	29.6	3.95
	.1875	3⁄16	†24.50	7.21	114.	22.9	3.98
8 × 8	.6250	⅝	56.98	16.8	142.	35.5	2.91
	.5000	½	47.35	13.9	124.	31.1	2.99
	.3750	⅜	36.83	10.8	102.	25.4	3.06
	.3125	5⁄16	31.24	9.19	88.1	22.0	3.10
	.2500	¼	‡25.44	7.48	73.4	18.4	3.13
	.1875	3⁄16	†19.41	5.71	57.2	14.3	3.17
7 × 7	.5000	½	40.55	11.9	79.2	22.6	2.58
	.3750	⅜	31.73	9.33	65.6	18.8	2.65
	.3125	5⁄16	26.99	7.94	57.4	16.4	2.69
	.2500	¼	22.04	6.48	48.1	13.7	2.72
	.1875	3⁄16	†16.85	4.96	37.7	10.8	2.76
6 × 6	.5000	½	34.48	10.1	48.6	16.2	2.19
	.3750	⅜	27.04	7.95	40.5	13.5	2.26
	.3125	5⁄16	23.02	6.77	35.5	11.8	2.29
	.2500	¼	18.82	5.54	29.9	9.95	2.32
	.1875	3⁄16	‡14.41	4.24	23.5	7.83	2.35
5 × 5	.5000	½	27.68	8.14	25.7	10.3	1.78
	.3750	⅜	21.94	6.45	22.0	8.80	1.85
	.3125	5⁄16	18.77	5.52	19.5	7.81	1.88
	.2500	¼	15.42	4.54	16.6	6.64	1.91
	.1875	3⁄16	11.86	3.49	13.2	5.28	1.95
4 × 4	.5000	½	20.88	6.14	11.4	5.70	1.36
	.3750	⅜	16.84	4.95	10.2	5.10	1.44
	.3125	5⁄16	14.52	4.27	9.23	4.61	1.47
	.2500	¼	12.02	3.54	8.00	4.00	1.50
	.1875	3⁄16	9.31	2.74	6.47	3.24	1.54
3½ × 3½	.2500	¼	10.50	3.09	5.29	3.02	1.31
	.1875	3⁄16	8.14	2.39	4.29	2.45	1.34
3 × 3	.2500	¼	8.80	2.59	3.16	2.10	1.10
	.1875	3⁄16	6.86	2.02	2.60	1.73	1.13
2 × 2	.2500	¼	5.40	1.59	.766	.766	.694
	.1875	3⁄16	4.31	1.27	.668	.668	.726

* Outside dimensions across flat sides.
† Non-compact section for $F_y = 36$ ksi and $F_y = 46$ ksi, bending only.
‡ Non-compact section for $F_y = 46$ ksi, bending only.
Sections subjected to axial compression or compression due to bending should be checked for compliance with Specification Sect. 1.9.2.2.

STRUCTURAL TUBING
Rectangular
Dimensions and properties

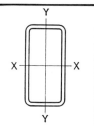

	DIMENSIONS				PROPERTIES					
					X - X AXIS			Y - Y AXIS		
Nominal* Size	Wall Thickness		Weight per Foot	Area	I_x	S_x	r_x	I_y	S_y	r_y
In.	In.		Lb.	In.²	In.⁴	In.³	In.	In.⁴	In.³	In.
12 × 8	.5000	½	60.95	17.9	337.	56.2	4.34	181.	45.2	3.18
	.3750	⅜	47.03	13.8	270.	45.0	4.42	145.	36.3	3.24
	.3125	5⁄16	39.74	11.7	232.	38.7	4.46	125.	31.3	3.27
	.2500	¼	‡32.23	9.48	192.	32.0	4.50	103.	25.9	3.30
	.1875	3⁄16	†24.50	7.21	148.	24.7	4.54	80.1	20.0	3.33
12 × 6	.5000	½	54.15	15.9	271.	45.2	4.13	92.0	30.7	2.40
	.3750	⅜	41.93	12.3	220.	36.6	4.22	75.0	25.0	2.47
	.3125	5⁄16	35.49	10.4	190.	31.6	4.26	65.1	21.7	2.50
	.2500	¼	28.83	8.48	157.	26.2	4.31	54.2	18.1	2.53
	.1875	3⁄16	‡21.96	6.46	122.	20.4	4.35	42.2	14.1	2.56
12 × 4	.5000	½	47.35	13.9	205.	34.2	3.84	35.2	17.6	1.59
	.3750	⅜	36.83	10.8	169.	28.1	3.95	29.5	14.7	1.65
	.3125	5⁄16	31.24	9.19	147.	24.5	4.00	25.9	13.0	1.68
	.2500	¼	25.44	7.48	123.	20.5	4.05	21.9	10.9	1.71
	.1875	3⁄16	‡19.41	5.71	96.0	16.0	4.10	17.3	8.63	1.74
12 × 2	.3750	⅜	31.73	9.33	118.	19.7	3.56	5.62	5.62	.776
	.3125	5⁄16	26.99	7.94	104.	17.4	3.62	5.14	5.14	.805
	.2500	¼	22.04	6.48	88.3	14.7	3.69	4.51	4.51	.834
	.1875	3⁄16	‡16.85	4.96	69.8	11.6	3.75	3.70	3.70	.863
10 × 8	.5000	½	54.15	15.9	215.	43.0	3.67	153.	38.1	3.10
	.3750	⅜	41.93	12.3	174.	34.7	3.75	123.	30.8	3.16
	.3125	5⁄16	35.49	10.4	150.	30.0	3.79	107.	26.7	3.20
	.2500	¼	‡28.83	8.48	124.	24.8	3.83	88.4	22.1	3.23
	.1875	3⁄16	†21.96	6.46	96.3	19.3	3.86	68.7	17.2	3.26
10 × 6	.5000	½	47.35	13.9	170.	34.0	3.49	76.9	25.6	2.35
	.3750	⅜	36.83	10.8	139.	27.8	3.58	63.1	21.0	2.41
	.3125	5⁄16	31.24	9.19	120.	24.1	3.62	55.0	18.3	2.45
	.2500	¼	25.44	7.48	100.	20.1	3.66	45.9	15.3	2.48
	.1875	3⁄16	‡19.41	5.71	78.3	15.7	3.70	35.9	12.0	2.51

* Outside dimensions across flat sides.
† Non-compact section for $F_y = 36$ ksi and $F_y = 46$ ksi, when bending occurs about X - X axis.
‡ Non-compact section for $F_y = 46$ ksi, when bending occurs about X - X axis.
Shapes subjected to combined axial load and bending may not be compact under Specification Sect. 1.5.1.4.1. Check all shapes for compliance with this section.
Shapes subjected to axial compression or compression due to bending should be checked for compliance with Specification Sect. 1.9.2.2.
Shapes subjected to bending about the Y - Y axis may not be compact under Specification Sect. 1.5.1.4.1.

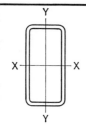

STRUCTURAL TUBING
Rectangular
Dimensions and properties

Nominal* Size	Wall Thickness		Weight per Foot	Area	X - X AXIS			Y - Y AXIS		
					I_x	S_x	r_x	I_y	S_y	r_y
In.	In.		Lb.	In.²	In.⁴	In.³	In.	In.⁴	In.³	In.
10 × 4	.5000	½	40.55	11.9	125.	24.9	3.23	29.0	14.5	1.56
	.3750	⅜	31.73	9.33	104.	20.8	3.34	24.5	12.3	1.62
	.3125	⁵⁄₁₆	26.99	7.94	91.2	18.2	3.39	21.7	10.8	1.65
	.2500	¼	22.04	6.48	76.6	15.3	3.44	18.3	9.17	1.68
	.1875	³⁄₁₆	16.85	4.96	60.2	12.0	3.48	14.5	7.26	1.71
10 × 2	.3750	⅜	27.04	7.95	72.4	14.5	3.02	4.74	4.74	.772
	.3125	⁵⁄₁₆	23.02	6.77	64.0	12.8	3.07	4.33	4.33	.800
	.2500	¼	18.82	5.54	54.2	10.8	3.13	3.80	3.80	.828
	.1875	³⁄₁₆	14.41	4.24	42.9	8.58	3.18	3.11	3.11	.857
8 × 6	.5000	½	40.55	11.9	96.2	24.1	2.84	61.7	20.6	2.27
	.3750	⅜	31.73	9.33	79.7	19.9	2.92	51.2	17.1	2.34
	.3125	⁵⁄₁₆	26.99	7.94	69.7	17.4	2.96	44.8	14.9	2.38
	.2500	¼	22.04	6.48	58.4	14.6	3.00	37.6	12.5	2.41
	.1875	³⁄₁₆	‡16.85	4.96	45.8	11.4	3.04	29.6	9.85	2.44
8 × 4	.5000	½	34.48	10.1	71.6	17.9	2.66	23.7	11.9	1.53
	.3750	⅜	27.04	7.95	59.9	15.0	2.74	20.1	10.0	1.59
	.3125	⁵⁄₁₆	23.02	6.77	52.6	13.1	2.79	17.7	8.87	1.62
	.2500	¼	18.82	5.54	44.2	11.1	2.83	15.0	7.52	1.65
	.1875	³⁄₁₆	14.41	4.24	34.8	8.71	2.87	11.9	5.96	1.68
8 × 3	.5000	½	31.08	9.14	57.6	14.4	2.51	11.6	7.74	1.13
	.3750	⅜	24.49	7.20	49.0	12.3	2.61	10.1	6.74	1.18
	.3125	⁵⁄₁₆	20.90	6.15	43.3	10.8	2.65	9.05	6.04	1.21
	.2500	¼	17.12	5.04	36.7	9.18	2.70	7.77	5.18	1.24
	.1875	³⁄₁₆	13.13	3.86	29.1	7.28	2.74	6.24	4.16	1.27
8 × 2	.3750	⅜	21.94	6.45	38.1	9.52	2.43	3.73	3.73	.760
	.3125	⁵⁄₁₆	18.77	5.52	34.1	8.52	2.48	3.43	3.43	.788
	.2500	¼	15.42	4.54	29.2	7.31	2.54	3.03	3.03	.817
	.1875	³⁄₁₆	11.86	3.49	23.4	5.85	2.59	2.49	2.49	.845
7 × 5	.5000	½	34.48	10.1	60.8	17.4	2.45	35.9	14.3	1.88
	.3750	⅜	27.04	7.95	50.7	14.5	2.52	30.0	12.0	1.94
	.3125	⁵⁄₁₆	23.02	6.77	44.4	12.7	2.56	26.4	10.6	1.97
	.2500	¼	18.82	5.54	37.4	10.7	2.60	22.3	8.90	2.00
	.1875	³⁄₁₆	14.41	4.24	29.4	8.40	2.63	17.6	7.02	2.04

* Outside dimensions across flat sides.

‡ Non-compact section for $F_y = 46$ ksi, when bending occurs about X - X axis.

Shapes subjected to combined axial load and bending may not be compact under Specification Sect. 1.5.1.4.1.   Check all shapes for compliance with this section.

Shapes subjected to axial compression or compression due to bending should be checked for compliance with Specification Sect. 1.9.2.2.

Shapes subjected to bending about the Y - Y axis may not be compact under Specification Sect. 1.5.1.4.1.

STRUCTURAL TUBING
Rectangular
Dimensions and properties

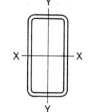

	DIMENSIONS				PROPERTIES					
Nominal* Size	Wall Thickness		Weight per Foot	Area	X - X AXIS			Y - Y AXIS		
					I_x	S_x	r_x	I_y	S_y	r_y
In.	In.		Lb.	In.²	In.⁴	In.³	In.	In.⁴	In.³	In.
6 × 4	.5000	½	27.68	8.14	33.4	11.1	2.02	17.6	8.79	1.47
	.3750	⅜	21.94	6.45	28.6	9.54	2.11	15.2	7.58	1.53
	.3125	5⁄16	18.77	5.52	25.4	8.46	2.14	13.5	6.74	1.56
	.2500	¼	15.42	4.54	21.6	7.19	2.18	11.5	5.76	1.59
	.1875	3⁄16	11.86	3.49	17.2	5.72	2.22	9.20	4.60	1.62
6 × 3	.5000	½	24.28	7.14	25.8	8.60	1.90	8.44	5.63	1.09
	.3750	⅜	19.39	5.70	22.7	7.56	1.99	7.51	5.01	1.15
	.3125	5⁄16	16.65	4.90	20.3	6.77	2.04	6.79	4.52	1.18
	.2500	¼	13.72	4.04	17.4	5.82	2.08	5.88	3.92	1.21
	.1875	3⁄16	10.58	3.11	14.0	4.66	2.12	4.76	3.17	1.24
6 × 2	.3750	⅜	16.84	4.95	16.7	5.57	1.84	2.72	2.72	.741
	.3125	5⁄16	14.52	4.27	15.3	5.08	1.89	2.53	2.53	.770
	.2500	¼	12.02	3.54	13.3	4.44	1.94	2.25	2.25	.799
	.1875	3⁄16	9.31	2.74	10.8	3.61	1.99	1.87	1.87	.827
5 × 3	.5000	½	20.88	6.14	15.5	6.21	1.59	6.86	4.57	1.06
	.3750	⅜	16.84	4.95	14.0	5.58	1.68	6.21	4.14	1.12
	.3125	5⁄16	14.52	4.27	12.6	5.06	1.72	5.65	3.77	1.15
	.2500	¼	12.02	3.54	11.0	4.38	1.76	4.93	3.29	1.18
	.1875	3⁄16	9.31	2.74	8.87	3.55	1.80	4.02	2.68	1.21
5 × 2	.2500	¼	10.50	3.09	8.48	3.39	1.66	1.92	1.92	.789
	.1875	3⁄16	8.14	2.39	6.89	2.75	1.70	1.60	1.60	.816
4 × 3	.2500	¼	10.50	3.09	6.45	3.23	1.45	4.10	2.74	1.15
	.1875	3⁄16	8.14	2.39	5.23	2.62	1.48	3.34	2.23	1.18
4 × 2	.2500	¼	8.80	2.59	4.69	2.35	1.35	1.54	1.54	.770
	.1875	3⁄16	6.86	2.02	3.87	1.93	1.38	1.29	1.29	.798
3 × 2	.2500	¼	7.10	2.09	2.21	1.47	1.03	1.15	1.15	.742
	.1875	3⁄16	5.59	1.64	1.86	1.24	1.06	.977	.977	.771

* Outside dimensions across flat sides.

Shapes subjected to combined axial load and bending may not be compact under Specification Sect. 1.5.1.4.1. Check all shapes for compliance with this section.

Shapes subjected to axial compression or compression due to bending should be checked for compliance with Specification Sect. 1.9.2.2.

Shapes subjected to bending about the Y - Y axis may not be compact under Specification Sect. 1.5.1.4.1.

1-2 R_A = 16 KIPS
BC = 12 KIPS TENS.
R_D = 20 KIPS

1-4

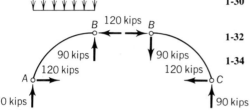

1-6

1-8 $R_A \downarrow$ = 1.40 P
$R_{BH} \rightarrow$ = 0.80 P
$R_{BV} \downarrow$ = 2.15 P

1-10 X = 2 m

1-12 R_E = 6.33 kN $\leftarrow$; R_A = 10.20 kN
AB = 6.33 kN comp; BC = 9.50 kN comp.
DC = 11.41 kN comp.
ED = 0; DB = 6.33 kN comp.

1-14 F_{TIE} = 450 kN (TENS)
R_A = 725.6 kN $\searrow$; R_C = 810 kN $\uparrow$

1-16 $\sigma = \dfrac{AL\gamma}{A} = L\gamma$
L_{max} = 7940 ft

1-18 A = 3.72 in.²

1-20 t = 0.459 in.

1-22 d_{link} = 0.874 in.
d_{bolt} = 0.874 in.

1-24 σ_{AB} = 1639.95 psi (tens.)
σ_{BC} = 1421.28 psi (tens.)

1-26 L = 4 in.

1-28 (a) d_{PIN} = 0.313 in.
(b) d_{PIN} = 0.402 in.

1-30 A_{AC} = ½ in.²
A_{BC} = 0.808 in.²

1-32 A_{TIE} = 6 in.²

1-34 ϵ_1 = 0.20 in./in.
ϵ_2 = 0.26 in./in. ϵ_{avg} = 0.295 in./in.
ϵ_3 = 0.40 in./in.
ϵ_4 = 0.32 in./in.

1-36 ϵ_{BC} = 0.00786 in./in.

1-38 $\gamma = \delta$

1-40 P = 6010 lb

1-42 (a) x = 0.857 l + 160.7l^2/P
(b) x = 0.462 l − 432.7l^2/P

1-44 (a) P = 2944 lb
(b) δ = 0.0209 ft

1-46

Memb.	Stress (psi)	δ (inches)
AB	28,100 Tens.	0.225 Elong.
BC	18,800 Tens.	0.150 Elong.
CD	15,000 Comp.	0.096 Comp.
BD	11,200 Tes.	0.054 Elong.
DE	15,000 Comp.	0.096 Comp.
BE	9,400 Comp.	0.075 Comp.
AE	5,600 Tens.	0.054 Elong.

1-48 δ_{tot} = 0.00428 ft

1-50 δ_C = 5.10 × 10⁻³m ↓

1-52 δ_{BV} = 98 × 10⁻⁴m

1-54 δ_C = 0.0017 in.

1-56 ϵ_A = 8.33 × 10⁻⁶in./in.; ϵ_C = 5 × 10⁻⁶in./in.
δ_B = 179.4 × 10⁻⁶in.

1-58 δ_C = 0.021 in.

1-60 δ_{CD} = 0.167 $\dfrac{P}{AE}$

δ_{BC} = 0.75 $\dfrac{P}{AE}$

1-62 t = 1.25 in.

1-64 $\epsilon_z = -\mu\epsilon_x - \mu\epsilon_y = -\mu(\epsilon_x + \epsilon_y)$

but $\epsilon_x = \sigma_x/E$ and $\epsilon_y = \sigma_y/E$

therefore, $\epsilon_z = \dfrac{-\mu}{E}(\sigma_x + \sigma_y)$

1-66 $\sigma_t = 37,900$ psi
$\sigma_y = 26,400$ psi

1-68 $P = 197,133$ lb

1-70 $x = {}^5/_6 b$

1-72 $W = 75,400$ lb

1-74 $n = 0.199$ turns

1-76 $A = 1.8$ in.²

1-78 $\sigma_s = 10,300$ psi
$\sigma_w = 1290$ psi

1-80 $\delta_D = 0.0252$ in.

1-82 $x = 8$ in.
$P = 23,550$ lb

1-84 $d_{min} = 13.9$ mm

1-86 $\sigma_1 = 2313$ psi
$\sigma_2 = 1205$ psi
$\sigma_3 = 648$ psi

1-88 $\delta = 6.95 \times 10^{-3}$ in.

1-90 $A_{min} = 331 \times 10^{-6}$ m²

1-92 $P = 10,400$ lb

1-94 $\sigma = 145.6 \cdot \Delta t$ psi

1-96 $\sigma_A = 9700$ psi
$\sigma_S = 19,350$ psi

1-98 $\sigma_A = 4270$ psi
$\sigma_S = 3780$ psi

1-100 $\sigma_S = \sigma_A = 166.2$ MN/m²
$\sigma_C = 83.1$ MN/m²

1-102 $\Delta T = 19.2°$F
$\sigma = 3744$ psi (tens)

1-104 $\sigma_1 = \sigma_3 = 19,850$ psi
$\sigma_2 = 300$ psi

1-106 $\sigma_s = 3700$ psi
$\sigma_A = 7400$ psi

1-108 $\sigma_s = 12,660$ psi
$\sigma_A = 8670$ psi

1-110 Temp. range $= T_0 \pm 96.9°$F

1-112 (a) $\delta_{load} = 54.10 \times 10^{-6}$m
(b) $\delta_{LD+temp} = 123.52 \times 10^{-6}$m

1-114 $\sigma_n = 7500$ psi
$\tau = 7500$ psi

1-116 $P = 10,200$ lb

1-118 $\sigma_n = 36$ MN/m²
$\tau = 48$ MN/m²

1-120 $\sigma = 9000$ psi
$\tau = 3000$ psi

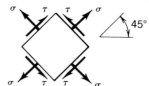

1-122

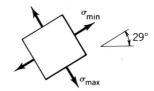

1-124 $\sigma_n = -1044$ psi
$\tau = 5934$ psi

1-126 $\sigma_n = -2160$ psi
$\tau = 6880$ psi

1-128

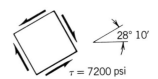

1-130

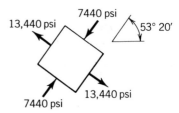

1-132

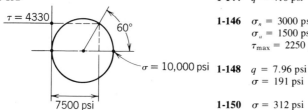

τ = 4330

60°

σ = 10,000 psi

7500 psi

1-134

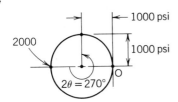

← 1000 psi

2000

1000 psi

$2\theta = 270°$

O

1-136

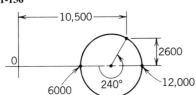

← 10,500 →

0

2600

6000   240°   12,000

1-138

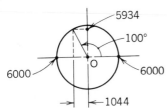

5934

100°

6000   O   6000

1044

1-140

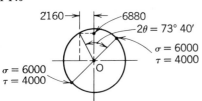

2160 →| |← 6880

$2\theta = 73° 40'$

σ = 6000
τ = 4000

σ = 6000
τ = 4000

O

1-142

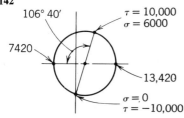

106° 40'

τ = 10,000
σ = 6000

7420

13,420

σ = 0
τ = −10,000

1-144   q = 416 psi

1-146 σ_n = 3000 psi
σ_a = 1500 psi
τ_{max} = 2250 psi

1-148   q = 7.96 psi
σ = 191 psi

1-150   σ = 312 psi

1-152   σ = 27,250

1-154

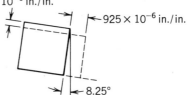

125×10^{-6} in./in.

925×10^{-6} in./in.

8.25°

1-156

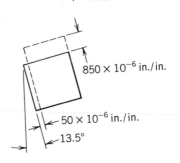

850×10^{-6} in./in.

50×10^{-6} in./in.

13.5°

1-158 $\epsilon_x = \epsilon_a$
$\epsilon_y = \epsilon_c$
$\gamma_{xy} = \epsilon_a + \epsilon_c - 2\epsilon_b$

1-160 $\epsilon_{max} = \dfrac{\epsilon_a + \epsilon_c}{2} + \frac{1}{2}\sqrt{2\epsilon_a(\epsilon_a - 2\epsilon_b)^2}$

$+ \frac{1}{2}\sqrt{2\epsilon_c(\epsilon_c - 2\epsilon_b) + 4\epsilon_b^2}$

2-2   d = 2.6 in.

2-4 T_s = 1.69T_t
W_s = 2.78W_t

2-6   (H.P.)$_s$ = 1.33 (H.P.)$_t$

2-8 τ_{AB} = 101.86 MN/m²
τ_{BC} = 117.89 MN/m²
θ_{AC} = 149.69 × 10⁻³ Rad

2-10   (a) T = 1.71 kN-m
(b) T = 0.966 kN-m

2-12 H.P. $= 588$
$\tau_{AB} = 6000$
$\tau_{BC} = 4330$

2-14 $T_1 = 15{,}700$ in.-lb
$T_2 = 68{,}700$ in.-lb
$T_3 = 178{,}600$ in.-lb

2-16 $T_1 = 88{,}000$ in.-lb
$T_2 = 150{,}000$ in.-lb

2-18 (a) $T_B = 60$ kN-m
$\tau_{AB} = 199$ MN/m²
(b) $\tau_{BC} = 203.73$ MN/m²
$\tau_{CD} = 235.85$ MN/m²

2-20 $(H.P.)_A = 1200$
$(H.P.)_B = 2800$
$(H.P.)_C = 800$

2-22

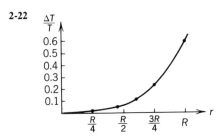

2-24 $\delta_C = 71.62 \times 10^{-2}$ in.

2-26 $\tau_{(2\text{-}5)} = 24{,}250$ psi
$\tau_{(2\text{-}6)} = 26{,}650$ psi
$\delta = 2.92$ in.

2-28 $\delta_C = 27.2$ in.

2-30 $P = 935$ lb
$\delta = 6.38$ in.

2-32 $(n_S/n_{P.B.}) = 2$
$(K_S/K_{P.B.}) = 1$

2-34 $x = {}^2\!/_5\, l$

2-36 $\delta = 18.9$ in. $\sigma_S = 50{,}100$ psi
$P_{P.B.} = 49.6$ percent $\sigma_{P.B.} = 40{,}000$ psi
$P_S = 50.4$ percent

2-38 $\tau = 283$ ksi
$\delta_A = 100.7$ in.

2-40 $P = 170$ lb in each, comp. on left and tens. on right.

2-42 $\tau_S = 7060$ psi
$\tau_A = 3100$ psi
$\phi = 0.0186$ Rad.

2-44 $\tau_S = 2235$ psi
$\tau_A = 436$ psi
$\phi_{A\text{-}S} = 0.0140$ Rad.

2-46 $\tau_{AL} = 1963$ psi
$\tau_{BR} = 1507$ psi

2-48 $\sigma_A = 508$ MN/m²
$\sigma_B = 507$ MN/m²
$\theta_C = 0.127$ Rad.

2-50 Steel 75 percent
Aluminum 25 percent

2-52 $T_B = 159{,}300$ in.-lb

2-54 $L_{BC} = 25$ in.
$T_B = 176{,}000$ in.-lb

2-56 $T_B = 2{,}891{,}000$ in.-lb

2-58 $P_a = 233$ lb
$P_b = 334$ lb
$P_c = 433$ lb
$\delta_{1000} = 2.48$ in.

2-60 $T = 4500$ in.-lb.

2-62 $\tau = 5100$ psi
$\phi = 0.00103$ Rad./ft

2-64 $\tau = 2540$ psi
$\phi = 0.000383$ Rad./ft

2-66 $\tau = 1450$ psi
$\phi = 0.000129$ Rad./ft.

2-68 *Circle* is 1.27 times as strong as square.

2-70 $T = 30{,}200$ in.-lb
$\phi = ?$

2-72 $\tau = 3080$ psi

2-74 (a) $\tau_\square = 7200$ psi $\tau_0 = 8480$ psi
(b) $\tau_\square = 900$ psi $\tau_0 = 1060$ psi
(c) $\tau_\square = 462$ psi $\tau_0 = 545$ psi

2-76 (a) $\tau_\Delta/\tau_0 = 15.7$
(b) $\phi_\Delta/\phi_0 = 0.376$
(c) $W_\Delta/W_0 = 0.546$

3-2 $V_A = \dfrac{+wl}{2}$ $V_{/} = \dfrac{+wl}{8}$
$M_A = \dfrac{-wl^2}{6}$ $M_{/} = \dfrac{-wl^2}{48}$

3-4 $V_A = 14$ kips $V_B = 8$ kips
$M_A = -75$ kips-ft $M_B = -9$ kips-ft

3-6 $V_x = \dfrac{w_B l}{6} - \dfrac{w_B x^2}{2l}$

$\quad\quad M_x = \dfrac{w_B l x}{6} - \dfrac{w_B x^3}{6l}$

3-8 $V_C = 0 \quad\quad\quad V_{12\,ft} = 0$

$\quad\quad M_C = +19.2$ kips-ft $M_{12ft} = +19.2$ kips-ft

3-10 $x = 0.25 \cdot l$

3-12 $P = 10$ kips

$\quad\quad V_B = -5.8$ kips

3-14 $V_5 = +1.95$ kips $\quad V_C = 4$ kips

$\quad\quad M_5 = +12.87$ kips-ft

Problems 3-16 to 3-30 are construction problems.

3-32 $L = 7.8$ ft

3-34 8 WF @ 28 lb/ft is one possibility.

3-36 $\sigma_{max} = 5271$ psi

3-38 $\sigma_t = 6314\,w$

$\quad\quad \sigma_c = 9876\,w$

3-40 $\sigma_{max} = 17.4$ MN/m²

3-42 $\sigma = 15,000$ psi

3-44 $P = 1020$ lb

3-46 8 WF @ 17 lb/ft is one possibility

3-48 $d = 126.5$ mm

3-50 $M_{(a)} = 24,000$ in.-lb

$\quad\quad M_{(b)} = 8000$ in.-lb

3-52 $\sigma_{max} = 26.6$ MN/m²

3-54 $l = 5.9$ ft

$\quad\quad \tau = 168$ psi

$\quad\quad \sigma = 850$ psi

3-56 $P = 1970$ lb

$\quad\quad \tau = 544$ psi

$\quad\quad \sigma = 20,000$ psi

3-58 14 WF @ 34 lb/ft is one possibility.

3-60 $\tau_{max} = 965$ psi

3-62 $w_B = 2110$ lb/ft

3-64 Force/in. = 1140 lb/in.

3-66 $P = 1020$ lb

3-68 $l = 19.91$ ft

3-70 $c_1 = 8.22$ in. (from left)

3-72 Proof

3-74 $\dfrac{M_P}{M_e} = 1.70$

3-76 $M_{PS} = 460,000$ in.-lb $\left(\dfrac{M_P}{M_e}\right)_s = 1.44$

$\quad\quad M_{PR} = 480,000$ in.-lb $\left(\dfrac{M_P}{M_e}\right)_r = 1.50$

3-78 $P = 3333$ lb

3-80 $w = 9620$ lb/ft

3-82 $M_{(a)} = 168,000$ in.-lb

$\quad\quad M_{(b)} = 630,000$ in.-lb

3-84 $M_{(a)} = 42,600$ in.-lb

$\quad\quad M_{(b)} = 37,200$ in.-lb

3-86 $\sigma_w = 682$ psi

$\quad\quad \sigma_s = 8182$ psi

3-88 $w = \dfrac{21,000}{l^2}$ lb/ft

3-90 $w_{LL} = 2030$ lb/ft

3-92 $P = 18,050$ lb

3-94 $b = 8.4$ in.

$\quad\quad A_s = 1.44$ in.²

3-96

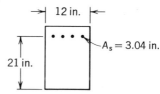

one possibility

3-98 $\sigma_t = 12,762$ psi

$\quad\quad \sigma_c = 7300$ psi

3-100 $\sigma_t = 14,800$ psi

$\quad\quad \sigma_c = 8570$ psi

3-102 $\sigma_t = 15{,}400$ psi
$\sigma_c = 17{,}050$ psi

3-104 $P = 1525$ lb

3-106 $\sigma_1 = 5800$ psi comp.
$\sigma_2 = 23{,}400$ psi tens.
$\tau_{max} = 14{,}600$ psi

3-108 $\sigma_1 = 20$ psi comp.
$\sigma_2 = 170$ psi comp.
$\tau = 95$ psi

3-110 $P = 41{,}000$ lb

3-112 $\sigma_1 = 245$ psi comp.
$\sigma_2 = 3860$ psi tens.
$\tau_{max} = 2060$ psi

3-114 $\sigma_{min} = 207$ psi
$\sigma_{max} = 4340$ psi
$\tau_{max} = 2280$ psi

3-116 $\sigma_{min} = 12.5$ psi comp.
$\sigma_{max} = 579$ psi tens.
$\tau_{max} = 296$ psi

3-118 $d = 1.69$ in

3-20 $d = 3.55$ in.

3-122 $l = 9.72$ ft
$\sigma_{max} = 11{,}000$ psi
$\tau_{min} = 1000$ psi

4-2 $\delta_1 = \dfrac{Px^3}{6EI} - \dfrac{Pax^2}{2EI} \quad 0 < x < a$

$\delta_2 = \dfrac{Pa^3}{3EI} - \dfrac{Pa^2}{2EI}(x - a) \quad a < x < l$

$\delta_{max} = \dfrac{Pa^2}{EI}\left(\dfrac{a}{3} + \dfrac{b}{2}\right)$

4-4 $\delta = \dfrac{wx^2}{120EI}\left(10l^2 + 10lx - 5x^2 + x^3/l\right)$

$\delta_{max} = \dfrac{wl^4}{30EI}$

4-6 $\delta_{max} = \dfrac{-71}{384}\dfrac{wl^4}{EI}$

4-8 $\delta_B = 0.80$ in.
$\delta_C = 4.43$ in.
$\delta_D = 7.75$ in.

4-10 $\delta = \dfrac{Mx^3}{6EIl} - \dfrac{Mlx}{6EI}$

$\delta_{max} = \dfrac{0.064\,Ml^2}{EI}$

4-12 $\delta_{max} = \dfrac{2.82 \times 10^5}{EI}$, ft

4-14 $\delta_{max} = \dfrac{3.44 \times 10^5}{EI}$, ft

4-16 $\delta_{max} = \dfrac{7.74 \times 10^4}{EI}$, ft

4-18 $\delta_{max} = \dfrac{19.5Pl^3}{EI}$, ft

4-20 $\delta_{max} = \dfrac{2.32l^4}{EI}$, ft

4-22 $I_{min} = 344$ in.^{4}

4-24 $I_{min} = 38.7$ in.^{4}

4-26 $I_{min} = 26.5$ in.^{4}

4-28 $\delta_{max} = \dfrac{7wl^4}{160\,EI}$

4-30 $\delta_{max} = \dfrac{wl^4}{512\,EI}$

4-32 $\theta_B = \dfrac{Pl^2}{2EI}$

$\delta_B = \dfrac{Pl^3}{3EI}$

4-34 $\theta_B = \dfrac{wl^3}{6EI}$

$\delta_B = \dfrac{wl^4}{8EI}$

4-36 $\theta_B = \dfrac{Ml}{EI}$ $\qquad \theta_C = \dfrac{Ml}{2EI}$

$\delta_B = \dfrac{Ml^2}{2EI}$ $\qquad \delta_C = \dfrac{Ml^2}{8EI}$

4-38 $\delta_{max} = \dfrac{Ml^2}{9V_3EI}$

4-40 $\theta_B = \dfrac{2.72 \times 10^4}{EI}$ $\qquad \theta_C = \dfrac{3.11 \times 10^4}{EI}$

$\delta_B = \dfrac{9.26 \times 10^4}{EI}$ $\qquad \delta_C = \dfrac{2.75 \times 10^5}{EI}$

4-42 $\theta_B = \dfrac{5.30 \times 10^3}{EI}$

$\delta_B = \dfrac{3.65 \times 10^4}{EI}$

4-44 $\delta_c = \dfrac{2.6 \times 10^5}{EI}$, ft

4-46 $\theta_D = \dfrac{9.7 \times 10^4}{EI}$

      $\delta_D = \dfrac{7.10 \times 10^3}{EI}$, ft

4-48 $\delta_C = \dfrac{31\,wl^4}{4096\,EI}$

4-50 $\theta_A = \dfrac{Pl^2}{6\,EI}$

      $\theta_B = \dfrac{Pl^2}{3EI}$

4-52 $\delta_B = \dfrac{3.66 \times 10^4}{EI}$

4-54 $\delta_C = \dfrac{7.75 \times 10^4}{EI}$

4-56 $\delta_D = \dfrac{4.16 \times 10^4}{EI}$

4-58 $\delta_B = \dfrac{6.1 \times 10^4}{EI}$

4-60 $\delta = \dfrac{31\,l^4}{1920EI}$

4-62 $\delta_E = \dfrac{1.25 \times 10^5}{EI}$

4-64 $\theta_{ABC} = \dfrac{2 \times 10^4}{EI}$

      $\theta_{CDE} = \dfrac{-2.36 \times 10^4}{EI}$

4-66 $R_S = 3.096$ kips

4-68 $\delta_C = \dfrac{93\,wl^4}{384\,EI}$

4-70 $\dfrac{7}{24}\dfrac{wl^4}{EI}$

4-72 $\dfrac{5}{3}\dfrac{Pl^3}{EI}$

4-74   2.70 in.

4-76 $\dfrac{wl^4}{16\,EI}$

4-78 $\dfrac{833\,wl^4}{5760\,EI}$

4-80 $\delta_B = \dfrac{7\,Pl^3}{96\,EI}$

      $\theta_B = \dfrac{Pl^3}{2\,EI}$

4-82 $\delta_D = \dfrac{Pl^3}{EI}$

      $\theta_D = \dfrac{4}{3}\dfrac{Pl^2}{EI}$

4-84 $\delta_{bend} = \dfrac{7.56 \times 10^4}{EI}$

      $\delta_{shear} = \dfrac{112.5}{GA}$

      Percentage = 2.91 percent

4-86 $\delta = \dfrac{29.04}{E}$

      Percentage = 2.2 percent

5-2 $R_A = \dfrac{23}{27}P$

      $R_B = \dfrac{4}{27}P$

      $M_A = \dfrac{-5Pl}{27}$

5-4 $R_1 = \dfrac{3\,P \cdot a}{2\,l}$

      $R_2 = \dfrac{3\,P \cdot a}{2\,l} + P$

5-6 $R_A = {}^3/_2\,P \downarrow$

      $R_B = {}^5/_2\,P \uparrow$

      $M_A = \dfrac{\overset{\frown}{Pl}}{4}$

5-8 $R_A = \dfrac{9W_0l}{40}$

      $R_B = \dfrac{11\,W_0l}{40}$

      $M_A = \dfrac{7\,W_0l^2}{120}$

5-10 $R_A = 3000 \downarrow$ (downward)

      $R_B = 3000 \uparrow$

      $M_A = +10{,}000$ ft-lb (clockwise)

5-12 $R_A = \dfrac{w(l-a)^2}{8\,l^3}(5l^2 + 2al - a^2)$

      $R_B = w(l-a) - \dfrac{w(l-a)^2}{8}(5l^2 + 2al - a^2)$

      $M_A = w(l-a)^2\left[\dfrac{3}{l^2}(5l^2 + 2al - a^2) - \tfrac{1}{2}\right]$

2

5-14 $R_A = \dfrac{21\,w_o l}{64}$

$R_B = \dfrac{11\,w_o l}{64}$

$M_A = \dfrac{5 w_o l^2}{64}$

5-16 $R_A = \dfrac{19\,w_o l}{64}$

$R_B = \dfrac{13\,w_o l}{64}$

$M_A = \dfrac{3 w_o l^2}{64}$

5-18 $\delta = \dfrac{Pl^2 a}{32 EI} - \dfrac{wa^3}{8 EI}(l + a)$

5-20 $R_A = \dfrac{Pa^2}{(a+b)^2}\left[\dfrac{2a}{a+b} + 1\right]$

$R_B = P - \dfrac{Pb^2}{(a+b)^2}\left[\dfrac{2a}{a+b} + 1\right]$

$M_A = \dfrac{Pb^2 a}{(a+b)^2}$

5-22 $R_A = R_B = \dfrac{w_o l}{4}$

$M_A = M_B = \dfrac{5 w_o l^2}{96}$

5-24 $R_A = \dfrac{w(l-a)^3(l+a)}{2l^3}$

$R_B = w(l-a)\left[1 - \dfrac{(l-a)^2(l+a)}{2l^3}\right]$

$M_A = \dfrac{w(l-a)^3}{l}\left[\dfrac{l+a}{4l} - \dfrac{1}{6}\right]$

$M_B = \dfrac{w(l-a)^2}{2}\left[1 - \dfrac{(l-a)(l+a)}{l^2}\right.$

$\left. + \dfrac{2(l-a)}{l}\left(\dfrac{l+a}{4l} - \dfrac{1}{6}\right)\right]$

5-26 $R_A = \dfrac{21 wl}{44}$

$R_B = \dfrac{23 wl}{44}$

$M_A = \dfrac{13 wl^3}{176}$; $M_B = \dfrac{17 wl^2}{176}$

5-28 $\delta = \dfrac{w_o}{960\,EI}\left[-25 l^2 x^2 + 40 l x^3 - \dfrac{16 x^5}{l}\right]$

$0 < x < \dfrac{l}{2}$

$\delta_{max} = \dfrac{3 w_o l^4}{1280\,EI}$

5-30 $R_A = \dfrac{3Pa}{2l}$

$R_B = \dfrac{P(2l + 3a)}{2l}$

$M_A = \dfrac{-Pa}{2}$

5-32 $R_A = R_B = \dfrac{3M}{2l}$

$M_A = \dfrac{M}{2}$

5-34 $R_A = \tfrac{3}{4}P \downarrow$
$R_B = \tfrac{3}{4}P \uparrow$
$\delta_C = \dfrac{5}{48}\dfrac{Pl^3}{EI} \rightarrow$

5-36 $R_1 = \dfrac{9}{8} wl \downarrow$

$R_2 = \dfrac{17}{8} wl \uparrow$

$M_1 = \dfrac{wl^2}{8}$ ↷

5-38 $R_1 = \dfrac{P}{8} \downarrow$
$R_2 = \tfrac{3}{4}P \uparrow$
$R_3 = \tfrac{5}{8}P \downarrow$

5-40 $R_a = \tfrac{3}{8} wl \downarrow$
$R_b = \tfrac{5}{8} wl \uparrow$
$M_a = \dfrac{wl^2}{8}$ ↷

5-42 $R_A = \tfrac{3}{5} wl \downarrow$
$H_C = 0$
$M_C = \dfrac{wl^2}{10}$ ↷

5-44 $R_A = \tfrac{4}{3}P \uparrow$
$R_B = \tfrac{2}{3}P \uparrow$
$M_A = \dfrac{pl}{3}$ ↷

5-46 $R_A = R_B = \dfrac{W_o l}{4}$

$M_A = M_B = \dfrac{5 w_o l^2}{96}$

5-48 $R_A = \dfrac{w}{2l^3}(l^4 + 2a^3 l - 2al^3 - a^4)$

$R_B = \dfrac{w}{2l^3}(l^4 - 2a^3 l + a^4)$

$M_A = \dfrac{w}{12 l^2}(l^4 - 6l^2 a^2 + 8a^3 l - 3a^4)$

$M_B = \dfrac{w}{12 l}(l^4 - 4a^3 l + 3a^4)$

5-50 $R_A = \dfrac{49\,wl}{80}$

$R_B = \dfrac{31\,wl}{8}$

$M_A = \dfrac{9wl^2}{80}$

5-52 $R_A = \dfrac{21\,wl}{44}$

$R_B = \dfrac{29\,wl}{44}$

$M_A = \dfrac{13wl^2}{176}$; $M_B = \dfrac{17wl^2}{176}$

. 5-54 $R_1 = 2.5$ kips ↑
$R_2 = 10.625$ kips ↑
$R_3 = 3.75$ kips ↓
$R_4 = 0.625$ kips ↑

5-56 $R_A = 4.376\,\dot{w}$
$R_B = 6.248\,w$
$R_C = 0.624\,w$

$\delta_c = \dfrac{3.52w}{E}$

5-58 $R_A = \dfrac{Pa^3}{l^3 + a^3}$

$R_B = R_C = \dfrac{Pl^3}{l^3 + a^3}$

$M_A = \dfrac{Pa^3 l}{l^3 + a^3}$; $M_c = \dfrac{Pl^3 a}{l^3 + a^3}$

5-60 $R_A = R_C = \frac{2}{3}P$ $\qquad R_B = R_D = P/3$
$M_A = \frac{2}{3} \cdot Pl$

5-62 $R_A = 0.36\,wl$
$R_B = 1.32\,wl$
$R_C = 1.74\,wl$
$R_D = 0.58\,wl$

5-64 $R_A = \dfrac{3Pa}{2l}$

$R_B = \dfrac{P}{2l}(3a + 2l)$

$M_A = \dfrac{Pa}{2}$

5-66 $R_A = \dfrac{11 M_A}{4l}$ kips ↑

$R_B = \dfrac{9\,M_A}{2\,l}$ kips ↓

$R_C = \dfrac{7 M_A}{4\,l}$ kips ↑

5-68 $R_A = R_B = \dfrac{3M}{2l}$

$M_A = \dfrac{M}{2}$

5-70 $R_A = \dfrac{21 w_0 l}{64}$

$R_B = \dfrac{11}{64}\,w_0 l$

$M_A = \dfrac{5 w_0 l^2}{64}$

5-72 $R_A = \dfrac{19}{64}\,w_0 l$

$R_B = \dfrac{13}{64}\,w_0 l$

$M_A = \dfrac{3 w_0 l^2}{64}$

5-74 $R_A = \dfrac{11P}{16} - \dfrac{3wa^2}{4l}$

$R_B = \dfrac{5P}{16} + \dfrac{3wa^2}{4l} + wa$

$M_A = \dfrac{3Pl}{16} - \dfrac{wa^2}{4}$

5-76 $R_A = P - \dfrac{Pa}{(a + b)}\left[1 + \dfrac{ba - b^2}{(a + b)^2}\right]$

$R_B = \dfrac{Pa}{(a + b)}\left[1 + \dfrac{ab - b^2}{(a + b)^2}\right]$

$M_A = \dfrac{Pab^2}{(a + b)^2}$; $M_B = \dfrac{Pa^2 b}{(a + b)^2}$

5-78 $R_A = R_B = \dfrac{w_0 l}{4}$

$M_A = M_B = \dfrac{5 w_0 l}{96}$

5-80 $R_A = \dfrac{w}{2l^3}\left[l^4 - 2l^3 a + 2a^3 l - a^4\right]$

$R_B = \dfrac{w}{2l^3}\left[a^4 - 2a^3 l + l^4\right]$

$M_A = \dfrac{w}{12l^2}\left[-3a^4 + 8a^3 l - 6a^2 l^2 + l^4\right]$

$M_B = \dfrac{w}{12l^2}\left[3a^4 - 4a^3 l + l^4\right]$

5-82 $R_A = \dfrac{147\,wl}{308}$

$R_B = \dfrac{161\,wl}{308}$

$M_A = \dfrac{91wl^2}{1232}$; $M_B = \dfrac{17wl^2}{176}$

5-84 $R_1 = 13.38$ kips $\uparrow$ $\quad M_1 = -32$ kips-ft
$R_2 = 2.93$ kips $\uparrow$ $\quad M_2 = -2.42$ kips-ft
$R_3 = 9.08$ kips $\uparrow$ $\quad M_3 = -10.65$ kips-ft
$R_4 = 12.61$ kips $\uparrow$ $\quad M_4 = -18.00$ kips-ft

5-86 $M_A = M_C = 0$
$M_B = -26.77$
$R_A = 8.66 \uparrow$
$R_B = 12.68 \uparrow$
$R_C = 1.34 \downarrow$

5-88 $M_A = -25$ kips $- 1$
$M_B = -12.5$ kips -1; $\quad M_C = 0$
$R_A = 10.63$ kips $\uparrow$
$R_B = 4.99 \uparrow$
$R_C = 0.625 \downarrow$

5-90 $M_A = 0$; $\quad M_B = -100$ kips-1;
$M_C = 25$ kips $- 1$; $\quad M_D = 0$
$R_A = 5$ kips $\uparrow$; $\quad R_B = 27.5$ kips $\uparrow$;
$R_C = 15$ kips $\downarrow$
$R_D = 2.5$ kips $\uparrow$

5-92 $R_1 = 0.473$ kips $\uparrow$ $\quad M_1 = 0$
$R_2 = 3.54$ kips $\downarrow$ $\quad M_2 = 5.68$ kips-ft
$R_3 = 3.068$ kips $\uparrow$ $\quad M_3 = -25$ kips-ft

5-94 $R_1 = 7.31$ kips $\uparrow$ $\quad M_1 = 0$
$R_2 = 17.21$ kips $\uparrow$ $\quad M_2 = -24.3$ kips-ft
$R_3 = 9.84$ kips $\uparrow$ $\quad M_3 = -8.63$ kips-ft
$R_4 = 8.04$ kips $\uparrow$ $\quad M_4 = -24.0$ kips-ft

5-96 $R_a = \dfrac{wl}{8} \downarrow$

$R_b = \dfrac{3}{8} wl \uparrow$

$M_a = \dfrac{wl^2}{24}$

5-98 $R_1 = 3.6$ kips $\uparrow$ $\quad M_1 = -5.9$ kips-ft
$R_2 = 5.7$ kips $\uparrow$ $\quad M_2 = -7.0$ kips-ft
$R_3 = 4.8$ kips $\downarrow$ $\quad M_3 = -15.0$ kips-ft

5-100 $R_1 = 1.4$ kips $\uparrow$ $\quad M_1 = -0.5$ kips-ft
$R_2 = 7.8$ kips $\uparrow$ $\quad M_2 = -11.4$ kips-ft
$R_3 = 10.8$ kips $\uparrow$ $\quad M_3 = -25.0$ kips-ft

5-102 $R_1 = 1.1$ kips $\downarrow$ $\quad M_1 = +4.5$ kips-ft
$R_2 = 8.2$ kips $\uparrow$ $\quad M_2 = -9.0$ kips-ft
$R_3 = 2.3$ kips $\uparrow$ $\quad M_3 = -1.4$ kips-ft
$R_4 = 3.2$ kips $\uparrow$ $\quad M_4 = -30.0$ kips-ft

5-104 $P_u = 6000$ lb

5-106 $W_u = \dfrac{23.81 M_P}{l^2}$

5-108 $W_u = \dfrac{13.51 M_P}{l^2}$

5-110 $P_u = \dfrac{3 M_P}{l}$

5-112 $P_u = \dfrac{6 M_P}{l}$

5-114 $w_u = 2420$ lb/ft

6-2 $U = \dfrac{6750 A l^3}{E}$ in.-lb

6-4 $\sigma_{\max} = 89{,}150$ psi
$\delta = 0.962$ in.
$U = 6320$ ft-lb

6-6 $U = \dfrac{17.8 P^2}{AE}$ ft-lb

6-8 $U = \dfrac{254 P^2}{AE}$ ft-lb

6-10 $U = \dfrac{1507 P^2}{32 AE}$ ft-lb

6-12 $\delta_C = 0.91$ in.

6-14 (a) $\dfrac{U_{AB}}{U_{BC}} = \dfrac{\delta^4{}_{AB}\, l_{BC}}{\delta^4{}_{BC}\, l_{AB}}$

(b) $\dfrac{U_{AB}}{U_{BC}} = \dfrac{\delta^4{}_{AB}\, G_{AB}\, l_{BC}}{\delta^4{}_{BC}\, G_{BC}\, l_{AB}}$

6-16 $\dfrac{U_{AB}}{U_{BC}} = 3$

6-18 $U_S = 16$ in.-lb
$U_B = 136$ in.-lb

6-20 $\theta = \dfrac{Ml}{EI}$

6-22 $U_a = \dfrac{w^2 l^5}{40EI}$

$U_B = \dfrac{w^2 l^5}{6EI}$

6-24 $\delta_i = \delta_{St} + \sqrt{\delta^2{}_{St} + 2\delta_{St} h}$

6-26 $\delta_i = \delta_{St} + \sqrt{\delta^2{}_{St} + \dfrac{2\delta_{St} Eh}{l}}$

6-28 $U = \dfrac{7P^2 l^3}{24EI}$

6-30 $k = \dfrac{33}{13}$

$R_A = \dfrac{7P}{13}$

6-32 $P_u = M_u \left(\dfrac{2a + b}{l} \right)$

6-34 $\delta_A = \dfrac{wl^4}{8EI}$

$\theta_A = \dfrac{wl^3}{6EI}$

6-36 $\delta_{max} = \dfrac{0.64Ml^2}{EI}$

$\theta_A = \dfrac{Ml}{6EI}$

6-38 $\delta_c = \dfrac{wl^4}{120EI}$

6-40 $\delta_c = 7.78 \times 10^{-2}$ in.

6-42 $\delta_A = \dfrac{13\,Pl^3}{6EI}$

$\theta_A = \dfrac{19\,Pl^2}{12}$

6-44 $\delta_D = \dfrac{9\,wl^4}{8\,EI}$

$\theta_D = \dfrac{7\,wl^3}{6\,EI}$

6-46 $\delta_{BH} = 9 \times 10^{-2}$ in.

6-48 $\delta_{CH} = 2.26 \times 10^{-2}$ in.

6-50 $\delta_{BH} = \dfrac{5}{6}\dfrac{Pl}{AE}$

6-52 0.783 in.

6-54 $\delta_{BH} = 1.92 \times 10^{-2}$ in. $\rightarrow$

6-56 $\delta_{CV} = 42.84 \times 10^6 / AE$
$\delta_{CH} = 12.96 \times 10^6 / AE$

6-58 $\delta_V = 0.113$ in. $\downarrow$
$\delta_H = 0.017$ in. $\leftarrow$

6-60 $\delta_H = 0.092$ in. $\rightarrow$

6-62 $\delta_A = \dfrac{wl^4}{8E +}$

$\theta_A = \dfrac{wl^3}{6\,EI}$

6-64 $\delta_c = \dfrac{5wl^4}{8EI}$

$\theta_c = \dfrac{2wl^3}{3EI}$

6-66 $\delta_{CH} = 2.26 \times 10^{-2}$ in.

6-68 $\delta_{CV} = 42.84 \times 10^6 / AE$ in.
$\delta_{CH} = 12.96 \times 10^6 / AE$ in.

6-70 $\delta_V = 0.109$ in. $\downarrow$
$\delta_H = 0.024$ in. $\rightarrow$

6-72 $R_A = \dfrac{3M}{2l}$

6-74 $R_A = \dfrac{P}{9}$

$R_B = \dfrac{8P}{9}$

7-2 $P = 1480$ lb
$\sigma_{cr} = 1970$ psi

7-4 $l = 7.94$ ft

7-6 $P_{cr} = 7040$ lb
$\sigma_{cr} = 3360$ psi

7-8 $\sigma_{cr} = 66,400$ psi

7-10 $P = 52$ kips

7-12 $M = 76.4$ in.-kips

7-14 S.F. = 1.39

7-16 $P = 120$ kips

7-18 S.F. = 2.29

7-20 S.F. = 2.33

8-2 $b = 3.2$ in.

8-4 $t = 0.283$ in.

8-6 $b = 7.66$ in.

8-8 $t = 0.564$ in.

8-10 $b = 6.45$ in.

8-12 $P = 39,780$
eff. $= 45.20$ percent

8-14 $d = 0.907$ in.

8-16

$R_1 = 6.67$ kips ↑

$R_2 = 2.67$ kips ↑

$R_3 = 1.33$ kips ↓

8-18 $d = 0.82$ in.

8-20 $d = 0.62$ in.

8-22 $e = 7.57$ in.

8-24 $d = 0.818$ in.

8-26 $l = 5.21$ in. each side

8-28 $l_1 = 10.6$ in.
$l_2 = 5.4$ in.

8-30 $q = 416$ psi

8-32 $(a)\, l_1 = 12.5$ in. $(b)\, l_1 = 15.3$ in.
$l_2 = 6.8$ in. $l_2 = 4.0$ in.

INDEX